AF616409

Integrating Hydrology, Ecosystem Dynamics, and Biogeochemistry in Complex Landscapes

Goal of this Dahlem Workshop:

To assess integrative experimental, observational, and modeling approaches that describe hydrology, biogeochemical cycles, and ecosystem dynamics and their feedbacks in complex landscapes.

Dahlem Workshop Report

Held and published on behalf of the
Freie Universität Berlin

Sponsored by:
Stiftung Deutsche Klassenlotterie

Integrating Hydrology, Ecosystem Dynamics, and Biogeochemistry in Complex Landscapes

Edited by

J.D. TENHUNEN and P. KABAT

Report of the Dahlem Workshop on
Integrating Hydrology, Ecosystem Dynamics, and
Biogeochemistry in Complex Landscapes
Berlin, January 18–23, 1998

Program Advisory Committee:
J.D. Tenhunen and P. Kabat, Chairpersons
H.-R. Bork, I.C. Burke, W. Mauser, J.F. Reynolds,
W.L. Steffen, and R. Valentini

JOHN WILEY & SONS
Chichester • New York • Weinheim • Brisbane • Singapore • Toronto

Baffins Lane, Chichester,
West Sussex PO19 1UD, England
National 01243 779777
International (+44) 1243 779777
e-mail (for orders and customer service enquiries): cs-books@wiley.co.uk

Visit our Home Page on http://www.wiley.co.uk or http://www.wiley.com

Other Wiley Editorial Offices

John Wiley & Sons, Inc., 605 Third Avenue,
New York, NY 10158-0012, USA

WILEY-VCH Verlag GmbH, Pappelallee 3,
D-69469 Weinheim, Germany

Jacaranda Wiley Ltd, 33 Park Road, Milton,
Queensland 4064, Australia

John Wiley & Sons (Asia) Pte Ltd, 2 Clementi Loop #02-01,
Jin Xing Distripark, Singapore 129809

John Wiley & Sons (Canada) Ltd, 22 Worcester Road,
Rexdale, Ontario M9W 1L1, Canada

Library of Congress Cataloging-in-Publication Data
A catalog record for this book is available from the Library of Congress

British Library Cataloguing in Publication Data
A catalogue record for this book is available from the British Library

ISBN 0-471-98474-4

Editorial Staff of Dahlem Konferenzen: J. Lupp, C. Rued-Engel, G. Custance
Typeset in 10/12pt Times by Dahlem Konferenzen

Printed and bound in Great Britain by Biddles Ltd., Guildford, Surrey

This book is printed on acid-free paper responsibly manufactured from sustainable forestry, in which at least two trees are planted for each one used for paper production.

Contents

The Dahlem Konferenzen

In 1974, the Stifterverband für die Deutsche Wissenschaft[1] in cooperation with the Deutsche Forschungsgemeinschaft[2] founded the *Dahlem Konferenzen.* It was created to promote an interdisciplinary exchange of scientific ideas as well as to stimulate cooperation in research among international scientists. Dahlem Konferenzen proved itself to be an invaluable tool for communication in science, and so, to secure its long-term future, it was integrated into the Freie Universität Berlin in January, 1990.

As has been evident over recent years, scientific research has become highly interdisciplinary. Now, before real progress can be made in any one field, the concepts, methods, and strategies of related fields must be understood and able to be applied. Coordinated research efforts, scientific cooperation, and basic communication between the disciplines and the scientists themselves must be promoted in order for science to advance.

To meet these demands, Dahlem Konferenzen created a special type of forum for communication, now internationally recognized as the *Dahlem Workshop Model.* These workshops are the framework in which coherent discussions between the disciplines take place and are focused around a topic of high priority interest to the disciplines concerned. At a Dahlem Workshop, scientists are able to pose questions and solicit alternative opinions on contentious issues from colleagues from related fields. The overall goal of a workshop is not necessarily to reach a consensus but rather to identify gaps in knowledge, to find new ways of approaching controversial issues, and to define priorities for future research. This philosophy is implemented at every stage of a workshop: from the selection of the theme to its breakdown in the discussion groups, from the writing of the background papers to the formulation of the group reports.

Workshop topics are proposed by leading scientists and are approved by a scientific board, which is advised by qualified referees. Once a topic has been approved, a Program Advisory Committee of scientists meets approximately one year before the

---

[1] The Donors' Association for the Promotion of Sciences and Humanities, a foundation created in 1921 in Berlin and supported by German trade and industry to fund basic research in the sciences.

[2] German Science Foundation.

workshop to delineate the scientific parameters of the meeting, select participants, and assign them their tasks. Participants are invited on the basis of their scientific standing alone.

Each workshop is organized around four key questions, which are addressed by four discussion groups of approximately ten participants. Lectures or formal presentations are taboo at Dahlem. Instead, concentrated discussion — within a group and between groups — is the means by which maximum communication is achieved. To facilitate this discussion, participants prepare the workshop theme prior to the meeting through the "background papers," the themes and authors of which are chosen by the Program Advisory Committee. These papers specifically review a particular aspect of the group's discussion topic as well as function as a springboard to the group discussion, by introducing controversies or unresolved problem areas.

During the workshop week, each group sets its own agenda to cover the discussion topic. Cross-fertilization between groups is both stressed and encouraged. By the end of the week, in a collective effort, each group has prepared a report reflecting the ideas, opinions, and contentious issues of the group as well as identifying directions for future research and problem areas still in need of resolution.

A Dahlem Workshop initiates and facilitates discussion between a certain number — necessarily restricted — of scientists. Because it is imperative that the discussion and communication should continue after a workshop, we present the results to the scientific community at large in the form of this published volume. In it you will find the revised background papers and group reports, as well as an introduction to the workshop theme itself.

The difference between proceedings of many conventional meetings and this workshop report will be easily discernable. Here, the background papers have not only been reviewed by formal referees, they have been revised according to the many comments and suggestions made by *all* participants. In this sense, they are reviewed more thoroughly than scientific articles in most archival journals. In addition, an extensive editorial procedure ensures a coherent volume. I am sure that you, too, will appreciate the tireless efforts of the many reviewers, authors, and editors.

On their behalf, I sincerely hope that the spirit of this workshop as well as the ideas and controversies raised will stimulate you in your work and future endeavors.

Prof. Dr. Klaus Roth, Director
Dahlem Konferenzen der Freien Universität Berlin
Thielallee 66, D–14195 Berlin, Germany

List of Participants with Fields of Research

J.D. ABER Complex Systems Research Center, 454 Morse Hall, University of New Hampshire, Durham, NH 03824, U.S.A.

Biogeochemistry of forested ecosystems, especially field and modeling analyses of nitrogen deposition and climate change on carbon, nitrogen, and water balances

B. ACOCK U.S. Dept. of Agriculture, ARS, Remote Sensing and Modeling Lab., Building 007, Rm. 008, BARC-West, 10300 Baltimore Avenue, Beltsville, MD 20705–2350, U.S.A.

Crop models as farm decision aids

D.D. BALDOCCHI Atmospheric Turbulence and Diffusion Division, NOAA, P.O. Box 2456, Oak Ridge, TN 37831–2456, U.S.A.

Canopy micrometeorology, atmosphere–biosphere trace gas exchange

H.-J. BOLLE Stücklenstr. 18 C, D–81247 Munich, Germany

Assessment of land-surface processes specifically in semi-arid areas with the aid of satellite measurements; desertification in the Mediterranean area

H.K.M. BUGMANN Potsdam Institute for Climate Impact Research (PIK), Postfach 60 12 03, D–14412 Potsdam, Germany

Forest ecosystem dynamics, impacts of global change on vegetation, ecological modeling, systems analysis

I.C. BURKE Dept. of Forest Sciences, Colorado State University, Fort Collins, CO 80523, U.S.A.

Patterns of soil organic matter dynamics across complex landscapes and regions, through both field research and simulation modeling; N cycling; effects of land use on net ecosystem production

N. CARACO Institute of Ecosystem Studies, Box AB, Millbrook, NY 12545, U.S.A.

Nutrient and CO_2 dynamics in aquatic systems; nutrient and grazer control of phytoplankton

W. CRAMER Potsdam Institute for Climate Impact Research (PIK), Postfach 60 12 03, D–14412 Potsdam, Germany

Impacts of climate and other global changes on terrestrial ecosystems and the global C cycle

L. DÜMENIL Max-Planck-Institut für Meteorologie, Bundesstr. 55, D–20146 Hamburg, Germany

Parameterization of land-surface processes in GCMs, interannual variability of the monsoon, Amazon and Mediterranean deforestation

W. EUGSTER Institute of Geography, University of Bern, Hallerstr. 12, CH–3012 Bern, Switzerland

Micrometeorology (energy and trace gas fluxes)

R.J. GURNEY Environmental Systems Science Centre, University of Reading, P.O. Box 238, GB–Reading RG6 6AL, U.K.

Land-surface processes and remote sensing

G.M. HORNBERGER Institute of Arctic and Alpine Research, Campus Box 450, University of Colorado, Boulder, CO 80309–0450, U.S.A.

Catchment hydrochemistry, colloid transport

R.B. JACKSON Dept. of Botany, Duke University, Durham, NC 27708, U.S.A.

Plant and soil ecology, biogeochemistry, ecosystem studies and global change

C.A. JOHNSTON Natural Resources Research Institute, University of Minnesota, 5013 Miller Trunk Highway, Duluth, MN 55803, U.S.A.

Landscape ecology, wetland biogeochemistry, application of geographic information systems to hydrologic and nutrient flows in landscapes

P. KABAT SC–DLO, Winand Staring Centre for Integrated Land, Soil, and Water Research, P.O. Box 125, NL–6700 AC Wageningen, The Netherlands

Land-surface processes, landscape–hydrology–atmosphere interactions, climate hydrology

H. LANG Hydrology/Dept. of Geography, ETH-Zentrum, Winterthurer Str. 190, CH–8057 Zurich, Switzerland

Hydrology, river basin modeling, climate variations and the water cycle, land-surface processes, mountain hydrology

P. MARTIN Institute for Systems, Informatics, and Safety (ISIS), EC Joint Research Centre, TP 650, I–21020 Ispra (VA), Italy

Atmosphere–biosphere interactions, in particular, global terrestrial biosphere modeling for climate modeling purposes and impacts studies; CO_2 *and* H_2O *exchange between the atmosphere and plant canopies; biogeochemical cycles modeling*

W. MAUSER Institute of Geography, University of Munich, Luisenstr. 37, D–80333 Munich, Germany

Hydrologic modeling on different scales, remote sensing

F.X. MEIXNER Abt. Biogeochemie, Max-Planck-Institut für Chemie, Postfach 3060, D–55020 Mainz, Germany

Biosphere–atmosphere exchange, (micro-) meteorology, (measurement of) surface fluxes

J.-C. MENAUT Dept. R.E.D., ORSTOM, 213, rue Lafayette, F–75480 Paris Cedex 10, France

Links between ecosystem function, structure, and dynamics; spatially explicit ecosystem modeling

L. MENZEL Potsdam Institute for Climate Impact Research (PIK), Postfach 60 12 03, D–14412 Potsdam, Germany

Land-surface processses and interactions with the atmosphere; hydrological modeling

R. NEMANI School of Forestry, University of Montana, Missoula, MT 59812, U.S.A.

Remote sensing of vegetation; ecosystem modeling

I.R. NOBLE Ecosystem Dynamics, Research School of Biological Sciences, The Australian National University, Canberra, ACT 0200, Australia

Modeling vegetation and landscape dynamics, global change

J.A. OLEJNIK Agrometeorology Dept., Agricultural University of Poznan, Ul. Witosa 45, PL–60677 Poznan, Poland

Heat balance structure; water balance structure; landscape–ecological impact of climatic change; measuring and modeling land surface–atmospheric interactions

J.M. PARUELO Dpto. Ecologia, Facultad de Agronomia (UBA), Av. San Martin 4453, 1417 Buenos Aires, Argentina

Ecosystem ecology, remote sensing, grassland and shrubland ecology

J. PEÑUELAS CREAF, Facultat de Ciencies, Universitat Autonoma de Barcelona, E–08193 Bellaterra (Barcelona), Spain

Plant physiological ecology under global change; remote sensing at ground level

R.A. PIELKE, SR. Dept. of Atmospheric Science, Colorado State University, Fort Collins, CO 80523, U.S.A.

Influence of landscape, including land–atmosphere biophysics and biogeochemical influences on local, regional, and global climate and weather

S.D. PRINCE Laboratory for Global Remote Sensing Studies, Geography Dept., University of Maryland, College Park, MD 20742–8225, U.S.A.

Study of regional and global vegetation dynamics using satellite remote sensing

M.R. RAUPACH CSIRO Land and Water, Environmental Mechanics Laboratory, GPO Box 821, Canberra, ACT 2601, Australia

Wind, turbulence, and transport processes in the lower atmosphere; biosphere–atmosphere interactions; fluid mechanics; soil erosion and particle transport by wind

J.F. REYNOLDS Dept. of Botany, Phytotron Building, Box 90340, Duke University, Durham, NC 27708–0340, U.S.A.

Ecological modeling of plant and ecosystem response to elevated CO_2 *and climate change; plant, ecosystem, and landscape models of desertification*

O. RICHTER Institute of Geography and Geoecology, Technical University of Braunschweig, Langer Kamp 19c, D–38106 Braunschweig, Germany

Environmental fate modeling, ecological modeling, agricultural ecology, coupling environmental fate models with geographical information systems, population dynamics

S.W. RUNNING School of Forestry/NTSG, University of Montana, Missoula, MT 59812, U.S.A.

Regional ecosystem modeling

W.L. STEFFEN Division of Wildlife and Ecology, GCTE Core Project Office/CSIRO, P.O. Box 84, Lyneham, ACT 2602, Australia

Global change ecology: emphasis on (i) global vegetation models; (ii) terrestrial global carbon cycle; (iii) regional-scale impact studies

J.D. TENHUNEN Dept. of Plant Ecology II, Bayreuth Institute for Terrestrial Ecosystem Research, University of Bayreuth, D–95440 Bayreuth, Germany

Landscape heterogeneity in surface exchanges

R. VALENTINI Dept. of Forest Science and Resources, DISAFRI, Universita della Tuscia, Via S. Camillo De Lellis, I–01100 Viterbo, Italy

Biosphere–atmosphere exchanges, trace gas exchanges

R.H. WARING College of Forestry, Oregon State University, Corvallis, OR 97331, U.S.A.

Forest ecosystem analyses: stress indices, plant–water relations

C.A. WESSMAN CIRES, Campus Box 216, University of Colorado, Boulder, CO 80309–0216, U.S.A.

Terrestrial ecosystems, scaling of ecological processes, and the expression of ecosystem processes in the state and pattern of the landscape; the use of remote sensing in regional and global ecology

E.F. WOOD Dept. of Civil Engineering, EQUAD, Princeton University, Princeton, NJ 08544, U.S.A.

Hydrology and hydroclimatology: land–atmosphere interactions, scaling of terrestrial water and energy fluxes, remote sensing for terrestrial hydrology

J. WU Dept. of Life Sciences (2352), Arizona State University West, P.O. Box 37100, Phoenix, AZ 85069–7100, U.S.A.

Landscape ecology: landscape pattern analysis, modeling landscape pattern and ecological processes, hierarchical patch dynamics

1

Ecosystem Studies, Land-use Change, and Resource Management

J.D. TENHUNEN[1], R. GEYER[1], R. VALENTINI[2], W. MAUSER[3], and A. CERNUSCA[4]

[1]Dept. of Plant Ecology, University of Bayreuth, 95440 Bayreuth, Germany
[2]Dept. of Forest Science and Environment, University of Tuscia, 01100 Viterbo, Italy
[3]Institute of Geography, Dept. of Geographical Remote Sensing, University of Munich, 80333 Munich, Germany
[4]Institute of Botany, University of Innsbruck, Innsbruck, Austria

ABSTRACT

After early studies of the interrelationships between land use and hydrology, ecosystem studies were strongly influenced during the 1960s and 1970s by the International Biological Programme with a focus on regulation mechanisms and processes related to production, decomposition, and nutrient cycling in different biomes. In recent years, a spatial orientation in ecosystem research has been revived with investment in large land-surface exchange experiments and with interest in scale problems, mostly related to nested watershed studies or meso-scale climate and atmospheric tranport modeling. With the development of the International Geosphere–Biosphere Programme, this information has had priority usage at coarse grid resolutions and to support global-scale analyses. The development of techniques that permit an integrated analysis of water, carbon, and nitrogen budgets within areas of 100 to 100,000 km^2 as well as the effects on these budgets of ecosystem disturbance and land-use change have been less emphasized. However, an integrated application of knowledge about ecosystem processes linked to hydrology and climate is needed for resource management and evaluation of the consequences of global change for human concerns. The concensus view reported in this book suggests that progress on landscape level understanding of coupled water, C, and N budgets is limited more by a lack of commitment to a rigorous development and application of synthetic techniques (ecosystem modeling, remote sensing, and geographic information systems) than by basic site-level measurements in various disciplines. This chapter introduces the background to this overall problem and provides an overview of the organization of the book with respect to the key topics considered, which examine current limits to our understanding and suggest where new research foci should lie. These overview comments are made in relation to a conceptual

Integrating Hydrology, Ecosystem Dynamics, and Biogeochemistry in Complex Landscapes
Edited by J.D. Tenhunen and P. Kabat

framework in development for estimating water and carbon balances for European landscapes based on ecophysiological and ecosystem-oriented field studies and simulation modeling efforts.

INTRODUCTION

As an indirect result of population and economic growth in recent decades, the biosphere has become increasingly stressed, often beyond the point at which long-term sustainability of internal structure and function of ecosystems is maintained (cf. Christensen et al. 1996). We have experienced an intensified "exploitation" of natural system resources to support agricultural and forest production, to provide water for human consumption, to supply the needs of industrial processes, and to provide attractive, diverse landscapes for recreation and tourism. Exceeding thresholds of sustainability via anthropogenic disturbance is highly undesirable, since the consequences of system level feedbacks (e.g., the effects of soil erosion and degradation, famine, polluted drinking water, etc.) are dangerous.

Finding appropriate compromises in resource use that satisfy existing competitive interests but result in sound environmental management requires an improved understanding of the trade-offs that accompany changes in "exploitation" or altered resource allocation at regional and landscape scales (Turner et al. 1994, 1995; Meyer and Turner 1994). The trade-offs are complex because multiple linked ecosystems (forests, grazing land, agricultural fields, wetlands, urban systems, etc.) and multiple budgets (carbon, water, and nitrogen, etc.) are impacted and must simultaneously be considered at the landscape scale. Decision makers logically ask ecosystem scientists what type of information they can provide that is useful in planning for alternative types of land and resource use. Together with the development of landscape ecology as a science (Risser et al. 1984; Forman and Godron 1986) and the blossoming of the International Geosphere–Biosphere Programme, the critical need for new management methods will stimulate the development of rigorous quantitative methods for analyzing ecosystem processes in a 4-dimensional framework, e.g., as they are influenced by environmental gradients and spatial heterogeneity within areas of 100 to 100,000 km^2 and by disturbance and land-use change over decades (Chapter 17). While remaining exceedingly important even within this new context, an *overall* orientation of ecosystem studies toward plot level (ha size) and seasonal influences on processes is passé.

Emerging and traditional measurement techniques applied in ecosystem science provide databases for the analysis of coupled water, carbon, and nitrogen budgets that are essential in the study of environmental problems (Chapter 16). In many (but not all) respects, information already exists for the key processes at the plot scale upon which to construct sound ecosystem models. To understand water, carbon, and nitrogen budgets at the landscape scale, we must develop synthetic tools for that purpose, combining modeling applications at the ecosystem (or plot or "pixel") level with data from remote sensing (Chapters 2–5) and the use of geographic information systems (GIS). To examine scenarios relevant to management, the atmospheric and hydrological boundary conditions for ecosystem response must be defined for selected landscape sections (Chapters 6, 9, 10, and 13). Both atmospheric drivers and soil water status in the rooting zone must be obtained for each pixel, thus, raising the question whether and to what extent analyses developed in hydrology and the atmospheric sciences are able to support landscape-level assessments of water, carbon, and nitrogen budgets (Chapters 15–17). A premise of the Dahlem Workshop entitled "Integrating Hydrology, Ecosystem Dynamics, and Biogeochemistry in Complex Landscapes," examined from several perspectives in the following chapters, is that progress on landscape-level understanding of coupled water, C, and N budgets is limited more by a lack of commitment to a rigorous development and application of synthetic techniques (ecosystem modeling, remote sensing, and GIS) than by basic site-level measurements in various disciplines (Burke et al. 1994; Pickett et al. 1994; Chapters 15, 16, and 17). Closely associated with this premise is the question whether current experimental designs for field studies support with optimal efficiency the coupling and closing of water, C, and N budgets at the landscape level.

A "complex landscape" is difficult to define. Complexity is seen differently with each emphasis in perspective, i.e., complexity of patterns in remotely sensed data, in topographic effects on transport or energy exchange, in management of defined land parcel mosaics, etc. This book struggles with several aspects of defining complexity and at least begins a process of examining how current information and techniques may be utilized for assessments in nonideal "real" landscapes. The limitations of current techniques are pointed out and suggestions are made as to where future emphases should be placed in order to utilize scientific observations more fully within a management context. The Dahlem group paid serious attention to considering how we may more strongly reorient thinking among natural scientists toward the analysis of landscape- and regional-level problems, to defining the most promising approaches for such analyses, and to suggesting how it might be possible to obtain greater understanding by working toward products defined from an interdisciplinary perspective.

RELATING "BOTTOM-UP" PROCESS INFORMATION TO "TOP-DOWN" REGIONAL AND LANDSCAPE PERSPECTIVES

The earliest observations of water cycles in terrestrial ecosystems (Lyell 1835; Boussingault 1845; Marsh 1864) recognized interactions between stream water yield, erosion, sedimentation, lake water levels, vegetation composition and production in lowland areas surrounding forests, and the degree of forestation. In these early descriptions by physical geographers, the landscape ecological perspective that is intimately tied to the study of water cycles was apparent. The consequences of deforestation were particularly pronounced in mountainous regions, leading Engler (1919) to initiate a paired watershed study to demonstrate the beneficial effects of forests in the Emmental in 1902 (see also Keller 1988; Burger 1934, 1945, 1954). Concerns about the influence of forests on streamflow and flooding initiated cooperative manipulations of hydrology, which were undertaken by the U.S. Weather Bureau and U.S. Forest Service in Colorado, based on clear-cutting of forested watersheds (Bates and Henry 1928). Clear definitions of watershed units, procedures for watershed standardization, and plans for organized studies within watersheds including the perspectives of meteorology, plant physiology, soil science, geology, and hydrology were laid out for Coweeta basin in the Southern Appalachian Mountains during this same time period (Swank and Crossley 1988). The long-term monitoring and manipulation studies at Coweeta provide an extremely significant landmark in ecosystem studies, and they continue to gain in value as new challenges requiring ecosystem analysis and understanding arise due to global change and anthropogenic disturbances.

In contrast, during the period of the International Biological Program (IBP), these landscape-oriented approaches lost ground as ecologists focused strongly on mechanisms and processes related to production, decomposition, and nutrient cycling. Detailed plot level investigations were promoted by technological advances in the development of field instrumentation. The products of IBP research were detailed one-dimensional "average representations" of process linkages in different biome types, an approach and shortcoming that is still often apparent in current global modeling efforts.

The degree to which ecosystems have been influenced by human activities and the nature of these disturbances make it inevitable that watersheds, or collections of watersheds in various shapes and forms, must provide a continuing focus of research (Chapters 10 and 13). The import of materials via atmospheric transport occurs at the landscape level and must be quantified for a ground area that is appropriate to link with hydrological processes (Chapter 8). Movement of these materials as well as their chemical transformations depend on water fluxes and water availability. The output of materials from terrestrial ecosystems to stream systems provides an integrated measure of overall ecosystem function (Bormann and Likens 1967; Chapter 12). This output is a function of the landscape mosaic, being influenced by soil and vegetation processes occurring in landscape patches encountered during transport along toposequences (Peterjohn and Correll 1984; Chapter 15). Water balance per se and transport potentials are a function of landscape and topography, being influenced by the spatial distribution of the vegetation canopy conductance, by topographic effects on energy input, by ground water distribution and availability to plants, and by spatial aspects of boundary layer climate (BAHC 1993).

These two historical perspectives on ecosystem research as related to environmental problems have been characterized as either "top-down" approaches, when the tendency to obtain information on entire system function is strongly represented, e.g., synthetic measures of watershed response or deposition over large spatial areas; or "bottom-up," when the focus has been on examining the inter-linkage of processes controlling system response. The purpose of our deliberations must be seen in the desire to include simultaneously both of these perspectives in the study of environmental problems and to bridge the gap between the two approaches. To illustrate this point, time and space scales relevant to different types of ongoing Central European forest

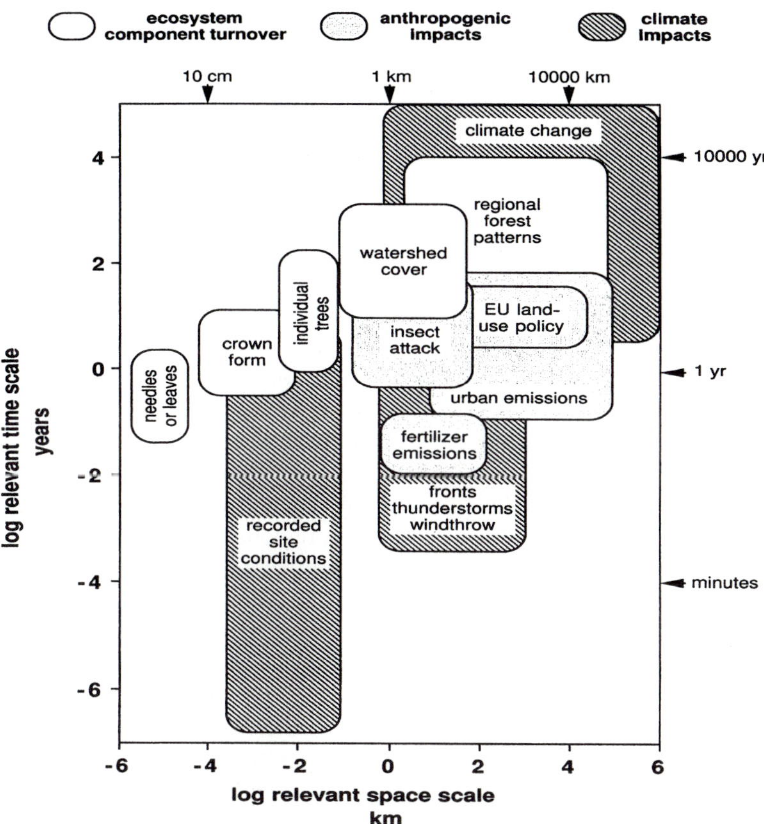

Figure 1.1 Relevant temporal and spatial scales of natural ecosystem processes and anthropogenic disturbances affecting temperate forests. EU = European Union.

ecosystem studies are shown in Figure 1.1. Current interests are increasingly to examine those problems clustered at scales indicated in the upper right of the figure, such as fertilization through emissions or patterns in the spread of insect infestations or the significance of change in land-use policies, but to understand as well how these are influenced by the planting of individual trees and stands and by processes that occur within trees and stands as indicated toward the lower left in the figure. Obviously, a merging of "top-down" and "bottom-up" approaches is needed and the challenge is to determine the extent to which this is possible, practical, and usable for the purposes of resource management.

During the initial phase of IGBP in recent years, many new insights have been gained on the integrated function of the biosphere and these insights are already documented in many books and publications (Walker et al. 1996, 1998). The progress made in large-scale experiments related to land-surface energy exchange, e.g., HAPEX-MOBILHY (coniferous forests of southwestern France), FIFE (grasslands of the U.S. Great Plains), HAPEX-SAHEL (Sahel region of northwest Africa), NOPEX and BALTEX (countries surrounding the Baltic Sea), EPHEDRA (east-central Spain), BOREAS (boreal forests of Canada), and the ongoing LBA (Amazon Basin), among others, has modified our views on ecosystem function in these regions as well as demonstrated continuing existing limitations on our knowledge (see Chapter 9). However, these experiments have been designed with a "top-down" perspective and have focused on "upscaling" information from point measurements in large regions (as surrogates for biomes) to the global scale. What has been less strongly represented in such experiments is the potential to interpret influences of temporal and spatial heterogeneity on observed energy exchange and other processes, i.e., an inability to determine fully the ranges in ecosystem response to climate perturbation that occur regionally (Chapters 6 and 7). It is specifically this heterogeneity and its management that must be understood with respect to socio-economic problems and environmental impacts.

To merge "top-down" and "bottom-up" approaches, integrated investigations of the biosphere must be carried out at continental, regional, and landscape scales. This may be visualized by examining the planning for two coupled European project networks (EUROFLUX and MEDEFLU; Figure 1.2; Chapter 7; Tenhunen et al. 1998; Baldocchi et al. 1996) designed to monitor net ecosystem exchange of CO_2 (NEE) of forests over long-term periods. The locations of sites for these networks are shown on a satellite composite view of the European continent in Figure 1.2; information on climate gradients and change in forest types is provided in Tables 1.1 and 1.2. Despite relatively high costs to operate such networks, difficulties in finding comparable sites, and the existing distribution of expertise to run such sites, one has, nevertheless, the impression that a new continental-scale view of the regulation of forest NEE can be obtained.

However, this provides a "top-down" perspective on forest ecosystem function that includes only a single tower site for intensive studies at each numbered location. Imagine now that interdisciplinary research, which includes biologists, forest managers, agriculturalists, hydrologists, atmospheric scientists and others, might be clustered and intensified within the circles surrounding these tower sites (see also Koch et al. [1995] and Steffen et al. [1998]). The information would become extremely valuable, since, on the one hand, a continental perspective is supported with links to global circulation models (GCMs) and global assessments, while the local heterogeneity and complex landscape function relevant at each location is examined at another appropriate scale. In addition to the single tower site, intensive research areas determined according to topography, land use, and remote sensing criteria should provide process information specific for the type of landscape at that location.

Networks such as that illustrated in Figure 1.2 can play a key role in bridging the gap between "top-down" and "bottom-up" perspectives, having a dual role in influencing both policy decisions related to large-scale global issues and building potentials for examining the implications of particular landscape management scenarios

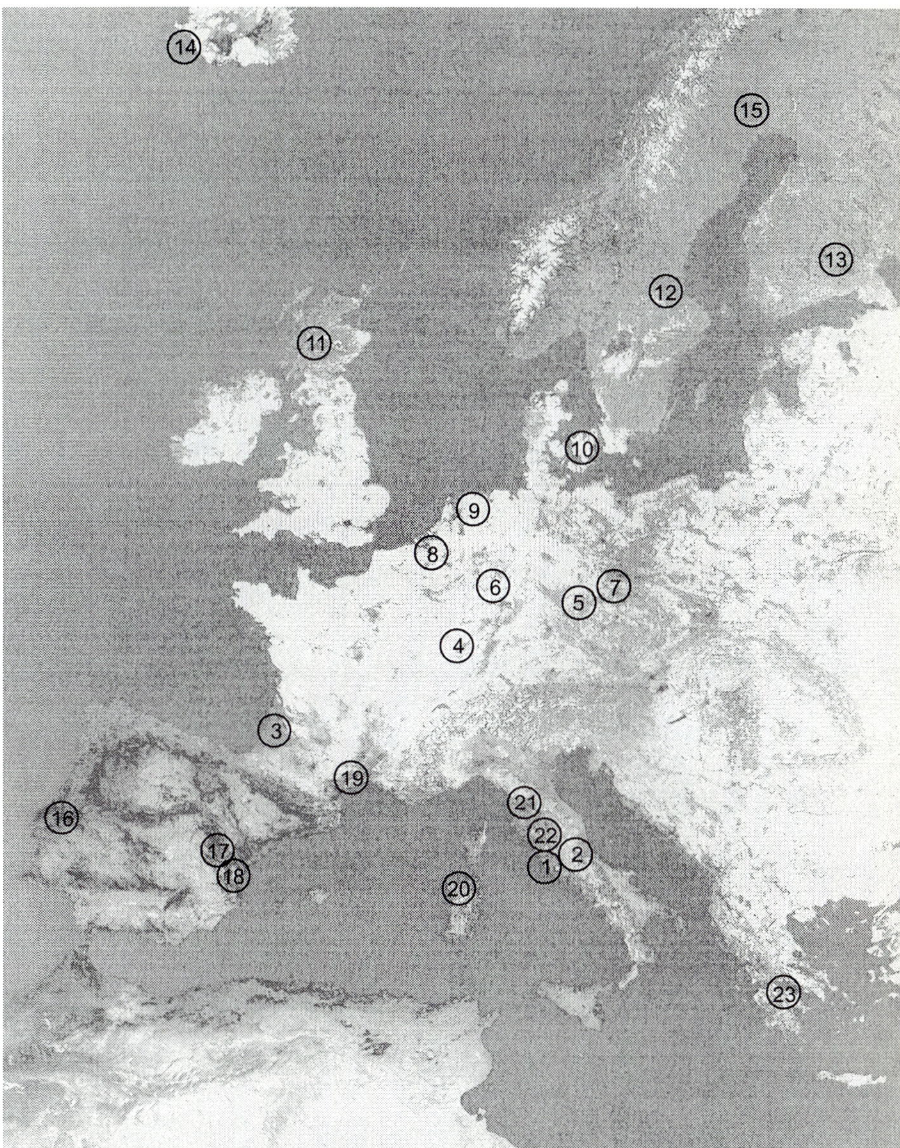

Figure 1.2 Sites of the European Union networks EUROFLUX and MEDEFLU designed to determine local forest net ecosystem CO_2 exchange (NEE) via eddy covariance techniques (cf. Chapter 7). Each numbered site indicates the location of a tower anchor station of the network. The climate as well as forest type and structure for each tower site are given in Tables 1.1 and 1.2. Circles are referred to in the text as indicating landscapes in the proximity of the anchor stations that might be used to examine the impacts of climate change or other disturbances at scales relevant to human dimensions and which might also provide a weighting of land-surface exchanges according to typical land use.

Table 1.1 Characteristics of the forest stands currently being studied by the project EUROFLUX arranged from south to north within Europe. Numbering of sites corresponds to the locations shown in Figure 1.2.

No.	Site	Position	Elevation (m a.s.l.)	Overstory species	Understory species	Mean T (°C)	Precipitation (mm)	LAI (m^2 m^{-2})	Canopy height (m)	Stand age (years)	Density (n ha^{-1})	Wood increment (m^3 ha^{-1} yr^{-1})
1	Italy	41°45′N 12°22′E	3	*Quercus ilex*	evergreen shrubs	15.3	770	3.5	12.5	50	1500	3.5
2	Italy	41°52′N 13°38′E	1550	*Fagus sylvatica*	herbs, cf. *Gallium*	7	1100	4.5	22	100	890	7
3	France	44°42′N 0°46′W	60	*Pinus pinaster*	*Molinia caerulea*	13.5	900	3	18	35	500	18
4	France	48°40′N 7°05′E	300	*Fagus sylvatica*	*Carpinus betulus*	9.2	820	5.5	13	30	4000	no data
5	Germany	50°09′N 11°52′E	780	*Picea abies*	*Deschampia flexuosa*	5.8	890	6.5	19	45	1000	5
6	Belgium	50°18′N 6°00’E	450	*Picea abies*	mosses	7	1000	4.5	27	60–90	200	7
7	Germany	50°58′N 13°38′E	380	*Picea abies*	*Deschampia flexuosa*	7.5	820	5	28	105	650	11
8	Belgium	51°18′N 4°31′E	10	*Pinus sylvestris*	herbs	10	750	3	22	70	540	7

Table 1.1 *continued.*

No.	Site	Position	Elevation (m a.s.l.)	Overstory species	Understory species	Mean T (°C)	Precipitation (mm)	LAI ($m^2\ m^{-2}$)	Canopy height (m)	Stand age (years)	Density ($n\ ha^{-1}$)	Wood increment ($m^3\ ha^{-1}\ yr^{-1}$)
9	Netherlands	52°10′N 5°44′E	25	*Pinus sylvestris*	*Deschampia flexuosa*	12	800	3	15	100	360	6.3
10	Denmark	55°29′N 11°38′E	40	*Fagus sylvatica*	herbs, cf. *Anemone*	8	600	5	25	80	430	no data
11	U. K.	56°37′N 3°48′E	340	*Picea sitchensis*	mosses	8	1400	8	6	15	2500	14
12	Sweden	60°05′N 17°28′E	45	*Pinus sylvestris Picea abies*	*Vaccinium*, mosses	5.5	530	5	25	100	600	5
13	Finland	61°51′N 24°17′E	170	*Pinus sylvestris*	*Vaccinium spp.*	3.5	640	3	12	30	2500	10
14	Iceland	63°50′N 20°13′W	78	*Populus trichocarpa*	grass and mosses	3.6	1120	2.5	1	7	10,000	no data
15	Sweden	64°07′N 19°27′E	225	*Picea abies*	—	1	570	2	8	30	2100	2.6

Table 1.2 Characteristics of the forest stands currently being studied by the project MEDEFLU arranged from west to east within the European Mediterranean region. Numbering of sites corresponds to the locations shown in Figure 1.2.

No.	Site	Position	Elevation (m a.s.l.)	Overstory species	Mean T (°C)	Precipitation (mm)	LAI ($m^2\ m^{-2}$)	Soil type	Features
16	Portugal	38°38′N 08°36′W		*Quercus suber* *Quercus rotundifolia*	14	450	2	sandy	managed woodland
17	Spain	38°46′N 00°15′E		*Pinus halepensis* *Quercus coccifera*	14	505	1	calcareous	macchia after fire
18	Spain	39°20′N 00°19′E		*Pinus halepensis* *Quercus coccifera*	15.1	422	3-5	sandy	macchia
19	France	43°44′N 03°35′E		*Quercus ilex*	13.4	778	3	calcareous	coppice forest
20	Italy	41°05′N 09°10′E		*Quercus ilex* Mediterranean shrubs	16.8	480	3	calcareous	nature reserve
21	Italy	43°20′N 10°30′E		*Quercus ilex* Mediterranean shrubs	14.5	520	1	sandy	macchia after grazing
22	Italy	42°30′N 11°50′E		*Quercus cerris*	14.8	800	5	clay	coppice forest
23	Greece	38°00′N 22°37′E		*Quercus conferta*	16.6	492	4	calcareous	managed forest

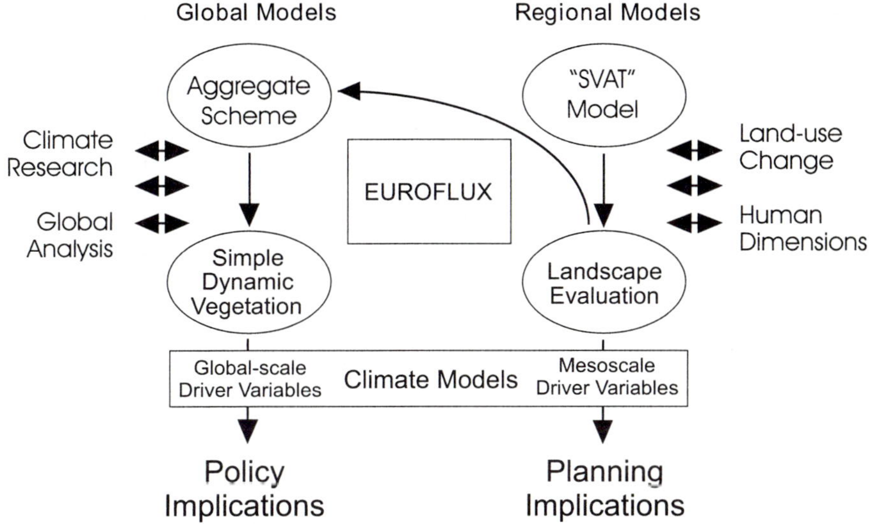

Figure 1.3 Schematic representation of the dual applications for information gained in the projects EUROFLUX and MEDEFLU (i.e., from coordinated comparative studies of forest gas exchange). Long-term observations of H_2O and CO_2 exchange permit development of a new generation of stand level "SVAT" models based on ecosystem physiology and biogeochemistry. Since both H_2O and CO_2 exchange are quantified with these models, they provide important interfacing with growth and production models and link via these to landscape assessments, evaluation of land-use change, and, thus, to the core projects of the IGBP and the International Human Dimensions Programme (IHDP). Such landscape evaluations as well as the flux measurements provide a new and better basis for parameterization of "SVAT" models designed for large-scale applications, i.e., in global models. "Aggregate SVATs" that include landscape characterization in their parameterization may be related abstractly to vegetation dynamics at large scales. Landscape level "SVAT" models for specific elements of the vegetation can provide information useful in resource management and planning when coupled with mesoscale climate models. More abstract "Aggregate SVATs" will provide information relevant to broad policy implications when coupled to global circulation models. The design of a "realistic" but simplified and efficient "SVAT" model for landscape applications is one goal being pursued within the EUROFLUX and MEDEFLU projects. From Tenhunen et al. (1998).

(Figure 1.3). Sound ecosystem models at the landscape scale are essential for the study of socioeconomic problems and for resource management (Cairns 1990; Slocombe 1993; Reynolds and Tenhunen 1996). Since it has already been recognized that GCMs must be improved in order to represent disturbance effects on biosphere processes, such a nested experimental strategy provides information for parameterization of dynamic global vegetation model (DGVM) schemes (Figure 1.3), e.g., concepts developed for specific landscape mosaics may be *correctly* projected to regional, continental, and global scales (Chapters 9 and 15).

A CONCEPTUAL FRAMEWORK FOR LINKING ECOSYSTEM PROCESSES AND ECOPHYSIOLOGY WITH LANDSCAPE FUNCTION

Analyses of ecosystem processes at catchment and landscape scales are extremely important, since such studies are carried out at the largest scale utilized to date for "ground truth" verification of ecosystem-related concepts (Risser 1990). Thus, they provide the only solid basis for formulating ecosystem models at the regional scale. To fit within an integrative scheme such as that described above and which is designed to increase the relevance of ecosystem studies, ecosystem scientists have a responsibility to adapt their methods to (a) focus on spatial heterogeneity, especially as revealed by remote sensing, (b) identify key functional properties and consciously prioritize efforts with respect to two foci, e.g., to examine key properties but to remain alert for detection of "surprise" properties, and (c) to develop a new relationship with ecosystem modelers in order to contribute effectively to the construction of landscape models. The limitations imposed by the dimensions at which ecosystem-level experiments may be conducted as well as the nonacceptability of large-scale ecosystem manipulations force us to rely on "bottom-up" techniques when formulating models that will be useful under conditions of climate change, for example at elevated atmospheric CO_2 levels (Jarvis 1987). Essential products of current ecosystem research must be so-called "phenomenological" or "aggregate models" that are simple, based on detailed process understanding, include appropriate "responsiveness" (Rastetter et al. 1992; Tenhunen, Siegwolf et al. 1994; Reynolds et al. 1996), and allow for effective application at landscape to regional scales.

A conceptual framework for including information from plot-scale "bottom-up" ecophysiological and ecosystem studies of water and carbon balance into landscape and regional-scale assessments may be visualized as follows: Figure 1.4 schematically depicts water fluxes and their regulation within an ecosystem patch of arbitrary spatial extent but with horizontal homogeneity (landscape functional units [LFUs] *sensu* Chapters 9, 14, and 15). We refer to the relatively simple model represented in the center column of the figure as a "Flux-PROXEL" (flux -process pixel model), emphasizing the need to generalize ecological processes in a pixel format for spatial applications. This Flux-PROXEL describes vegetation/atmosphere as well as vegetation/hydrological coupling in the vertical dimension for water and energy fluxes (CO_2 exchanges are relatively easily included as an extension of the model, e.g., Plate 1.1) and is in many ways equivalent with other "SVAT" (soil-vegetation-atmosphere transfer) models (Lee et al. 1993; Schädlich and Mauser 1996; Chapter 9). However, inherent in the PROXEL is (a) the desire to formalize methods for information transfer from more complex representations of the same processes, i.e., to link the PROXEL to "bottom-up hierarchies of process models," (b) the need to develop a focus on new ways for abstraction and simplification of ecological principles in this information transfer, (c) the intent to clearly define appropriate scales for relating "top-down" data inputs to LFUs of real spatial dimensions and defined heterogeneity, and (d) the

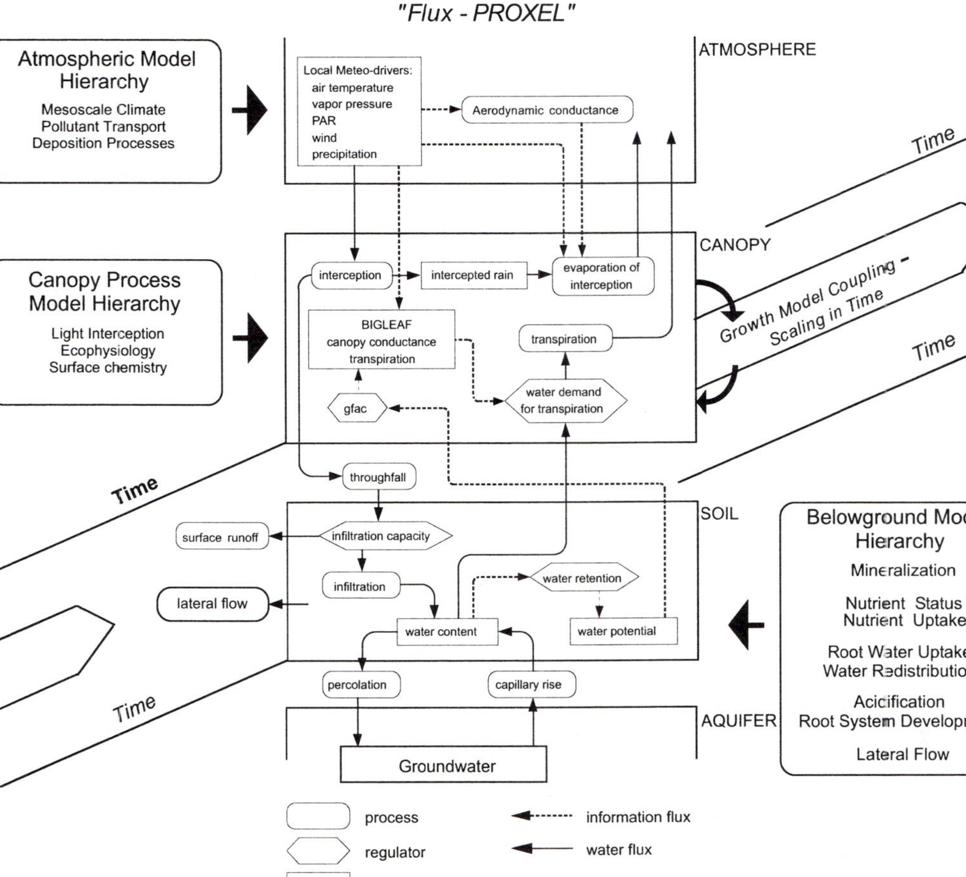

Figure 1.4 Diagram of process coupling, water fluxes, and information flows within a simple Flux-PROXEL (Process Pixel) and suggested relationship of the PROXEL to hierarchies of models for the atmospheric, canopy, and belowground compartments as well as to scaling over time. A hierarchy of models for describing canopy processes is discussed in the text and illustrated in Figure 1.5. BIGLEAF is a single layer representation of canopy processes derived as a simplification and equivalent model from either 3-D or 1-D layered representations. Since it is a derived model, it is based on biological principles rather than empirically measured relationships. gfac is a parameter that conveys information from the roots about water status to the leaves of the canopy (cf. Tenhunen et al. 1989, 1990; Tenhunen, Hanano et al. 1994; Sala and Tenhunen 1996; also Chapter 2).

preservation after maximal simplification of correct phenomenological responsiveness of LFUs to critical environmental variables (example given in Figure 1.5; see also Chapter 14).

One challenge in relating ecosystem research to effective management assessments at landscape and regional scales involves the design of "hierarchies of models" for the atmospheric, canopy, and belowground compartments (boxes indicated in Figure 1.4) that allow us to utilize our best process study information in simple landscape models (see also Chapters 9 and 14). An example demonstrates best what is meant here with the term "hierarchies of models." To quantify the water balance of landscapes found in Central Europe, the water use and gas exchange of coniferous spruce forest, deciduous beech forest, a variety of meadows, and wetland landscape elements must be simulated (conceptually illustrated in Figure 1.5). The up-scaling hierarchy of models for forest stand water and carbon balance includes modules for leaf physiology (Falge et al. 1996), light interception within a 3-D stand representation (Falge et al. 1997; Alsheimer et al. 1998), as well as techniques for simplifying the evaluated canopy function to either a 1-D layered or a single layered BIGLEAF representation (left side of Figure 1.5). We have carried out the parameterization of such models for sites within Germany, sites of EUROFLUX and MEDEFLU, and are working to include further sites as specific flux data become available (see also Plate 1.1). Determination of canopy fluxes via the more complex models, especially in the case of forests, is required due to the nonlinear response of gas exchange to light distribution within complex forest canopies, especially if we attempt to evaluate process rates for future conditions, e.g., with elevated atmospheric CO_2. With the use of such methods, compatibility between single leaf response or process regulation and measured canopy fluxes via eddy covariance techniques may be achieved. The derived simple or aggregate models, when appropriately calibrated, calculate "equivalent fluxes" but also provide an improved potential for extrapolating current knowledge beyond the limits of experience, i.e., at larger spatial scales or for altered climate conditions. In the case of grasslands or wetlands (right side of Figure 1.5), canopies may be more homogeneous in the horizontal dimension. In this case, 3-D information is seldom gathered and modeling concerns only the last two alternatives of the hierarchy for canopy models, e.g., 1-D layered or single layered BIGLEAF representations. Our grassland models have been constructed and parameterized for meadow and wetland sites in Germany, Switzerland, and Austria.

Figure 1.5 Conceptual diagram relating field studies and complex stand flux models to the regulation of vegetation/atmosphere exchange at the landscape level, i.e., illustrating the hierarchies of models for assessing water use and NEE of the vegetation as described in text. 3-D light interception and physiology modules illustrated for spruce forest are discussed in detail by Falge et al. (1996, 1997). 1-D layered models applied to grasslands and wetlands are discussed by Tenhunen, Siegwolf et al. (1994). For explanation of BIGLEAF, see caption to Figure 1.4. The landscape model illustrated in Plate 1.1 uses a C++ program library for manipulating spatial data (Ostendorf and Boyns 1996) to examine the effects of topography and vegetation distribution on water use and carbon exchange.

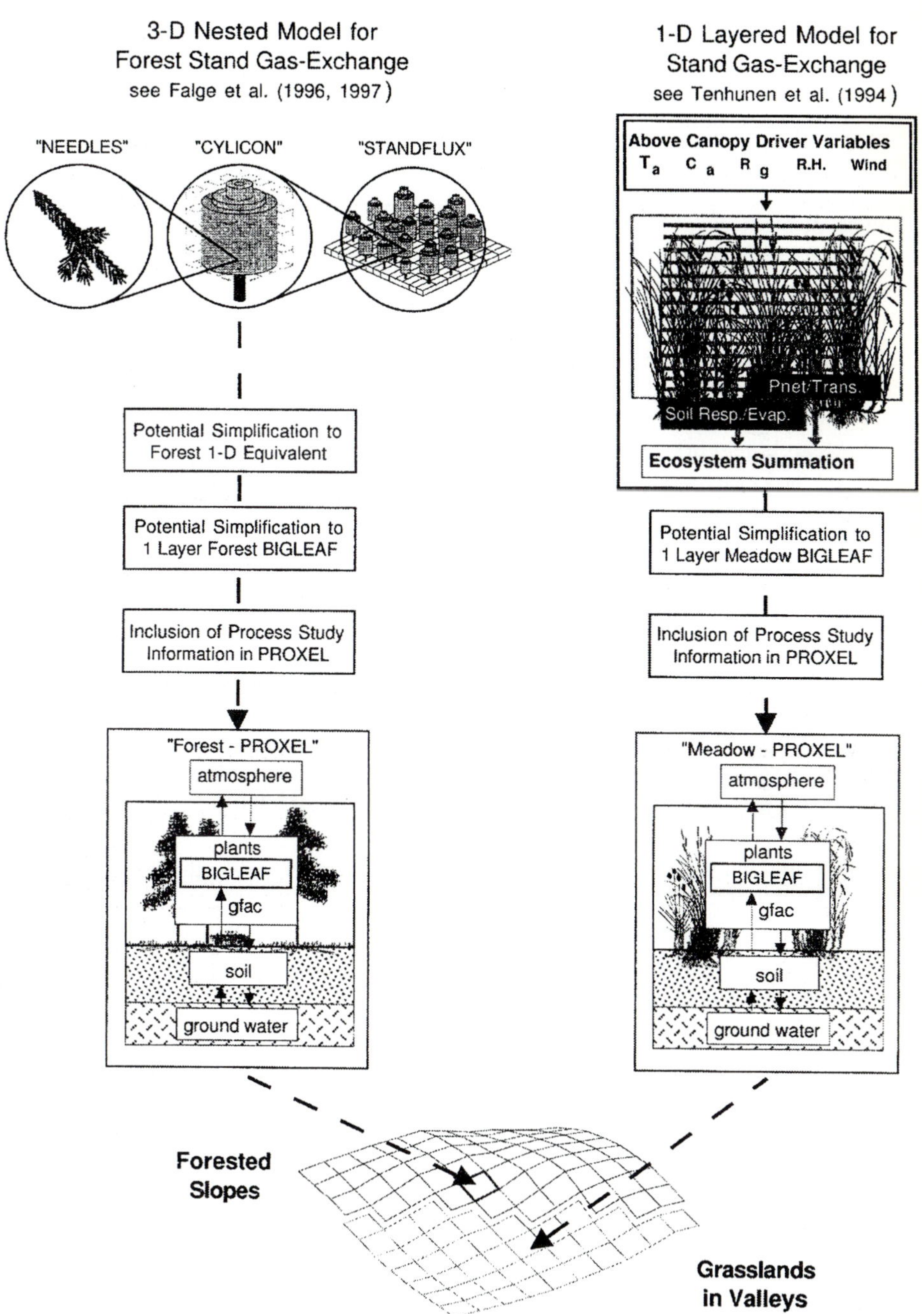
3-D Nested Model for
Forest Stand Gas-Exchange
see Falge et al. (1996, 1997)
"NEEDLES"
"CYLICON"
"STANDFLUX"
Potential Simplification to
Forest 1-D Equivalent
Potential Simplification to
1 Layer Forest BIGLEAF
Inclusion of Process Study
Information in PROXEL
"Forest - PROXEL"
atmosphere
plants
BIGLEAF
gfac
soil
ground water
1-D Layered Model for
Stand Gas-Exchange
see Tenhunen et al. (1994)
Above Canopy Driver Variables
Ta Ca Rg R.H. Wind
Pnet/Trans.
Soil Resp./Evap.
Ecosystem Summation
Potential Simplification to
1 Layer Meadow BIGLEAF
Inclusion of Process Study
Information in PROXEL
"Meadow - PROXEL"
atmosphere
plants
BIGLEAF
gfac
soil
ground water
Forested
Slopes
Grasslands
in Valleys
Linkage with Landscape Models

Important to the theme of this book is that the information from typical ecophysiological and ecosystem field studies is simplified and obtained in parallel form for a variety of LFUs within the PROXEL carbon and water balance simulator (lower part of figure), which is then used together with GIS and the spatial distribution of climate to estimate landscape gas exchange (see also Chapter 16). Results obtained in this manner with a prototype landscape model linking the canopies of 10 × 10 m pixels for six landscape elements at Monte Bondone in Trento, Italy, to the variation in radiation due to slope and aspect are illustrated in Plate 1.1. Correct PROXEL parameterization depends in this case on organized field studies of another European Union project, ECOMONT, which is assessing the effects of land-use changes in composite landscapes on water balance and vegetation/atmospheric exchanges (Cernusca et al. 1998). Broader applications in order to develop regional views on process regulation require an even more extensive information exchange and the development of both formal and informal research networks (see also Chapter 17). Interdisciplinary long-term collaborations are the key to understanding shifts in landscape function associated with land-use change at regional and continental scales.

A Flux-PROXEL (Figure 1.4) is undoubtedly one of the easiest PROXEL modules to design. Nevertheless, water fluxes may be quickly linked to transfer of other compounds both in gaseous and dissolved forms as suggested in Figure 1.5. Thus, specific applications of even a Flux-PROXEL require coupling of the PROXEL to mesoscale climate simulators or coupling with atmospheric transport models (Chapters 6, 8, and 9), i.e., require a closer collaboration between ecologists and atmospheric scientists. Models for belowground processes currently provide the greatest problems in landscape approaches, since budgets for carbon, nutrients, and water must be estimated for several soil layers, distribution of root activities is critical but mostly unknown at this scale, decomposition and mineralization are influenced by many populations of organisms, and lateral flows of water and dissolved substances are subject to a high degree of belowground heterogeneity. Improvement of our knowledge in these areas and the development of robust pixel models for landscape studies will require an intensified cooperation between ecologists and hydrologists (Chapters 10, 11, and 13). An additional challenge is to reconcile the PROXEL approach with identification of LFUs via remote sensing (Chapters 2–5).

While the Flux-PROXEL has a considerable value of its own, PROXEL-based assessments must also consider time-dependent changes in ecosystem structure and function (diagonal arrow in Figure 1.4) over time periods that can be clearly related to regional-scale environmental problems (Figure 1.1; see also Chapters 2, 9, 14, and 15). Water status and fluxes determine stand microclimate, N-mineralization, nutrient availability, nutrient uptake, emission of trace gases (Chapter 8), soil chemical equilibria, and transport of materials to groundwater (Chapter 13). As a framework for process integration, the Flux-PROXEL represents initial progress from one particular perspective (see also Chapters 7 and 9) toward "fully coupled ecosystem models" (Lauenroth et al. 1993), e.g., those which would predict dynamic changes in forest and crop growth at landscape and regional scales (cf. also SREMs [Simple Response

Ecosystem Models] Pielke et al. 1993; Chapters 6 and 14) or predict the potential for nutrient sequestering or nutrient release to stream systems (Peterjohn and Correll 1984; Wendland et al. 1993; Chapter 12). "Fully coupled ecosystem models" that are specifically designed to summarize our understanding of processes within landscapes will offer new tools for the analysis and solution of regional environmental problems. In this context, an entire family of PROXEL-type models will certainly be used in the future to examine alternatively questions related to water balance, forest growth, pollution damage, critical loads, nutrient disharmony, forest dieback, and socioeconomic impacts.

THE CHALLENGE OF WORKING OUR WAY UP- AND DOWN-SCALE

Due to the ubiquitous influence of humans on land use in all parts of the globe (Turner et al. 1995), the need for dynamic vegetation models that evaluate the vegetation mosaic and, thus, achieve a reasonable representation of the heterogeneity in vegetation/atmosphere exchange and a basis for relating these fluxes to carbon and nutrient balances, i.e., to currencies relevant to human concerns, is clearly recognized (Woodward and Steffen 1996; Turner et al. 1995; Pickett et al. 1994). In this new generation of global, regional, and landscape models, parameterization of the spatial differentiation in ecosystem function must be derived either from remote sensing (Potter et al. 1993; Schädlich and Mauser 1996; Martin and Aber 1997; Chapters 2–5) or for global models by upscaling and simplifying landscape-level estimates that approximate well the corresponding processes occurring at grid square scales (Woodward et al. 1995). Both research efforts focus attention on the understanding of aggregation or process integration within real landscapes. Sound ecosystem models at landscape and regional scales provide the crucial link between land-use change and socioeconomic problems, will aid resource management, and will allow us to test the assumptions of global-scale models (Chapters 16 and 17).

How do we define the appropriate structure of landscapes that should be modeled in order to learn the most from our efforts? How large an area should landscapes encompass that are used to build "fully coupled" ecosystem models sensitive to land-use patterns? While the 2 km^2 study area at Monte Bondone (Plate 1.1) proved adequate for developing techniques to quantify variation in ecosystem water balance, must such methods be applied to watersheds encompassing 100 km^2, 1000 km^2, or 100,000 km^2 before relevant management problems may be addressed? How much detail, how many ecological interactions must be eliminated and ignored, and what changes in pixel size are required in our landscape modeling efforts as we move across these scales? Is it possible to derive a formula that indicates the proper scale for ecosystem and landscape analysis, depending on complexity of ecological interactions involved in particular environmental concerns? These are all unanswered questions that must be addressed to utilize information on ecosystem processes in the context of resource management (see also Chapter 15).

In short, research efforts both at global or continental scale (Figure 1.2) and at the scale of small landscape sections are based on the scientific method of hypothesis testing. At intermediate scales, we have defined neither the type of hypothesis that can be successfully tested nor the appropriate methods for doing so. An analogy for this dilemma is provided by the old story where several blind men attempt to describe concisely an elephant: while the first man bases his description on a "groping and feeling" of the leg, the second examines the body, and a third gains an impression from the trunk of the beast. We can liken these differing views to the perspectives gained on landscape function when viewing it solely from the standpoint of ecosystems, or hydrology, or atmospheric/land-surface interactions. Until better communication and a multidisciplinary appreciation of natural systems is achieved, landscape-oriented approaches will not improve and we will not be able to formulate the most appropriate hypotheses. The quantification of ecosystem function at landscape and larger scales requires a crossing of traditional boundaries between scientific disciplines in a very practical sense. We must question whether the scientific support infrastructure that has developed in recent decades is up to the task of supporting landscape and regional research, or whether it only works to support a reductionist application of the scientific method where complexity is limited and linear relationships between cause and effect can be recognized. Otherwise stated, do existing funding agencies support synthesis or disciplinary survival? There are a number of examples on the side of the synthetic viewpoint, most related to the current IGBP, e.g., IGBP itself and its core projects as part of the ICSU, the IGBP Global Analysis, Integration, and Modeling core project (GAIM), efforts to establish the international FLUXNET Program (Chapter 7; Baldocchi et al. 1996), the U.S. TECO Program, etc. However, we must determine whether barriers to synthesis increase as the scales under consideration approach local relevance and have potentially a higher degree of influence on the activities of institutions established along disciplinary lines.

Thus, it is important to the goals of resource management that we consider (a) how to shift our focus to emphasize landscape approaches, (b) how to develop new experimental designs and technical methods for landscape studies, (c) how to develop new research funding structures, and (d) how to break through old barriers and raise the flag for cooperation and synthesis efforts at the landscape level. A number of suggestions to this end, including new creative use of electronic media, are presented in this book. The Dahlem Workshop entitled "Integrating Hydrology, Ecosystem Dynamics, and Biogeochemistry in Complex Landscapes" is a step in the direction of developing concensus views related to these themes. It was planned in order to produce a status report on those variables thought to be essential for conducting resource management assessments at the landscape scale. While it does not provide answers to all of our questions, it does provide a multiplicity of viewpoints on the phrasing of essential environmental management questions.

ACKNOWLEDGMENTS

Financial support for this Dahlem Workshop was received from the Stiftung Deutsche Klassenlotterie and is gratefully acknowledged. The enthusiastic participation of leaders from the IGBP core projects GCTE and BAHC as well as from the IGBP secretariat in Stockholm was an important stimulus for the overview achieved in submitted chapters. In particular, I would like to thank Prof. H.-J. Bolle for early input on the organization of this meeting and his continued efforts to link global and landscape perspectives. Finally, the director of Dahlem, Prof. K. Roth, and its staff, Ms. Julia Lupp, Caroline Rued-Engel, and others, contributed greatly to the success of our meeting by creating both an exciting and challenging atmosphere and an extremely well-organized meeting and processing of materials. The research reported here was financed in part by the EU projects EUROFLUX (ENV4–CT95–0078), MEDEFLU (ENV4–CT97–0455), ECOMONT (ENV4–CT95–0179), and by BMBF Grant No. BEO 51–0339476A.

REFERENCES

Alsheimer, M., B. Köstner, and J.D. Tenhunen. 1998. Temporal and spatial variation in transpiration of Norway spruce stands within a forested catchment of the Fichtelgebirge, Germany. *Ann. Sci. For.* **55**:103–123.

BAHC. 1993. Biospheric Aspects of the Hydrological Cycle. The Operational Plan. Stockholm: IGBP.

Baldocchi, D., R. Valentini, S. Running, W. Oechel, and R. Dahlman. 1996. Strategies for measuring and modeling carbon dioxide and water vapour fluxes over terrestrial ecosystems. *Glob. Change Biol.* **3**:159–168.

Bates, C.G., and A.J. Henry. 1928. Forest and streamflow experiments at Wagon Wheel Gap, Colorado. U.S. Dept. of Agriculture, Weather Bureau Monthly Weather Review, Suppl. No. 30. Washington, D.C.: GPO.

Bormann, F.H., and G.E. Likens. 1967. Nutrient cycling. *Science* **155**:424–429.

Boussingault, J.B. 1845. Rural Economy. London: Bailliere.

Burger, H. 1934. Einfluß des Waldes auf den Stand der Gewäßer. II. Mitteilung. Der Wasserhaushalt im Sperbel- und Rappengraben von 1915/16 bis 1926/27. *Mitt. Eidg. Anst. Forstl. Versuchswes.* **18**:311–416.

Burger, H. 1945. Der Wasserhaushalt im Valle di Melera von 1934/35 bis 1943/44. *Mitt. Eidg. Anst. Forstl. Versuchswes.* **24**:133–218.

Burger, H. 1954. Einfluß des Waldes auf den Stand der Gewässer. 5. Mitteilung. Der Wasserhaushalt im Sperbel- und Rappengraben von 1942/43 bis 1951/52. *Mitt. Eidg. Anst. Forstl. Versuchswes.* **31**:9–58.

Burke, I.C., W.K. Lauenroth, W.J. Parton, and C.V. Cole. 1994. Interactions of land use and ecosystem structure and function: A case study in the Central Great Plains. In: Integrated Regional Models, ed. P.M. Groffman and G.E. Likens, pp. 79–95. New York: Chapman and Hall.

Cairns, J., Jr. 1990. Lack of theoretical basis for predicting rate and pathways of recovery. *Env. Manag.* **14**:517–526.

Cernusca, A., M. Bahn, C. Chemini, W. Graber, R. Siegwolf, U. Tappeiner, and J. Tenhunen. 1998. ECOMONT: A combined approach of field measurements and process-based modelling for assessing effects of land-use changes in mountain landscapes. Ecol. Mod., in press.

Christensen, N.L., A.M. Bartuska, J.H. Brown, S. Carpenter, C. D'Antonio, R. Francis, J.F. Franklin, J.A. MacMahon, R.F. Noss, D.J. Parsons, C.H. Peterson, M.G. Turner, and R.G. Woodmansee. 1996. The report of the Ecological Society of America committee on the scientific basis for ecosystem management. *Ecol. Appl.* **6**:665–691.

Engler, A. 1919. Untersuchungen über den Einfluß des Waldes auf den Stand der Gewäßer. *Mitt. Eidg. Anst. Forstl. Versuchswes.* **12**:1–626.

Falge, E., W. Graber, R. Siegwolf, and J.D. Tenhunen. 1996. A model of the gas exchange response of *Picea abies* to habitat conditions. *Trees* **10**:277–287.

Falge, E.M., R.J. Ryel, M. Alsheimer, and J.D. Tenhunen. 1997. Effects of stand structure and physiology on forest gas exchange: A simulation study for Norway spruce. *Trees* **11**:436–448.

Forman, R.T.T., and M. Godron. 1986. Landscape Ecology. New York: Wiley.

Jarvis, P.G. 1987. Water and carbon fluxes in ecosystems. In: Potentials and Limitations of Ecosystem Analysis, ed. E.-D. Schulze and H. Zwölfer, pp. 50–67. Ecological Studies 61. Heidelberg: Springer.

Keller, H.M. 1988. European experiences in long-term forest hydrology research. In: Forest Hydrology and Ecology at Coweeta, ed. W.T. Swank and D.A. Crossley, pp. 407–414. Ecological Studies 66. Heidelberg: Springer.

Koch, G.W., P.M. Vitousek, W.L. Steffen, and B.H. Walker. 1995. Terrestrial transects for global change research. *Vegetatio* **121**:53–65.

Lauenroth, W.K., D.L. Urban, D.P. Coffin, W.J. Parton, H.H. Shugart, T.B. Kirchner, and T.M. Smith 1993. Modeling vegetation structure-ecosystem process interactions across sites and ecosystems. *Ecol. Mod.* **67**:49–80.

Lee, T.J, R.A. Pielke, T.G.F. Kittel, and J.F. Weaver. 1993. Atmospheric modeling and its spatial representation of land surface characteristics. In: Environmental Modeling with GIS, ed. M.F. Goodchild, B.O.Parks, and L.T. Steyaert, pp. 108–122. Oxford: Oxford Univ. Press.

Lyell, C. 1835. Principles of Geology, vol. III. London: Murray.

Marsh, G.P. 1864. Man and Nature. New York: Scribner.

Martin, M.E., and J.D. Aber. 1997. High spectral resolution remote sensing of forest canopy lignin, nitrogen, and ecosystem processes. *Ecol. Appl.* **7**:431–443.

Meyer, W.B., and B.L. Turner, II. 1994. Changes in Land Use and Land Cover: A Global Perspective. Cambridge: Cambridge Univ. Press.

Ostendorf, B., and M. Boyns. 1996. Spatial Information System SIS. Documentation for a C++ toolkit for linking environmental models and GIS. Bayreuther Forum Ökologie 26. Bayreuth: Bayreuth Institute for Terrestrial Ecosystem Research.

Peterjohn, W.T., and D.L. Correll. 1984. Nutrient dynamics in an agricultural watershed: Observations on the role of a riparian forest. *Ecology* **65**:1466–1475.

Pickett, S.T.A., I.C. Burke, V.H. Dale, J.R. Gosz, R.G. Lee, S.W. Pacala, and M. Shachak. 1994. Integrated models of forested regions. In: Integrated Regional Models, ed. P.M. Groffman and G.E. Likens, pp. 120–141. New York: Chapman and Hall.

Pielke, R.A., D.S. Schimel, T.J. Lee, T.G.F. Kittel, and X. Zeng. 1993. Atmosphere–terrestrial ecosystem interactions: Implications for coupled modeling. *Ecol. Mod.* **67**:5–18.

Potter, C.S., J.T. Randerson, C.B. Field, P.A.Matson, P.M. Vitousek, H.A. Mooney, and S.A. Klooster. 1993. Terrestrial ecosystem production: A process model based on global satellite and surface data. *Glob. Biogeochem Cyc.* **7**:811–841.

Rastetter, E.B., A.W. King, B.J. Cosby, G.M. Hornberger, R.V. O'Neill, and J.E. Hobbie. 1992. Aggregating fine-scale ecological knowledge to model coarser-scale attributes of ecosystems. *Ecol. Appl.* **2**:55–70.

Reynolds, J.F., and J.D.Tenhunen, eds. 1996. Landscape Function and Disturbance in Arctic Tundra. Ecological Studies 120. Heidelberg: Springer.

Reynolds, J.F., J.D. Tenhunen, P. Leadley, H. Li, D.L. Moorhead, B. Ostendorf, and F.S. Chapin, III. 1996. Patch and landscape models of arctic tundra: Potentials and limitations. In: Landscape Function and Disturbance in Arctic Tundra, ed. J.F. Reynolds and J.D. Tenhunen, pp. 293–324. Ecological Studies 120. Heidelberg: Springer.

Risser, P.G. 1990. Landscape pattern and its effects on energy and nutrient distribution. In: Changing Landscapes: An Ecological Perspective, ed. I.S. Zonneveld and R.T. Forman, pp. 45–56. Heidelberg: Springer.

Risser, P.G., J.R. Karr, and R.T.T. Forman. 1984. Landscape ecology: Directions and approaches. Special Publication 2. Champaign, IL: Illinois Natural History Survey.

Sala, A., and J.D. Tenhunen. 1996. Simulations of canopy net photosynthesis and transpiration in *Quercus ilex L.* under the influence of seasonal drought. *Agr. For. Meteorol.* **78**:203–222.

Schädlich, S., and W. Mauser. 1996. Spatial evapotranspiration calculation on a microscale test site using the GIS-based PROMET-model. HydroGIS 96: Application of geographic information systems in hydrology and water resources management. *IAHS Publ.* **235**: 649 657.

Slocombe, D.S. 1993. Implementing ecosystem-based management. *BioScience* **43**: 612–622.

Steffen, W.L., C. Valentin, R.J. Scholes, X.-S. Zhang, and J.-C. Menaut. 1998. The IGBP Terrestrial Transects. In: Global Change and the Terrestrial Biosphere: Implications for Natural and Managed Ecosystems, A Synthesis of GCTE and Related Research, ed. B. Walker, W. Steffen, J. Canadell, and J. Ingram, pp. 66–87. IGBP Book Series 4. Cambridge: Cambridge Univ. Press.

Swank, W.T., and D.A. Crossley, eds. 1988. Forest Hydrology and Ecology at Coweeta. Ecological Studies 66. Heidelberg: Springer.

Tenhunen, J.D. 1999. Model hierarchies for relating vegetation structure, ecosystem physiology, and plant community distribution to landscape water use. In: ECOMONT: Ecological Effects of Land-use Changes in European Terrestrial Mountain Ecosystems, ed. A. Cernusca, U. Tappiner, and N. Bayfield. Oxford: Blackwell, in press.

Tenhunen, J.D., R. Hanano, M.A.I. Abril, E.W. Weiler, and W. Hartung. 1994. Above- and belowground environmental influences on leaf conductance of *Ceanothus thyrsiflorus* growing in a chaparral environment: Drought response and the role of abscisic acid. *Oecologia* **99**:306–314.

Tenhunen, J.D., J.F. Reynolds, S. Rambal, R. Dougherty, and J. Kummerow. 1989. QUINTA: A physiologically-based growth simulator for drought adapted woody plant species. In: Biomass Production by Fast-Growing Trees, ed. J.S. Pereira and J.J. Landsberg, pp. 135–168. NATO ASI Series, Applied Science 166. Dordrecht: Kluwer Academic.

Tenhunen, J.D., A. Sala Serra, P.C. Harley, R.L. Dougherty, and J.F. Reynolds. 1990. Factors influencing carbon fixation and water use by mediterranean sclerophyll shrubs during summer drought. *Oecologia* **82**:381–393.

Tenhunen, J.D., R. Siegwolf, and S.F. Oberbauer. 1994. Effects of phenology, physiology, and gradients in community composition, structure, and microclimate on tundra ecosystem CO_2 exchange. In: Ecophysiology of Photosynthesis, ed. E.D. Schulze and M.M. Caldwell, pp. 431–460. Ecological Studies Series 100. Heidelberg: Springer.

Tenhunen, J.D., R. Valentini, B. Köstner, R. Zimmermann, and A. Granier. 1998. Variation in forest gas exchange at landscape to continental scales. *Ann. Sci. For.* **55**:1–11.

Turner, B.L., II, W.B. Meyer, and D. Skole. 1994. Global land-use/land-cover change: Towards an integrated study. *Ambio* **1**:91–95.

Turner, B.L., II, D. Skole, S. Sanderson, G. Fischer, F. Louise, and R. Leemans. 1995. Land-use and land-cover change. Science/research plan. IGBP Report 35. Stockholm: IGBP.

Walker, B., and W. Steffen, eds. 1996. Global Change and Terrestrial Ecosystems. IGBP Book Series 2. Cambridge: Cambridge Univ. Press.

Walker, B.H., W.L. Steffen, J. Canadell, and J.S.I. Ingram, eds. 1998. Global Change and the Terrestrial Biosphere: Implications for Natural and Managed Ecosystems. A Synthesis of GCTE and Related Research. IGBP Book Series 4. Cambridge: Cambridge Univ. Press.

Wendland, F., H. Albert, M. Bach, and R. Schmidt. 1993. Atlas zum Nitratstrom in der Bundesrepublik Deutschland. Heidelberg: Springer.

Woodward, F.I., T.M. Smith, and W.R. Emanuel. 1995. A global land primary productivity and phytogeography model. *Glob. Biogeochem. Cyc.* **9**:471–490.

Woodward, F.I., and W.L. Steffen, eds. 1996. Natural disturbances and human land use in dynamic global vegetation models. IGBP Report 38. Stockholm: IGBP.

2

Remote Sensing Requirements to Drive Ecosystem Models at the Landscape and Regional Scale

R.H. WARING[1] and S.W. RUNNING[2]

[1]College of Forestry, Oregon State University, Corvallis, OR 97331, U.S.A.
[2]School of Forestry, University of Montana, Missoula, MT 59812, U.S.A.

ABSTRACT

To evaluate ecosystem response at the landscape and regional scale requires a search for properties that are both rich in information and measurable from space. Although it is only possible to distinguish broad surface features (forests, grasslands, wetlands, agricultural fields, urban areas) at the regional scale, present satellites can monitor seasonal and annual changes in these surface features on a daily basis with 1 km spatial resolution. Additional information on vegetation structure is desirable at the landscape scale, but satellites sensors with finer spatial resolution pass over less frequently, so a suite of satellites would be required to obtain daily coverage.

At the regional scale, remote sensing can provide good estimates of incoming radiation, ambient temperature, and relative humidity. Rain radar and microwave radiometers carried on a newly launched satellite should greatly improve estimates of precipitation, but other important drivers of ecosystem processes, such as wind speed and aerosol transfers, are not yet measurable with present satellite technology. At the landscape level, improved estimates of incoming radiation can be obtained by accounting for differences in slope and aspect. Finer resolution coverage can also provide detailed information on the spatial and seasonal distribution of water in its liquid or solid state. Present remote sensing technology has the additional capability of documenting the state of vegetation phenology, a property important in all ecosystem models. To improve the modeling of ecosystem processes across complex landscapes further, we need remote sensing techniques that provide more direct measures of water stress and biochemical status of vegetation. At the same time, we need to reconfigure ecosystem models so that they can be initialized, driven, and validated more readily with present remote sensing technology.

Integrating Hydrology, Ecosystem Dynamics, and Biogeochemistry in Complex Landscapes
Edited by J.D. Tenhunen and P. Kabat

INTRODUCTION

Over the last decade, much structural and functional detail has been incorporated into water, carbon, and mineral cycling models. As a result, current models can, in principle, account for changes in climate, land use, and natural disturbances (Waring and Running 1998). There is a problem, however, in providing the detailed information required to apply these complicated models. In this chapter we attempt to distill from a variety of ecosystem models those variables that might be monitored from satellites and thus allow the analysis of complex landscapes ranging from small watersheds to the regional scale.

In the following presentation, the driving variables and internal components of water, carbon, and mineral cycling models are first reviewed and then synthesized into a simplified form that can be extrapolated across areas where detailed information on soils, vegetation, and climate are lacking. In redesigning ecosystem models for use in complex landscapes, we must take into account both the capabilities and limits of remote sensing.

Remote sensing technology is admirably designed to provide up-to-date classification of vegetation assemblages and to map their distribution. To a lesser extent, remote sensing provides a measure of canopy structure and the presence of water in vegetation, or on the surface in a liquid or frozen state. With continuous coverage, remote sensing also provides a mean of recording variation in snowmelt, flooding, drought, leaf phenology, and disturbance. Remote sensing is unable to provide direct estimates of gaseous exchanges from ecosystems or any below-ground properties. These "hidden variables" can only be inferred from surface properties and through modeling.

SURFACE METEOROLOGY

Most water, carbon, and mineral cycling models are driven by a common set of five meteorological variables: (1) incoming solar radiation, (2) ambient temperature, (3) humidity, (4) precipitation, and (5) wind speed. At present, an inadequate network of weather stations is available to serve as a basis for extrapolating climatic data in many areas. As an alternative, optical sensors mounted on a series of geostationary platforms provide frequent measures of cloud cover from which incoming solar radiation may be accurately derived. Net radiation, important for solving the energy balance, is calculated based on surface reflectance (albedo) measured during cloud-free periods. With dual-band thermal sensors, air temperature and humidity may also be estimated (Goward et al. 1994). Radar offers a direct means of monitoring precipitation. At present, no technology provides a good estimate of wind speed or rates of atmospheric deposition, which makes the calculation of canopy aerodynamic properties and nutrient budgets difficult.

HYDROLOGIC MODELS

Hydrologic models that incorporate physiological and structural information about the vegetation contain similar features (Figure 2.1). In addition to a common set of meteorological driving variables, the state of the system must be defined. Whether there is standing water, ice, or snow cover is fairly easily determined with imaging radar (Waring et al. 1995). It is more difficult to determine the *amount* of water stored on foliage, in vegetation, litter, and snowpack, and nearly impossible to measure water stored in soils.

To account for physiological limitations on water vapor diffusion out of leaves into the atmosphere (via transpiration) and CO_2 diffusion into leaves (via photosynthesis),

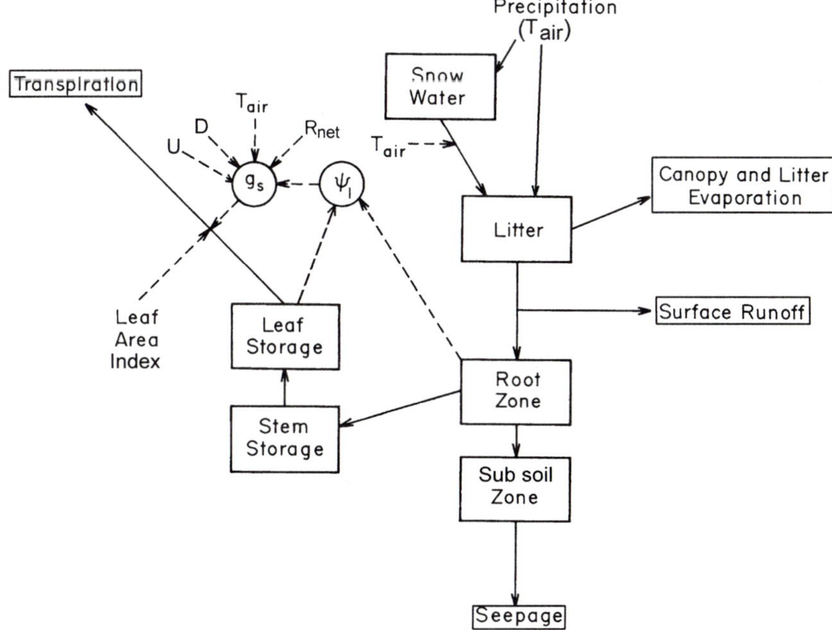

Figure 2.1 General structure of a forest water balance model that accounts for precipitation entering a snowpack, litter, soil surface, and subsoil horizons. Landscape models also include lateral transport through seepage and surface runoff. Water is eventually lost from the system through transpiration, evaporation, runoff, or seepage. Within trees and other woody vegetation, water is stored temporarily in the sapwood of stems and branches, and in the leaves. Evaporation from wet canopies or other surfaces depend on the driving variables (R_{net}), wind speed (U), and vapor pressure deficit (D), the latter is a function of air temperature and relative humidity. Structural variables that affect vapor transfer are associated with height and leaf area index of the vegetation. Calculation of transpiration requires an additional variable, stomatal conductance (g_s), which reflects all the hydraulic resistances in the path between soil and leaves. Maximum g_s is a function of leaf water potential (ψ_l) measured under nontranspiring conditions at night (after Running 1984).

hydrologic and carbon cycling models calculate an *intermediate variable,* average leaf stomatal conductance (g_S), which, when multiplied by leaf area index (LAI) provides a canopy conductance value. An important simplification comes from comparative studies that show the maximum integrated canopy conductance approaches a constant once LAI exceeds 3.0 for broad types of vegetation (Kelliher et al. 1995). No direct means is yet available to monitor (g_S) from space, although subtle shifts in the reflectance spectrum in visible wavelengths that relate to diurnal changes in photosynthetic efficiency also mirror changes in stomatal conductance (Gamon et al. 1992).

Over longer periods, where drought limits the degree to which stomata open during the day, some process models calculate an additional internal variable, leaf water potential (ψ_l), which at night, when plants are not transpiring, is functionally related to the maximum daily g_S. Measurement of predawn ψ_l is therefore an effective measure of soil drought (Running and Coughlan 1988). During sustained periods of drought, the relative water contents of foliage, branches, and woody stems decrease. These reductions in water content alter the dielectric constant of the vegetation, but whether imaging radar can detect these subtle changes from satellites is highly speculative (Waring et al. 1995). At present, sustained periods of drought are inferred from satellite-derived near-infrared and red reflectances, which provide a measure of landscape greenness (and LAI), and from estimates of surface temperature acquired with thermal sensors (Nemani and Running 1989).

CARBON BALANCE MODELS

Carbon balance models that are coupled to water and nutrient cycling operate by predicting carbon uptake and losses through a series of processes, starting with photosynthesis (Figure 2.2). About half the total incoming solar radiation is in the visible portion of the spectrum (400–700 nm) which represents photosynthetically active radiation (PAR). The fraction of PAR absorbed by the canopy is mainly a function of LAI. The upper limits for gross photosynthesis are set by the total amount of PAR absorbed (APAR) and the photosynthetic capacity of the foliage, which is a function of

---

Figure 2.2 A schematic diagram of the major components of advanced carbon balance models. All carbon dioxide uptake is associated with the photosynthetic process and represents gross primary production (GPP). Net assimilation (A) and net canopy exchange (NCE) take into account CO_2 respired by plants during the day and at night, respectively, as a function of maintenance respiration (R_m). Net primary production (NPP) represents the remaining carbon allocated to the synthesis of plant tissue during the year, taking into account additional respiration associated with growth of new tissue (R_s). Carbon allocation is highly dependent on environmental conditions. Plants allocate proportionally more carbon below ground as environmental conditions become less favorable. The detritus pool of dead organic matter is subject to decomposition as a result of the heterotrophic activities of animals and microbes (R_h). Net ecosystem production (NEP) is the net carbon sequestered in biomass and soil organic matter annually. After Waring and Running (1998).

chlorophyll or nitrogen concentrations per square meter of leaf area. Gross primary production (GPP) is further limited by other environmental variables affecting canopy stomatal conductance and the biochemical/biophysical processes associated with photosynthesis, in particular, atmospheric CO_2 concentrations, temperature of the air and soil, vapor pressure deficits, and nitrogen availability (Leuning 1995).

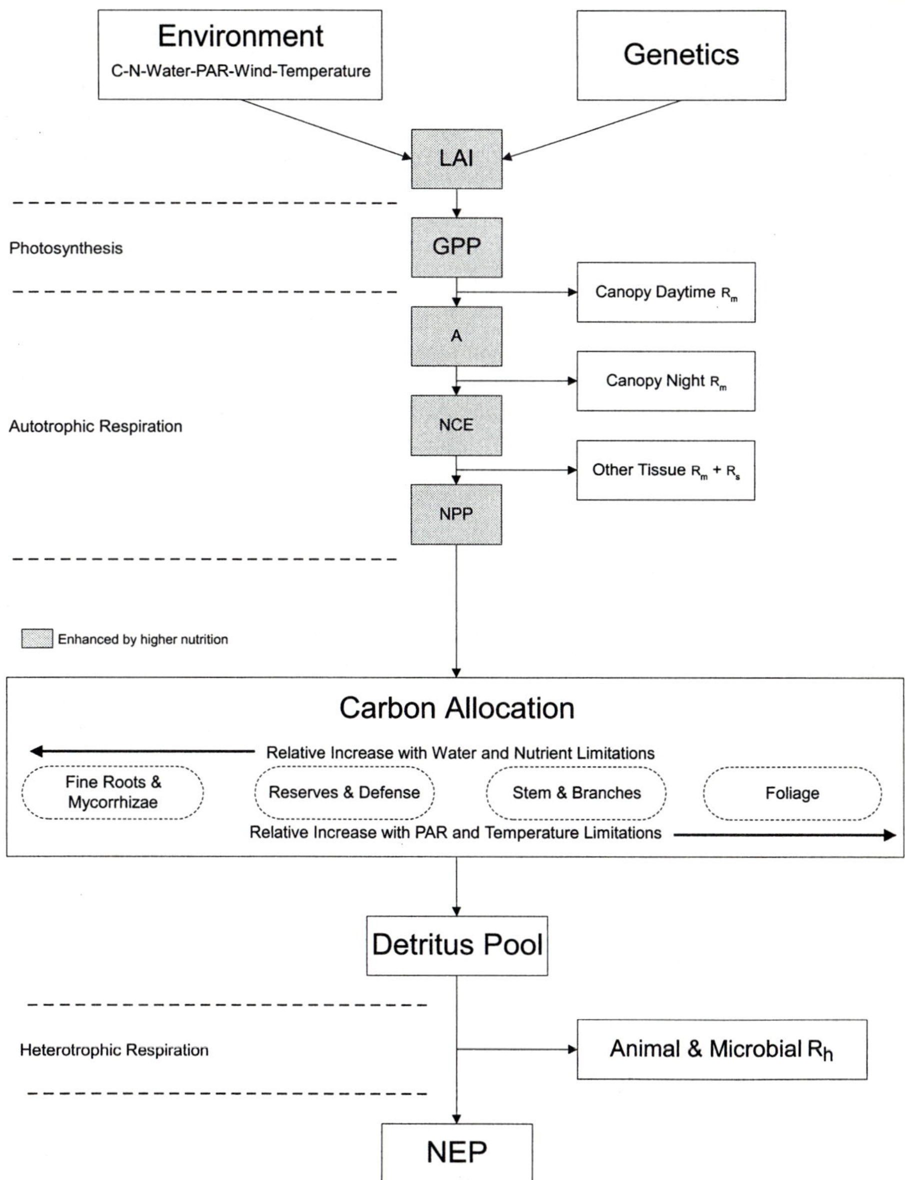

Deducting foliar maintenance respiration during the daylight hours provides an estimate of net assimilation (A). Including canopy metabolic respiration at night yields an estimate of daily net canopy exchange (NCE) for a 24–hr period. Net primary production (NPP) is calculated by accounting for additional autotrophic (plant) losses associated with synthesis of dry matter (R_s) and maintenance of living cells (R_m) throughout each day. NPP is partitioned into various components based on schemes associated with the relative availability of carbon (C) to nitrogen (N), which change with the availability of water and nutrients and absorbed PAR.

Leaf and fine-root turnover are the major contributors to litter on a seasonal basis; however, all biomass components eventually enter the detritus pool. The annual turnover of leaves and roots is correlated with seasonal variation in LAI, specific leaf area, and nitrogen content, factors that also affect the canopy photosynthetic capacity (Reich et al. 1997).

Decomposition of litter and release of CO_2 by heterotrophic (mostly microbial) organisms are a function of substrate quality (C:N or lignin:N ratio), temperature, and moisture conditions (Melillo et al. 1982). Net ecosystem production (NEP) represents the net amount of carbon sequestered annually into biomass or soil humus. In undisturbed systems, NEP is normally a small positive value, representing the residual amount of carbon available after subtracting autotrophic respiration (R_a) and heterotrophic respiration (R_h) from GPP.

Most of the environmental variables driving carbon balance models are similar to those required for hydrologic models with the addition of ambient CO_2 concentrations and available nitrogen. We have good information on changing concentrations in ambient CO_2; however, the only obvious way to assess the available nitrogen from space is to keep track of changes in the canopy N content. The normalized difference vegetation index (NDVI), which is derived from reflectance of red and near-infrared radiation from the Earth's surface, is more indicative of the photosynthetic capacity of the canopy (Myneni et al. 1992) than to total nitrogen content (Yoder et al. 1994). With finer spectral resolution, however, the biochemical constituents of canopies (chlorophyll, protein nitrogen, lignin, and cellulose) can be assessed from space (Matson et al. 1994; Smith and Curran 1995; Martin and Aber 1997).

Rarely is vegetation undisturbed, so NEP may be negative if large amounts of biomass are harvested, consumed by fire, or microbial respiration is increased following changes in land use. Even in protected forests, windstorms transfer biomass to the ground surface. Therefore, it is critical, in predicting NEP, that changes in standing biomass can be monitored accurately. Radar can only reliably measure aboveground biomass up to about 100 Mg ha^{-1}, which is less than 10% of what some productive forests can accumulate (Waring et al. 1995). Newly developed near-infrared laser altimeters offer the promise of increasing the range in biomass that can be monitored from space (Weishampel et al. 1996). This technology also provides a means of estimating changes in canopy height and the vertical distribution of leaf area, important structural variables in many carbon cycling models.

Although belowground allocation of NPP cannot be directly measured, detailed carbon balance analyses suggest that NPP may be a constant fraction of annual GPP (Gifford 1994; Waring et al. 1998). This finding greatly simplifies the calculation of annual NPP, and if accurate measures of aboveground growth are available, the remaining fraction of NPP must go into root growth (Landsberg and Waring 1997).

Litter decomposition is another process than cannot be directly assessed from space, but may be estimated with accurate measurement of annual leaf litterfall. From studies in different climatic regions, Landsberg and Gower (1997) showed that the mean residence time (MRT), defined as the ratio of pool size to the rate that material is added or removed from it, per unit of time, can be estimated from the ratio of forest floor mass to annual litterfall, once forest floor accumulation stabilizes (Figure 2.3). A prediction of MRT under steady-state conditions can be derived from the relation depicted in Figure 2.3, purely as a function of annual litterfall (Waring and Running 1998).

The mean residence time of forest floor litter is an important ecosystem property that is likely to change substantially with climatic variation and with the type of vegetation present. Improvements in the accuracy by which seasonal changes in LAI are monitored from space, along with supplemental data on specific leaf mass would provide an opportunity to assess MRT across a wide range of spatial and temporal scales.

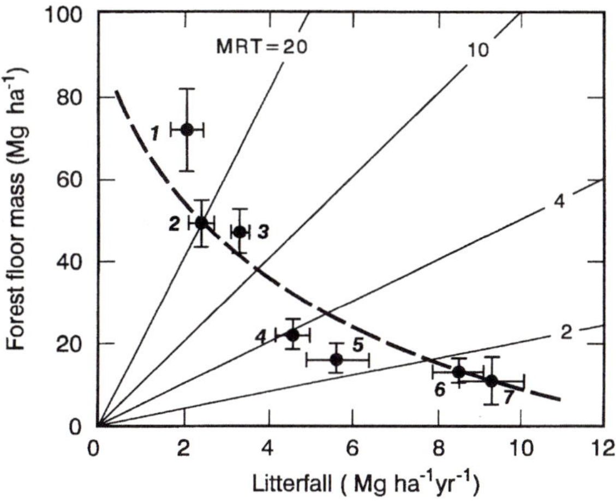

Figure 2.3 Average forest floor mass plotted against litterfall for (1) boreal needle-leaved evergreen (n = 16), (2) boreal broad-leaved deciduous (n = 7), (3) temperate needle-leaved evergreen (n = 73), (4) temperate broad-leaved evergreen (n = 11), (5) temperate broad-leaved deciduous (n = 2), (6) tropical broad-leaved evergreens (n = 31), and (7) tropical broad-leaved deciduous forests (n = 2). Assuming the forest floor is in steady state, the average mean residence time (MRT, years) can be calculated as forest floor mass/litterfall mass. The straight lines indicate these values. The curve line is fit to the mean values and is the basis for deriving a related equation that predicts MRT from equilibrium annual litterfall. From Landsberg and Gower (1997).

MINERAL CYCLING MODELS

The major processes that operate in the cycling of minerals through ecosystems are presented in Figure 2.4. Nutrients are deposited on the landscape in wet and dry fall or weathered from soils and rock. Plants modify the cycling of many elements through their selective uptake, internal redistribution, and the form of detritus returned

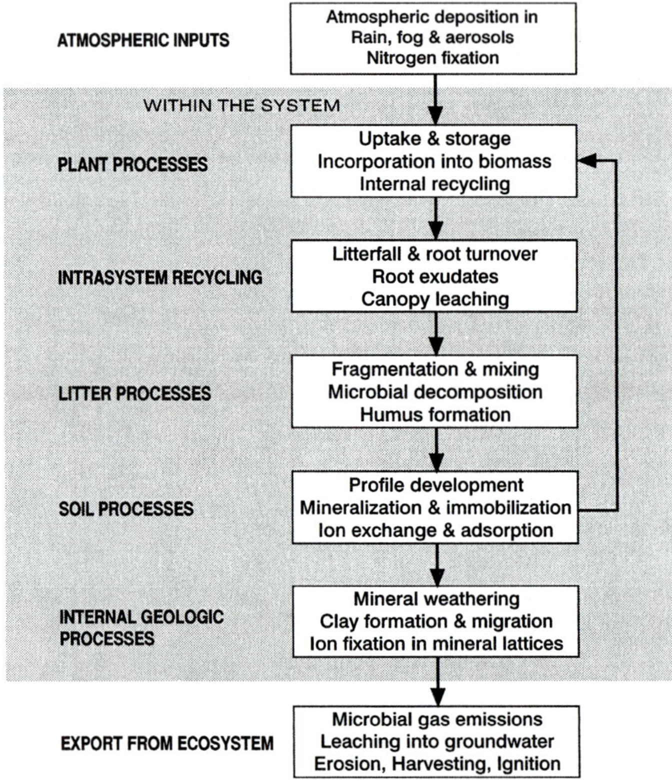

Figure 2.4 Schematic of a mineral cycling model. Atmospheric inputs of nutrients vary, depending on distance from sources, use of fertilizers, presence of nitrogen-fixing organisms within the ecosystem, and the absorbing capacity of the vegetation. Weathering of minerals from rock and soil particles provides an additional source of nutrients. Inorganic forms of nutrients are taken up by vegetation from the soil and incorporated into organic matter. As leaves and other organs die, the organic matter is broken down by the activity of a host of animals, fungi, and bacteria. In the decomposition process, inorganic forms of the elements are released (mineralized). Some mineralized elements are held on soil particles, others are released into solution. Once in solution the elements may be taken up again by plants if not leached below the rooting zone. Some elements are transferred into the groundwater and flow into streams. Large amounts of nutrients may be removed periodically from the system through erosion, harvesting of biomass, or ignition. A small amount is also converted in trace gases that produce haze and modify in other ways the energy balance of the Earth. After Waring and Running (1998).

annually to the soil. Organic matter is metabolically consumed by many soil organisms, so that only a small fraction of the detritus is eventually converted into soil humus. During the decomposition process, minerals are transformed from organic to inorganic forms. Whether the elements are immobilized in microbial biomass, made available on soil exchange sites, adsorbed on to clay surfaces, or fixed permanently into mineral lattices depends on a variety of soil and geologic processes that differ within the soil profile. Some minerals are again taken up by plants and recycled through the system, while others may be lost as gases or as leachate. When disturbed, ecosystems can lose large amounts of elements through erosion, harvesting, or by ignition but the losses from one system may be deposited in another.

Nutrient requirements differ among life forms; however, imbalances in the availability of critical nutrients cause a shift in the way that plants allocate resources and increase plant susceptibility to insects and diseases (Landsberg and Gower 1997; Waring and Running 1998). Leaf litterfall reflects in its elemental composition the relative availability of essential nutrients. Because microbes have less developed cell walls than vascular plants, calcium is normally mineralized faster than the organic substrate losses mass (Figure 2.5). Nutrients in short supply, such as N, P, and S in boreal pine forests, are selectively accumulated in microbial biomass (Staaf and Berg 1982). A remote sensing technology that could analyze the full complement of nutrients sequestered in live and senescence foliage would greatly improve our capabilities of extrapolating mineral cycling models.

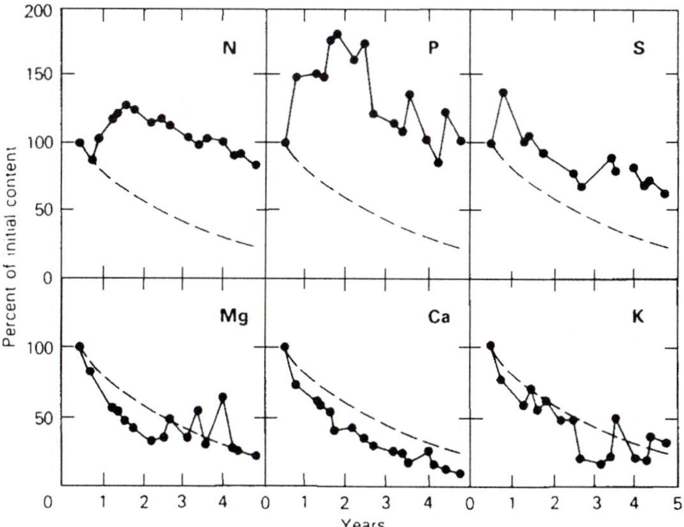

Figure 2.5 Loss of nutrient elements from the litter of Scot pine during the first five years of decomposition. For each nutrient, the solid line indicates the percentage of the initial content remaining at various intervals. The loss of Mg, Ca, and K are more rapid than the disappearance of the organic mass of litter (dashed line), whereas N, P, and S are retained during the period of litter decay. From Staaf and Berg (1982).

During transformations of nitrogen in the soil, a variety of nitrogen gases including NH_3, NO, N_2O, and N_2 are produced as products of microbial activity. The relative proportion of these gases shift, depending to a large extent on the moisture content of soils and the rates that nitrogen is mineralized (Schlesinger 1997; Meixner and Eugster, this volume). There is thus a close link between trace gas emissions of nitrogenous compounds (and sulfur compounds) and the hydrologic cycle. The carbon cycle is also strongly coupled to trace gas emission of methane, particularly in wetlands where CH_4 flux has been shown to increase linearly as a function of daily estimates of net ecosystem production (Schlesinger 1997).

INTEGRATIVE MODELS DRIVEN BY REMOTELY SENSED OBSERVATIONS

A diagram of a simplified but integrated ecosystem model is presented in Figure 2.6. This model has been widely applied in Australia with considerable success (Coops et al. 1998; Landsberg and Coops 1999). Remote sensing of the normalized difference vegetation index provides a measure of seasonal variation in the fraction of radiant

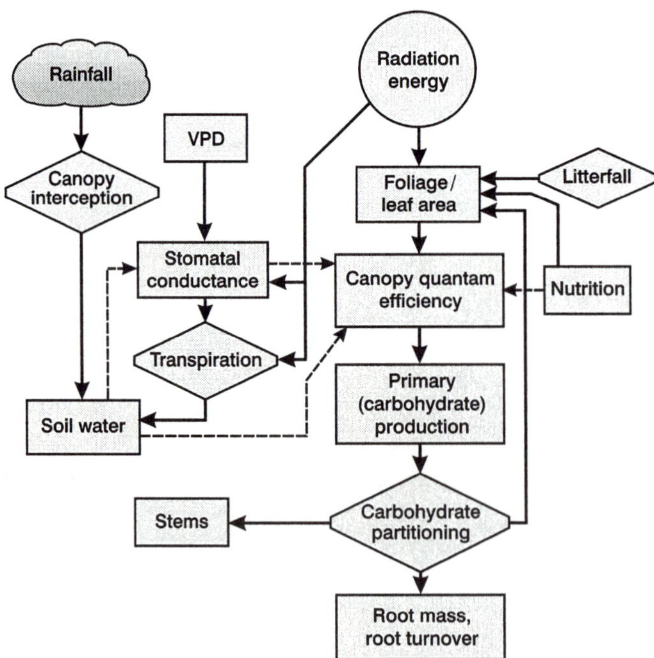

Figure 2.6 Diagram of a simplified ecosystem model that is driven at monthly time steps with remotely sensed information of canopy properties and climatic data. Nutrition and litterfall, together with some soil properties, are required to complete the computation of water, carbon, and mineral balances (Landsberg and Coops 1998). VPD = vapor presseure deficit.

energy absorbed by vegetation. Weather data, combined with an estimate of soil water holding capacity, allows the model to estimate monthly transpiration, evaporation, seepage, and runoff. If sustained drought causes litterfall to increase substantially, a reduction in NDVI occurs, signifying exhaustion of available soil water. If the original estimates of soil water storage capacity are incorrect, they may be adjusted based on the satellite observations of when drought induces massive reduction in LAI (Spanner et al. 1994).

Limitations in the availability of water or subfreezing conditions constrain the rates that water vapor and CO_2 can diffuse through leaf stomata and thus the rates of transpiration and gross photosynthesis. The maximum photosynthetic capacity (canopy quantum efficiency) is related to leaf chlorophyll content and specific leaf mass, variables widely reported in the literature (Gitelson and Merzlyak 1997; Reich et al. 1997) and potentially measurable from space. Important differences in soil fertility are indirectly expressed through the maximum seasonally observed LAI (Spanner et al. 1994).

The model partitions carbohydrates into stems, leafs, and roots based on the extent that soil water storage has been depleted or surface evaporation (associated with vapor presseure deficit constraints on photosynthesis) restricts nutrient uptake to lower, less fertile soil horizons (Coops et al. 1998). Remote sensing with LIDAR, which can provide accurate estimates of aboveground growth, will play a critical role testing the generality of such integrative models across diverse landscapes.

Alternative tests of linked ecosystem models are possible in areas where diverse landscapes include gauged streams. In such cases, ecosystem models must be linked to take into account lateral flow of water and nutrients (Wood, this volume). For landscape analysis to proceed efficiently, it is necessary to combine areas with similar physiography (hillslope partitioning; Figure 2.7). Seasonal changes in meteorological conditions and soil properties combine to affect the relative "wetness" and rate that water is laterally transported down slope. This information, when combined with knowledge of the spatial distribution of LAI provide a means of predicting streamflow, along with other components of water, carbon, and nutrient cycles (Figure 2.8; Nemani et al. 1993).

SUMMARY

Table 2.1 presents a list of variables that we seek to monitor from space. The majority of meteorological driving variables are acquired already from satellite-derived measurements; however, the data sets need to be well documented and more broadly distributed. Structural features such as LAI, canopy height, and standing biomass are also derivable from satellites, particular those equipped with LIDAR. Monitoring the presence of surface water, ice, and snow with optical or radar sensors can aid in extrapolating model predictions of trace gas emissions, snowmelt, and soil drainage. Although canopy biochemistry, nutrient balance, and stomatal conductance are important internal variables in many models, only well-calibrated, fine-resolution spectrometers

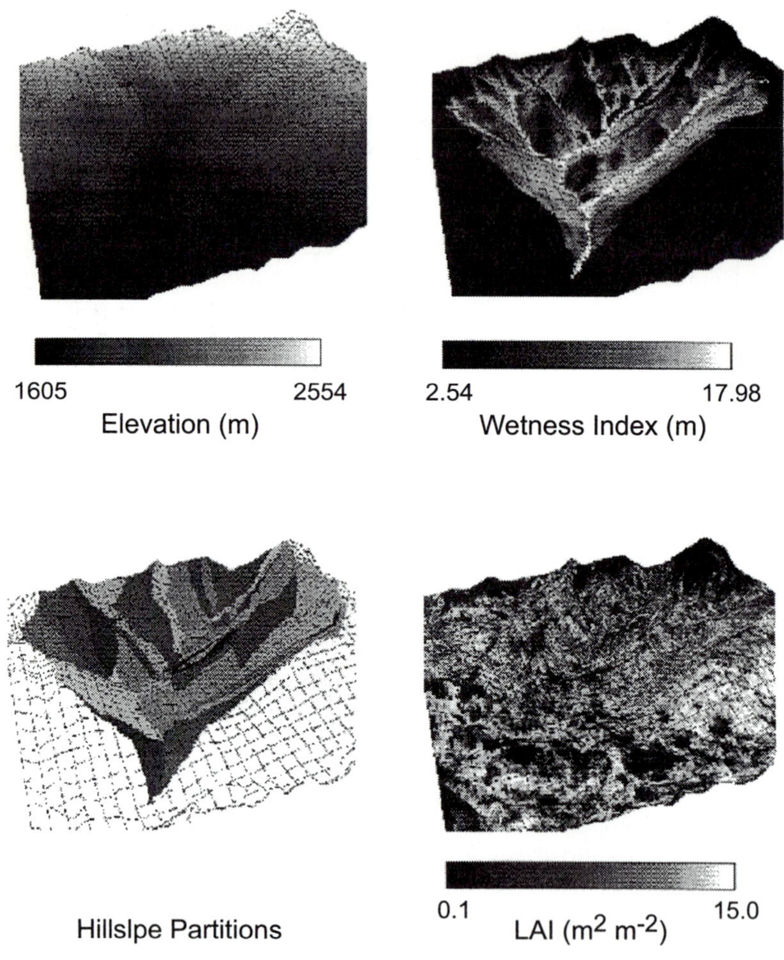

Figure 2.7 To apply ecosystem models to complex landscapes it is necessary to overlay a digital elevation map (upper left) with other data layers: hillslope partiions, a wetness index indicating the transfer rates of water down slope, and the spatial distribution of leaf area index (seasonally). Models can then predict streamflow and other variables that integrate multiple interactions across complex landscapes. After Waring and Running (1998).

(passive or laser powered) would appear to have the capacity to monitors these properties.

The remote sensing capability most useful for regional biogeochemical analyses clearly requires a combination of spatial, temporal, and spectral qualities. Ideally, we should like a geostationary satellite positioned above each region of interest. From such a platform we could obtain surface meteorological data diurnally. In addition, an

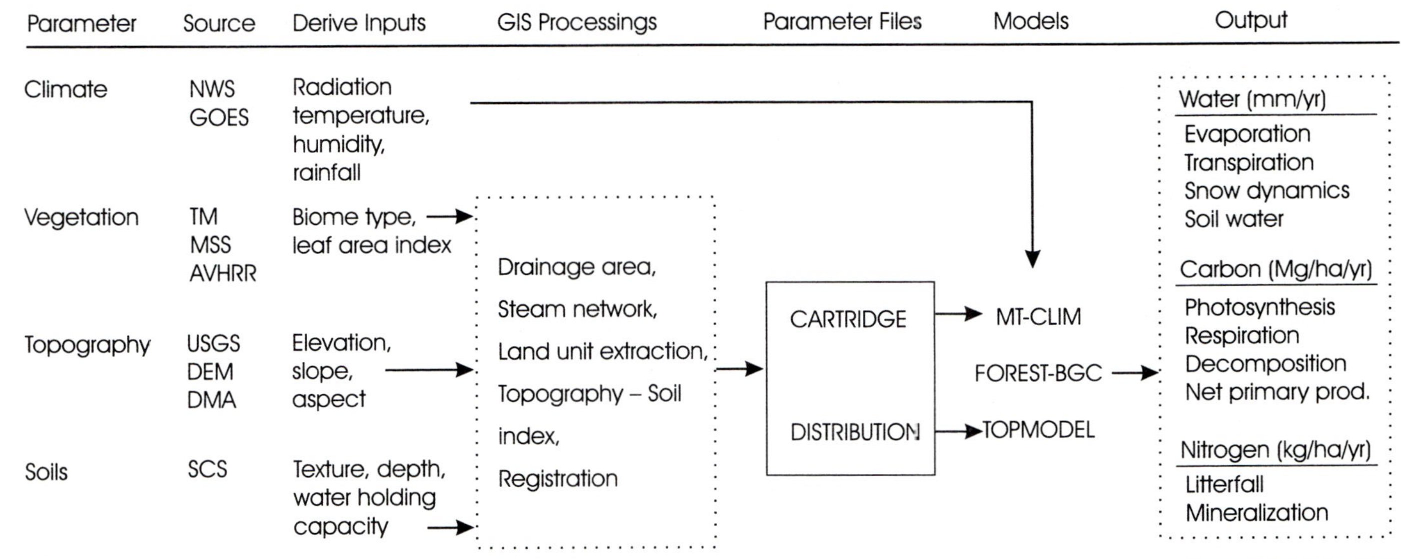

NWS: National Weather Service
GOES: Geostation Operational Environmental Satellite
TM: Lansat/Thematic Mapper
MSS: Landsat/Multispectral Scanner
AVHRR: NOAA/Advanced Very High Resolution Radiometer
USGS: United States Geological Survey
DEM: Digital Elevation Model
DMA: Digital Elevation Model
SCS: Soil Conservation Survey
CARTRIDGE: Land unit parameterization
DISTRIBUTION: Within land unit parameterization
MT-CLIM: Mountain Microclimate Simulator
FOREST-BGC: Forest Ecosystem Simulator
TOPMODEL: Hydrologic Routing Simulator

Figure 2.8 Integration of data layers of physical and biological characteristics of a landscape with process models to yield key ecosystem responses for the Regional Ecosystem Simulation System (RESSys). After Nemani et al. (1993) and Waring and Running (1998).

Table 2.1 Desired variables for extrapolating models predictions across landscapes.

Class	Variables
1. Meteorological	Solar radiation, air temperature, humidity, precipitation and (wind speed). Atmospheric composition and chemical deposition.
2. Physical	Topography: soil texture, fertility and depth. Vegetation height, density, leaf area, standing and fallen biomass. Surface water, ice, and snow.
3. Physiological	Canopy nutrient balance, lignin and nitrogen concentrations, stomatal conductance, photosynthetic capacity, water potential or content.

array of sensors capable of providing 10 m or better spatial resolution would be required on each platform to discriminate complex patterns in topography and vegetation. Spectral quality would need to be sufficient to separate plant functional units and to provide information on canopy water and biochemical status. Ideally, each geostationary satellite would provide data in real time in an interpretable form. Further, they would be cross-calibrated and periodically replaced to assure a continuous high quality data base for study of interannual variability of regional biogeochemistry associated with climatic and anthropogenic change.

A major advancement in extrapolating ecosystem models across complex landscapes will require a change in modeling perspective. We must strive to incorporate remote sensing data to drive, initialize, and validate our models with the technology available today so that we are in a position to take full advantage of future advances in remote sensing.

REFERENCES

Coops, N.C., R.H. Waring, and J.J. Landsberg. 1998. Assessing forest productivity in Australia and New Zealand using a physiologically-based model driven with averaged monthly weather data and satellite derived estimates of canopy photosynthetic capacity. *For. Ecol. Manag.* **104**:113–127.

Gamon, J.A., J. Penuelas, and C.B. Field. 1992. A narrow-waveband spectral index that tracks diurnal changes in photosynthetic efficiency. *Remote Sens. Env.* **41**:35–44.

Gifford, R.M. 1994. The global carbon cycle: A viewpoint on the missing sink. *Austral. J. Plant Physiol.* **21**:1–15.

Gitelson, A.A., and M.N. Merzlyak. 1997. Remote estimation of chlorophyll concentration in higher plant leaves. *Intl. J. Remote Sens.* **18**:2691–2697.

Goward, S.N., R.H. Waring, D.G. Dye, and J. Yang. 1994. Ecological remote sensing at Otter: Satellite macroscale observations. *Ecol. Appl.* **4**:322–343.

Kelliher, F.M., R. Leuning, M.R. Raupach, and E.-D. Schulze. 1995. Maximum conductances for evaporation from global vegetation types. *Agr. For. Meteorol.* **73**:1–16.

Landsberg, J.J., and N.C. Coops. 1999. Modelling forest productivity across large areas and long periods. *Nat. Res. Mod.*, in press.

Landsberg, J.J., and S.T. Gower. 1997. Applications of Physiological Ecology to Forest Production. San Diego: Academic.

Landsberg, J.J., and R.H. Waring. 1997. A generalised model of forest productivity using simplified concepts of radiation-use efficiency, carbon balance and partitioning. *For. Ecol. Manag.* **95**:209–228.

Leuning, R. 1995. A critical appraisal of a combined stomatal-photosynthesis model for C3 plants. *Plant Cell Env.* **18**:339–355.

Martin, M.E., and J.D. Aber. 1997. High spectral resolution remote sensing of forest canopy lignin, nitrogen, and ecosystem processes. *Ecol. Appl.* **7**:431–443.

Matson, P.A., L. Johnson, C. Billow, J. Miller, and R. Pu. 1994. Seasonal patterns and remote spectral estimation of canopy chemistry across the Oregon transect. *Ecol. Appl.* **4**: 280–298.

Melillo, J.M., J.D. Aber, and J.F. Muratore. 1982. Nitrogen and lignin control of hardwood leaf litter decomposition dynamics. *Ecology* **63**:621–626.

Myneni, R.B., B.D. Ganapol, and G. Asrar. 1992. Remote sensing of vegetation canopy photosynthetic and stomatal conductance efficiencies. *Remote Sens. Env.* **42**:217–238.

Nemani, R., and S.W. Running. 1989. Estimation of regional surface resistance to evapotranspiration from NDVI and thermal infrared AVHRR data. *J. Appl. Meteorol.* **28**:276–284.

Nemani, R., S.W. Running, L.E. Band, and D.L. Peterson. 1993. Regional hydrological simulation system: An illustration of the integration of ecosystem models in a GIS. In: Environmental Modeling with GIS, ed. M.F. Goodchild, R.O. Parks, and T. Stevaert, pp. 296–304. New York: Oxford Univ. Press.

Reich, P.B., M.B. Walters, and D.S. Ellsworth. 1997. From tropics to tundra: Global convergence in plant functioning. *Proc. Natl. Acad. Sci. USA* **94**:13,730–13,734.

Running, S.W. 1984. Documentation and preliminary validation of H20TRANS and DAYTRANS, two models for predicting transpiration and water stress in western coniferous forests. USDA Rocky Mountain Forest and Range Experiment Station Research Paper RM–252. Fort Collins, CO.

Running, S.W., and J.C. Coughlan. 1988. A general model of forest ecosystem processes for regional applications. I. Hydrologic balance, canopy gas exchange and primary production processes. *Ecol. Mod.* **42**:125–154.

Schlesinger, W.H. 1997. Biogeochemistry: An Analysis of Global Change (2d ed.). San Diego: Academic.

Smith, G.M., and P.J. Curran. 1995. The estimation of foliar biochemical content of a slash pine canopy from AVIRIS imagery. *Can. J. Remote Sens.* **21**:234–244.

Spanner, M., L. Johnson, J. Miller, R. McCreight, J. Freemantle, J. Runyon, and P. Gong. 1994. Remote sensing of seasonal leaf area index across the Oregon transect. *Ecol. Appl.* **4**:258–271.

Staaf, H., and B. Berg. 1982. Accumulation and release of plant nutrients in decomposing Scots pine needle litter. Long-term decomposition in a Scots pine forest II. *Can. J. Bot.* **60**:1561–1568.

Waring, R.H., J.J. Landsberg, and M. Williams. 1998. Net primary production of forests: A constant fraction of gross primary production? *Tree Physiol.* **18**:129–134.

Waring, R.H., and S.W. Running. 1998. Forest Ecosystems: Analysis at Multiple Scales. San Diego: Academic.

Waring, R.H., J. Way, E.R. Hunt, Jr., L. Morrissey, K.J. Ranson, J.F. Weishampel, R. Oren, and S.E. Franklin. 1995. Imaging radar for ecosystem studies. *BioScience* **45**:715–723.

Weishampel, J.F., D.J. Harding, and J.B. Blair. 1996. Remote sensing of forest canopies. *Selbyana* **7**:6–14.

Yoder, B.J., and R.H. Waring. 1994. The normalized difference vegetation index of small Douglas-fir canopies with varying chlorophyll concentrations. *Remote Sens. Env.* **49**:81–91.

3

What Practical Information about Land-surface Function Can Be Determined by Remote Sensing? Where Do We Stand?

S.D. PRINCE

Laboratory for Global Remote Sensing Studies, Geography Dept., University of Maryland, College Park, MD 20742–8225, U.S.A.

ABSTRACT

The attitude of the land-surface modeling community to remote sensing has often been one of disappointment that ground-based instrument measurements cannot be replaced. In fact, remote sensing replaces no widely used, ground measurement technique except, perhaps, some types of mapping; rather the measurements should be regarded as different and complementary to field observations. Each technique for the inference of a surface variable must be understood in terms of what it is, an interaction of radiation with a surface. Viewed from this perspective, several remote-sensing techniques (e.g., the fraction of incident photosynthetically active radiation absorbed by a vegetation canopy, FAPAR) emerge as more appropriate to meet the aims of the study of land-surface processes than are traditional measurements used at the field scale — measurements which remotely sensed data have sometimes been expected to mimic (e.g., leaf area index in certain applications, LAI). Currently available, remotely sensed measurements of land-surface properties are reviewed, emphasizing those aspects of the data that are under-utilized, such as the correlation structure of variables in adjacent fields of view. An emerging aspect of remote sensing that has enormous significance for land-surface studies of larger areas is that of data handling. This includes techniques for analysis of very large data sets including atmospheric and other corrections, processing of up to 25 years of continuous data acquisition, and innovative database designs that allow for simultaneous analysis of diverse data types. The rapid expansion of remote-sensing techniques and data archives in recent years has not been matched by greater understanding and use by the land-surface process community, rather the gap seems to have widened. A greater commitment of both communities to work together would be very beneficial at a time when many questions are being asked about the spatial extent of certain processes, their relative importance at a global scale, and the exact locations of sensitive areas.

Integrating Hydrology, Ecosystem Dynamics, and Biogeochemistry in Complex Landscapes
Edited by J.D. Tenhunen and P. Kabat

INTRODUCTION

The stability, repeat measurement capability, and global coverage of instruments mounted on Earth-orbiting and geostationary satellite platforms has led to widespread use of their measurements in studies of land-surface and atmospheric processes. These highly desirable characteristics of Earth observation from satellites outweigh the disadvantages of remote sensing which include (by its very nature) an inability to make direct measurements, interferences caused by the intervening atmosphere, and various biases such as variations in Earth-sensor geometry.

Significant advances in the inference of surface conditions from space have been made over the past ten years, and the improved satellite instruments that are often necessary to implement these developments are rapidly becoming available. At the same time, as new sensors and algorithms for their application are being developed, the archives of data supplied by earlier satellite observatories have continued to accumulate. These archives have provided new opportunities for analysis, most important of which is the capability to observe very large areas, including the entire Earth's surface, over tens of years. These archives themselves present significant technical challenges that require the attention of applications as well as measurement specialists.

TEMPORAL AND SPATIAL DIMENSIONS

Remote sensing is particularly suited to the study of spatial and temporal variation in the hydrology, primary production, and biogeochemistry of large segments of the Earth's surface. Remote sensing allows us to scale upwards owing to the multiple scales of observations and the stability of satellite platforms in space and time. Temporal scales of interest that can be addressed with these data sources include interannual, seasonal, the duration of transient weather systems, and diurnal. Change detection techniques coupled with the stability of measurements from remote-sensing instruments are very powerful and can answer questions about trends and cycles in the Earth's processes. For example, the controversy over large-scale desertification of the Sahel region of Africa has been largely dispelled by the use of spectral vegetation indices computed from the long-term advanced very high resolution radiometer (AVHRR) data record (Tucker et al. 1991; Prince, Brown de Colstoun et al. 1998). El Niño southern oscillation (ENSO) related changes in surface radiation and vegetation have been mapped (Dye 1992).

Nevertheless, remotely sensed data have limitations for longer-term trend detection, especially when these are drawn from multiple instruments (Myneni et al. 1997), the calibrations of which may not be fully established and, especially in the case of the AVHRR, for which changing sun-sensor geometry may have, as yet, poorly understood effects on the inference of surface properties, especially near the poles.

A wide range of spatial scales is represented in satellite observations, whereas field measurements made from the ground surface are typically limited to a few hectares or less. The use of direct measurements of larger areas is only possible, if the site is located in a larger, uniform domain (e.g., via eddy correlation techniques for fluxes). The derivation of fields of meteorological variables such as mean air temperature, rainfall and humidity from observations made in networks of ground measurement stations, often separated by hundreds of kilometers, remains as much an art as a science. This is partly because the variables of interest are poorly defined in relation to the process they will be used to model. For example, many physiological processes are dependent on extreme values of meteorological variables, which may be poorly correlated with means, and spatial interpolation of extremes is less accurate than that of means which, themselves, may be poorly sampled. Problems of deviations of local site conditions from the mean can only partly be studied using networks of ground measurement stations.

Remote sensing has a valuable contribution to make in optimizing the sampling schemes necessary for selecting the spatial locations for field measurements (Prince and Steininger 1999). In regions of complex land surfaces where, for example, small "hot spots" of activity in a landscape (e.g., trace gas emissions from wetlands) significantly affect or even dominate the processes of interest, remote sensing becomes an essential part of the sampling design. Studies of the land surface using satellite sensor data often indicate that physical structure, biodiversity, water, carbon, and nitrogen, for example, may each require a different sampling design.

SATELLITE REMOTE SENSING OF LAND-SURFACE PROPERTIES

Satellite remotely sensed data contribute to measurement and modeling of the land surface in three distinct ways, each of which is considered before turning to issues concerning data processing.

Land-surface Classification

Classification was the first major application of digital, multi-spectral remotely sensed data to be developed, building on a tradition of visual interpretation of aerial photographs, especially false-color infrared photography. The remotely sensed data that are classified are radiances, either the Sun's radiation reflected and measured by remote-sensing instruments ("passive" instruments), or from "active" remote-sensing systems that emit their own radiation such as radar and lidar. Several different wavebands or frequencies are often used at the same time and sometimes different polarizations as well. Each type of radiation is selected because of its specific interactions with materials and structures of interest in the scene. Some of these interactions are discussed in the next section.

Classification techniques are limited to the satellite data types that are available, and therefore explicit physical interactions cannot always be used to select input data sets, beyond those incorporated in the design of the original instrument. Satellite sensor channels are often selected empirically and later prioritized using statistical, covariance methods. Land-cover classification remains an important application of remotely sensed data, but more purpose-oriented methods are being adopted, particularly with the availability of many more types of data. These newer methods include the a priori specification of the scene classes or spectral end-members that are of interest and attempts to derive biophysically meaningful classes, e.g., by regression-tree and regression classification methods in which an explicit dependent variable is identified (Hansen et al. 1996; Prince and Steininger 1999).

Another productive approach has been the use of indices or indicators that can diagnose surface conditions such as albedo*temperature (Menenti and Bastiaanssen 1995), rain use efficiency (RUE = net primary production/rainfall) (Prince, Brown de Colstoun et al. 1998) the relationship between spectral vegetation indices and surface radiant temperature (Nemani et al. 1993), and foliar phenology using temporal profiles of spectral vegetation indices (Menenti et al. 1993). Using a maximum likelihood classification of AVHRR spectral vegetation index (SVI) temporal profiles, several land-cover classifications at scales from 1° × 1° to 1 km × 1 km with global coverage have been produced specifically for global change models (Defries and Townshend 1994; Defries et al. 1995). The future 250 m channel of MODIS is being considered as a means of developing a land cover and land-cover change indicator data set based on multitemporal and multispectral reflectance (Townshend, pers. comm.).

The success of classification in relation to land-surface process studies depends on two issues: (a) the degree of correlation of the classes with the relevant ranges of biophysical variables used in modeling and (b) the practicability of obtaining values of the desired biophysical variables to assign to each class. Clearly the first is much more likely using remotely sensed data closely related to the process of interest and a priori classes or regression-tree approaches, while the second can benefit from the use of remotely sensed data to develop a sampling design. In the HAPEX-Sahel campaign, field measurement sites were used to define the classes, and every grid cell of the entire study domain was assigned to one of the classes (Prince et al. 1997). An optimal field sampling design is an important asset when attempting to assign surface properties to remotely sensed classes and the regression-tree approach with remotely sensed data has proved to be very effective for this at both local and continental scales (Prince and Steininger 1999).

In addition to spatial and temporal classification, remote sensing can be useful in the assessment of the heterogeneity of land-surface classes, defined either by prior classification of remotely sensed data or by other means. Given the spatially comprehensive nature of most remote-sensing instruments, the variability of any class composed of more than a few pixels may be described using simple measures, such as within-class variance. This variance may be expressed in terms of radiances, or in terms of inferred biophysical variables. The advent of sensors with spatial resolutions as fine as 1m, a resolving power 900-fold greater than Landsat, will soon enable the detection and monitoring of larger, individual clumps of vegetation and even individual trees. Fine-scale patterns of land cover, bare ground, and pools of water will be detectable and boundaries will be more accurately measured, but only for small scenes owing to the very large data volumes and limited spatial coverage of the sensors at any one time. Intermediate resolution data have already been analyzed for very large areas, such as the 128 m land-cover maps for Amazonia developed by the Landsat Pathfinder Humid Tropical Deforestation Project (Townshend et al. 1995).

The coarse spatial resolution of satellite data has often been regarded as a disadvantage, although it has the advantage of providing greater spatial coverage. Spectral mixture decomposition techniques have been developed to gain information that would otherwise require data having finer spatial resolution. The spectral reflectance can be modeled as a linear sum of the products of area and reflectance of subpixel components, assuming no interactions. Using this approach it has been possible to unmix the SVIs (used to estimate FAPAR) of fine-scale components of semi-arid vegetation landscapes independent of the varying brightness of bare ground (Hanan et al. 1991). A study of coastal wetlands to determine losses caused by sea level rise more accurately (Rizzo 1996) has shown that significant areas of marsh surface are lost due to the formation of small ponds and creeks within a marsh area. These features are detected using a spectral-mixture decomposition technique. Although the causes of loss may be diverse, once interior ponds begin to form, the surrounding areas become subject to increased erosion. Customizable, global land-cover classifications for use in Earth system studies have been developed using mixture modeling approaches in which 8 km AVHRR Pathfinder data were used to derive characterizations of vegetation as continuous fields indicating proportional cover of important vegetation attributes (Defries et al. 1997). Each pixel was described in terms of the aereal proportions of the different growth forms (woody, herbaceous, or bare surfaces). Users of these classifications may therefore specify any one of a virtually infinite number of combinations of levels of the components, so the final land-cover classification is more in the hands of the modeler rather than the classifier.

Direct Inference of Surface and Atmospheric Biophysical Properties from Remotely Sensed Measurements

A major advance made possible by digital remote sensing over air photograph interpretation was the quantitative inference of surface, biophysical properties. Initially, empirical relationships were derived between measurements and surface properties using statistical comparisons of field and remotely sensed data sets (e.g., Prince 1991b) which have not infrequently, but erroneously, been referred to as "calibrations." The relationships sometimes turned out to be simple, robust, and transferable between sites and sensors, such as that between thermal radiation and surface temperature. In other cases, correlations have to be developed afresh for every new application, as is the case for that between spectral vegetation indices and biomass (e.g., Prince and Astle 1986).

Relationships between remotely sensed radiances and surface variables obtained by inversion of explicit physical models have many advantages over correlation. It is no longer always necessary to derive the parameters of the relationship for each new

Table 3.1 Selected, remotely sensed variables, principles of inference, range, and accuracy of measurement. For a more complete catalog of the more than 45 current satellites, including future firm and tentative missions, instruments complements, and mission details, listed according to type of measurement, see CEOS (1997). For one recent synthesis of remote sensing of the land surface with a modeling emphasis, see Sellers (1995).

Variable	Remote Sensing Technique	Principle	Range/Accuracy	Instruments/Platforms/Archives	Ref.
Surface albedo	Surface reflectance (broad or narrow spectral bands)	Spectral reflectance with known sun/sensor geometry. Spectral and spatial interpolation (using BRDF)	Typical satellite accuracy ± 0.03	METEOSAT, GOES AVHRR, Landsat, SPOT Future: AATSR, MISR, POLDER	1
Incident solar radiation (total and broad bands such as PAR)	Plane, horizontal surface: top of the atmosphere reflectance, surface albedo	Model cloud, aerosol, and surface reflectance	Total short wave at monthly time scale 280 km cells < 9 Wm^{-2}, RMSE = 18 Wm^{-2} PAR (monthly): error < 10%	http://eosweb.larc.nasa.gov http://metosrv2.umd.edu/~srb/	2
	Inclined surfaces: topographic correction	Modeled from spectral reflectance and digital elevation	Hourly in range 0–1000Wm^{-2}, RMSE 163Wm^{-2}	Algorithm available for GOES, METEOSAT	3
Leaf area index (LAI)	Red and near infrared reflectance	Correlation with SVIs. Canopy radiative transfer models	Typical satellite accuracy ± 0.74	Many satellite SVI data sets. Global LAI at 1° in Meeson et al. (1995)	4
Fraction of incident PAR absorbed (FAPAR)	Red and near infrared reflectance	Differential absorption and scattering in red and near infrared (SVIs)	Typical satellite accuracy 10–20%	Landsat, SPOT, AVHRR, SeaWiFS Future: MODIS; Meeson et al. (1995)	5
Phytoplankton	Spectral reflectance	Differential reflection in visible	Chlorophyll concentration ± < 20% aircraft, ± < 35% satellite	CZCS SeaWiFS Future: MODIS	6
Cover	FAPAR and reflectance of bare ground	SVIs and mixture decomposition	Depends on pixel size, typically 10%	Landsat, SPOT, AVHRR Future: MODIS	7
Biomass	C, L, P band synthetic aperture radar (SAR)	Radar backscatter	RMSE 4 kgm^{-2}. Range approx. 2–15 kg m^{-2}	Shuttle Imaging Radar, AIRSAR, JERS	8
	Visible reflectance	Geometric optical modeling of shadows in sparse canopies	Depends on application, typically ± 50%	Landsat, AVHRR	9
	Laser ranging	Lidar reflectance profiles of entire canopy return	Std error 2–30% of mean	SLICER, LVIS on aircraft. VCL-global biomass at 25 m spatial resolution	10

Variable	Remote Sensing Technique	Principle	Range/Accuracy	Instruments/Platforms/Archives	Ref.
Structure	Lidar laser ranging for canopy height only	Only first or last reflected return recorded	20 ± 10 cm	Aircraft laser profilers and scanners	11
	Laser ranging and backscatter analysis for height and internal structure	Full signal recorded from canopy top to ground	Vertical location < 0.5 m	Aircraft e.g., SLICER, LVIS (satellite VCL in 2000)	12
	Visible reflectance	Geometric-optical models	Qualitative	Landsat, SPOT	13
	Angular visible reflectance	BRDF	Can measure several structural variables	Landsat, SPOT, AVHRR Future: MISR	14
	Microwave backscatter	Radar	Qualitativ	RADARSAT, ERS 1/2, JERS	15
Roughness	Laser ranging	Lidar	Vertical structure < 0.5 m	Aircraft (VCL in 2000)	16
	Multispectral reflectance	Land-surface classification	Depends on accuracy of assigned value and intra-class uniformity	Landsat, SPOT	17
Topography	Laser ranging	Lidar	< 0.5 m	Aircraft; Future: VCL	18
	Stereo imagery	Parallax	Base-height ratio 0.45 (± 45°) –0.75 (equator)	SPOT; Future: ASTER	
	Microwave	Radar	< 5 cm	RADARSAT, ERS 1/2	
	Multispectral reflectance	Land-surface classification into topographic zones	Qualitative	Landsat, SPOT	19
Surface temperature	Thermal emission in 1 or 2 bands	Stefan-Boltzmann law with measured or estimated emissivity	Typical ± 3°C	AVHRR, Landsat TM, ATSR Future: MODIS, ASTER	20
	Thermal/microwave emission in many bands	Sounding	Approximately 1°C	TOVS HIRS/MSU	21
Air temperature	Thermal/microwave emission in many bands	Sounding	Approximately 1°C	TOVS HIRS/MSU	22
	SVI and surface temperature	TVX	RMSE 4°C	AVHRR; Future: MODIS	23

Variable	Remote Sensing Technique	Principle	Range/Accuracy	Instruments/Platforms/Archives	Ref.
Surface air humidity	Thermal emission in two bands	Thermal split window	RMSE 10 mb	AVHRR, TOVS	24
	Thermal/microwave emission in many bands	Sounding	RMSE 1–4 mb	TOVS	25
Soil moisture (near surface)	Passive microwave	Variation in emissivity	25%	Aircraft; Future: AMSR	26
	Synthetic aperture radar (SAR) C, X, L bands	Variation in emissivity		RADARSAT, ERS–1/2, JERS	27
	Slope of SVI and surface temperature relationship for adjacent pixels	TVX	Not known	AVHRR	28
Rainfall	Satellite radar	Reflectivity of water droplets	Daily totals	TRIMM latitude ± 35°	29
	Ground radar			Within range of a ground radar installation	30
	Cloud-top temperature	Cold cloud duration Local calibration needed	CV for monthly estimate for Niger 44%	METEOSAT, GOES, GMS, INSAT	31
Pattern; patch size distribution, boundary length, mean interpatch distance	Reflectance	Contextual analysis of adjacent pixel values	Currently 15 m, 1 m expected	SPOT HRV, Landsat TM and MSS	32
Fire	Temperature	Active fires and fire scars	Depends on fire size at satellite overpass time	AVHRR; Future: MODIS	33
	Visible and nir reflectance	Burn scars, low visible reflectance, and SVI	Depends on pixel size	Landsat, SPOT, AVHRR Future: MODIS	
Chemical composition	High spectral resolution reflectance	Absorption spectroscopy for N, lignin, cellulose	N std err 0.18%, lignin 1.43% dry leaf mass; not practical with current satellite sensors	Aircraft AVIRIS data	34

Table 3.1 References

[1]Myneni et al. (1995)	[18]Dubayah et al. (1998)
[2]Dye (1992); Whitlock et al. (1995); Frouin and Pinker (1995)	[19]Sellers et al. (1995)
[3]Dubayah and Loechel (1997)	[20]Prince, Goetz et al (1998)
[4]Gholz et al. (1997)	[21]Lakshmi et al. (1998)
[5]Hanan et al. (1997)	[22]Lakshmi et al. (1998)
[6]Harding et al. (1994)	[23]Prince and Goward (1995)
[7]Yang and Prince (1997)	[24]Prince, Goetz et al (1998)
[8]Waring et al. (1995); Ranson et al. (1997)	[25]Lakshmi et al. (1998)
[9]Nilson and Peterson (1994)	[26]Wang (1995)
[10]Means et al. (1998)	[27]Hall et al. (1995)
[11]Ritchie and Menenti (1997)	[28]Prince and Goward (1995)
[12]Means et al. (1998); Dubayah et al. (1998)	[29]Petty (1995)
[13]Yang and Prince (1997)	[30]NOAA (1990–1992)
[14]Walthall (1997)	[31]Petty (1995)
[15]Ranson et al. (1997)	[32]Skole and Tucker (1993)
[16]Dubayah et al. (1998)	[33]Malingreau et al. (1993); http://modarch.gsfc.nasa.gov/fire_atlas/
[17]Sellers et al. (1995)	[34]Martin and Aber (1997)

application. Moreover, the user is more likely to be aware of potential problems with the relationship, since the physical principles on which the model depends are generally better understood. For example, the use of shadowing models to infer biomass (Yang and Prince 1997) stands in marked contrast to the use of SVIs (Prince and Astle 1986). The relationship between variables inferred by inversion of models and field measurements should be specified by appropriate comparison statistics (e.g., Prince, Goetz et al. 1998), an aspect of applied statistics that is not always adequately addressed in validation studies.

The physical processes that enable surface variables to be inferred from remotely sensed radiances involve passive emission (microwave and thermal) and reflection (specular, lambertian, and non-lambertian) and scattering with selective absorption of specific wavebands, polarization, and shadowing. Active systems such as optical (lidar) and microwave (radar) are also used for ranging. Each of these phenomena may be measured for a single field of view or in a linear or rectangular array of fields of view. The first type of device is referred to as non-imaging, and the second as imaging, since the two-dimensional array of measurements has the properties of an image. The fields of view may be contiguous, overlapping, or separated by areas of the surface that are not sensed so that the remotely sensed measurements are samples of the surface, not continuous. In Table 3.1 some widely used surface variables are listed together with the principal, available techniques for measurement and inference, and the major archives of data.

The first remotely sensed variables to be used were meteorological (Gurney et al. 1993). The discovery that SVIs are linearly related to certain canopy variables (Goward and Dye 1996) led to the incorporation of remotely sensed measurements of canopy properties into physiological models (Tucker and Sellers 1986). These canopy models typically use land-surface maps to specify variables that are effectively static at the temporal scale of the models; however, studies of the impact of vegetation on climate require measurements of more slowly changing properties (Goward and Prince 1995). A number of recent developments in remote sensing, such as radar and lidar, are likely to make useful contributions to the measurement of structure.

The question of the accuracy and precision of most remotely sensed measurements is not easy to answer, for at least three reasons. First, calibrations of satellite measurements using field data inevitably include the error of the field measurement. Second, like any measurement technique, remote sensing measurements are affected by various interfering variables, but these are often more difficult to control in the case of remote sensing (e.g., atmospheric effects). Much greater accuracy is usually achievable under circumstances where the interferences are at least partly controlled. Third, satellite measurements are intrinsically measurements of patches of the land surface, whereas the conventional measurements are generally for points. Thus the two are not measures of the same entity and are often not directly comparable. This latter factor should cause the user of traditional measurements to carefully consider whether a point measurement is relevant and what the implications of using the inappropriate spatial resolution might be.

Indirect Inference of Biophysical Properties of the Surface Using Models Forced by Remotely Sensed Data

More complete analyses of the interaction between radiation and the surface are providing innovative ways of making use of the data contained in remotely sensed observations. Remotely sensed data contain spectral, spatial, and temporal information, all of which need to be considered if the information content is to be fully exploited. For example, the presence of vegetation foliage is the primary determinant of observed surface temperature patterns, and the relationship between the normalized difference vegetation index (NDVI) and surface temperature is diagnostic of aspects of the surface energy balance (Goward and Dye 1996). This relationship is defined by the correlation of the two variables in an adjacent set of pixels, information that is lost if only individual pixels are analyzed. These findings have led to the development of the temperature-vegetation index (TVX) contextual algorithms to infer air temperature from reflected and emitted measurements from AVHRR data. Surface temperature (T_S), air temperature (T_a), atmospheric perceptible water (U), and vapor pressure deficit (D) can now be derived from AVHRR measurements alone using the TVX approach. When compared with ground observations collected during FIFE, HAPEX-Sahel, and BOREAS field measurement campaigns, these results showed that T_a could be retrieved with a RMS error of 3.5° C for a range of 48° C, U with 0.6 cm over a range of 3.6 cm, and D with 10.9 mb over a range of 58 mb (Prince, Goetz et al. 1998).

The inference of net primary production (NPP), a key ecological variable that describes the fixation of atmospheric carbon by plants, and thus measures the potential supply of carbon to humans and animals for food, fuel, and fiber, takes this approach a step further. With the development of functional linkages between satellite observations and biophysical attributes of vegetation, multitemporal studies of NPP can be made using certain types of satellite data (Prince 1991a). Using estimates of the required environmental variables (surface and air temperature, vapor pressure deficit, and soil moisture) from satellite data, energy (incident photosynthetically active radiation and FAPAR) and autotrophic respiration, the global production efficiency model (GLO-PEM) provides NPP estimates at the spatial and temporal scales of the satellite data (typically 1 or 8 km globally and 1–10 d for 1981 to present) (Prince and Goward 1995). GLO-PEM represents the first attempt to infer potential plant production from a radiation absorption model forced with remotely sensed data and, at the same time, to infer the environmental factors that determine the actual production from remotely sensed measurements. The goal is to examine the variations in NPP from year to year associated with land-cover changes, interannual vegetation dynamics, and climate fluctuations. The basis of the model is the production efficiency approach in which the NPP is estimated from the amount of photosynthetically active radiation (PAR) absorbed by the plant canopy (APAR) and an efficiency of conversion of this energy into biomass (Prince 1991a). Results from GLO-PEM give a global, total NPP of 68.97 PgC for 1987. Examination of NPP values for different vegetation classes indicates that the greatest contribution is by broad-leaved evergreen forests (15.95 PgC yr^{-1}), and wooded grasslands (13.61 PgC yr^{-1}); the first mainly because of the high NPP per unit area, and the second because of its large area on a global scale. The volume of data processed to achieve these outputs is very large (~20 Gbytes per year) and the volume of products is an order of magnitude larger. Significant computing technological innovations are used including fast networks, arrays of processors, and task scheduling software (see below).

Comprehensive analysis of remotely sensed variables leads to mechanistic modeling of the interaction of radiation and the land surface. These studies are beginning to reveal the fact that remotely sensed data have sometimes been used in inappropriate ways. The most obvious case concerns the inference of leaf area index (LAI) from SVIs (Goward and Huemmrich 1992). A linear relationship between LAI and SVIs exists at low LAI, but it saturates at approximately an LAI of 4 — well within the range of interest in biophysical studies. The relationship of SVIs with FAPAR does not saturate and can be linear. Thus, the use of remotely sensed SVIs to estimate FAPAR is more logical and subject to less error if it is direct, rather than first estimating LAI from SVIs and then using a radiative transfer model to estimate FAPAR from LAI. The attempt to estimate the traditional variables used in models forced with direct field measurements when they are applied to remotely sensed measurements is often inappropriate. There is a need for changes in models to incorporate the new measures that are available from remote sensing (Gurney et al. 1993), rather than regarding remote sensing as a crude and fundamentally limited means of estimating traditional, but inappropriate variables.

DATA PROCESSING

The development of remote-sensing techniques has led to a rapid rise in expectations in the land-surface hydrology, primary production, and biogeochemistry communities. Whereas analysis of the spectral reflectance of a single pixel, involving geometric and atmospheric correction, was, until recently, the entire focus of investigations (e.g., Hanan et al. 1997), now entire Landsat scenes of 44×10^6 pixels are being processed (Liang and Townshend 1997) and this is only a prelude to entirely machine-based, multiple-scene studies.

An essential step in the use of remotely sensed data is atmospheric correction, which requires information about the composition of the atmosphere to recover surface radiances from top of the atmosphere observations. The atmospheric composition is normally obtained from solar transmissometer measurements in the field. Clearly this is not practical for more than a few ground locations, but the possibility of deriving atmospheric information from top of the atmosphere radiances has been exploited in new, very fast algorithms (Fallah-Adl et al. 1996). Whereas existing, operational correction schemes assume a standard atmosphere with zero or constant aerosol loading and a uniform, lambertian surface, future atmospheric correction of MODIS data will use aerosol and water vapor data derived from other MODIS channels and will also correct for adjacency effects and take into account the directional reflectance properties of the observed surface.

Corrections of top of the atmosphere radiances at a scene-scale for certain surface conditions are also possible. For example, Dubayah (1992) has developed a model that can recover net solar radiation using Landsat thematic mapper (TM) corrected for topographic and atmospheric effects, by combining digital elevation data with TM data sets. Application of this model over the Rio Grande River basin in Colorado, U.S.A., revealed that topographic variability is significant even at annual time scales (Dubayah and van Katwijk 1992). The radiation maps corrected for topographic effects generated from this research provide valuable inputs to hydroclimatic models.

Long-term data sets have a special value for studies of interannual variation, but they raise special challenges particularly when multiple sensors have been in operation. For example, four different sensors have contributed to the afternoon data AVHRR record since 1981. Not only are changing calibrations of the sensors a problem, but added complexity arises, for example with the widely used AVHRR data set, owing to temporal drift in equatorial crossing time as satellite platforms spend longer in orbit. Fundamental issues also arise, for example, in compositing techniques which are used to minimize atmospheric and geometric problems for optical and thermal data (Prince and Goward 1996). Research programs have been instituted to address these problems but advances will continue to be made in atmospheric physics and in algorithms for retrieval of variables, and so it is clear that repeated reprocessing of these very large data sets (approximately 0.25 petabytes for current AVHRR archive) will be necessary for the foreseeable future. AVHRR data have already been reprocessed more than 6 times (Townshend 1992). The Pathfinder programs (Smith et al. 1997) are particularly valuable in this respect.

The use of very large data sets having high spatial and temporal resolutions, in complex, integrated modeling (e.g., REHSSys Coughlan and Dungan [1996]), GLO-PEM Prince and Goward [1995]), means that data input/output and computation time are limiting factors in the application of remotely sensed data. These computational restrictions recall the early years of satellite remote sensing, when specially designed image-processing computers were necessary, although now the cause is the increase in spectral, spatial, and temporal dimensions. Several orders of magnitude improvement in execution times are possible using advanced algorithms, parallel processing, and new hardware configurations. As an illustration, a highly optimized global bidirectional reflectance distribution function (BRDF) retrieval algorithm has been developed and applied to the Pathfinder AVHRR land data covering a period of four years (1983 through 1986), using three widely different models (Fallah-Adl et al. 1996). The global BRDF retrieval of 30 GB took less than four minutes on a 16-node IBM SP2 array, including I/O time (Kalluri et al. 1997).

As remote-sensing data sets expand and large regions are being studied, it has become desirable to work with more than a single segment of an orbit or "scene." For example, the spatial extent and arrangement of forest disturbance is an important factor in estimating its influence on global biogeochemistry, biodiversity, and climate cycles

(Skole and Tucker 1993). The NASA Landsat Pathfinder Humid Tropical Deforestation Project (Townshend et al. 1995) is mapping global deforestation for 75% of the humid tropics within which it is believed that most deforestation takes place. This includes the entire Amazon Basin, Central Africa, and Southeast Asia. Deforestation occurs at a scale smaller than 100 m and in unpredictable locations and so these vast areas must be analyzed at fine resolution. A raster-based image processing system is used to classify the Landsat data into the seven classes. The classified data are then converted to vector form in a geographical information system. A human editor then visually interprets the vectorized, classified image and enters corrections manually. This combined procedure greatly reduces analysis time over that of hand digitization of an image and reduces the confusion of classes associated with purely digital processing techniques. Databases are being created for three epochs from the 1970s, 1980s, and 1990s for the parts of the tropical forest where deforestation is known to be most significant. Comparisons have been made with previous estimates of forest cover and deforestation in the Pan-Amazon and the rates of change made by the FAO sampling procedure may be up to half an order of magnitude too high on the basis of this comprehensive analysis. The unique contribution of Landsat to land-cover assessment is that it provides the combination of spectral integrity, spatial detail, and temporal repetitiveness which captures and delineates the spatio-temporal dynamics of the Earth's land areas in sufficient detail to assess not only the location but also the causes of land-cover change. To exploit these features fully, multiple scenes are needed; in the case of the Landsat Pathfinder Deforestation project, over 3000 Landsat scenes are being analyzed.

Large area studies place new demands on analytical procedures and also on operational satellite data acquisition technologies and policies. Adjacent scenes should, ideally, be acquired on as near as possible the same date and with the same viewing conditions; however, the trade-off of spatial resolution, repeat observation frequency, and data processing capacities necessary in current, single-satellite systems means that compromises have to be made. A tuned data acquisition, handling, and processing system is therefore a high priority for global-scale, land-cover/land-use assessment and one such is under development for Landsat (S.N. Goward, pers. comm.). The automated Landsat acquisition, processing, and analysis (LAPA) system aims to make acquisitions based on specified criteria that take account of quantitative measures of needs and opportunities for acquisition. Fully automated land-cover analysis will be undertaken through effective use of the spectral, spatial, and temporal information structure of the data. The LAPA system needs to handle between 20–40 gigabytes per day to employ Landsat as a global-scale, terrestrial, land-cover, monitoring system.

Studies of land-surface processes are much more powerful if multiple data layers can be incorporated so as to benefit from their different characteristics; however, different data sets, gathered by different sensors, may have quite different spatial sampling and geometry and so innovative database techniques are needed. For example, to create and validate global land-surface classifications based on coarse-resolution AVHRR data, Landsat scenes have been co-registered with AVHRR so that, using simple commands, images can be overlaid or exchanged. Co-registration has been carried out by identifying ground control points on topographic maps and then shifting Landsat MSS and AVHRR data to find the maximum correlation between the two (Defries et al. 1997). Similarly, the Coastal Marsh Project, which is undertaking an analysis of the surface condition of coastal marshes in the U.S.A. to detect areas that are at risk of rapid loss, uses topographic maps, the national wetlands inventory (NWI), and products developed from Landsat TM data (Kearney et al. 1995). The NWI classifications are used to determine marsh boundaries. Within those areas, the satellite imagery is used to determine the percentage of the marsh surface that is covered with water, using a mixture modeling technique (Rizzo et al. 1996). The marshes are categorized into one of four different classes, viz. healthy, moderately, heavily, and completely deteriorated (Kearney et al. 1995). So far, 14 satellite images covering the east coast of United States from Massachusetts to Georgia, and for a time period from the early 1980s to 1990s have been analyzed. The 1984 to 1993 change results for the Delaware Bay indicate that roughly 10% of the marshes show significant deterioration.

The need to co-register multiple, spatial data sets is particularly evident in the plans for the vegetation canopy lidar (VCL) mission (Dubayah et al. 1998), in which lasers will be mounted on an Earth-orbiting satellite. The mission's goals are land-cover characterization for terrestrial ecosystem and climate modeling, and generation of topographic ground control points. VCL data are intended to provide information on a variety of remote-sensing problems that can only be investigated with knowledge of the three-dimensional canopy structure. It is planned to co-register VCL measurements with data from other sensors such as Landsat-7, MODIS, ASTER, and other EOS sensors. The reasons for this are: (a) an understanding of quantities such as NDVI and LAI in terms of the structural components of the canopy can only be realized with co-incident lidar and reflectance measurements; (b) because VCL measurement tracks are not contiguous, it will be highly desirable to relate the measurements to those obtained from spatially continuous sensors to allow extrapolation. Thus VCL data need to be registered and fused with other data sets having different resolutions, geometry, and frequencies. This is no simple task given the unique nature and the amount of VCL data that will be generated (total data volume will be approximately 5 Terabytes). A typical task will be to query a database to provide VCL return profiles together with observations from other sensors such as Landsat-7 ETM+ and MODIS. This is a daunting task, which will require much computing power, and more importantly, database innovation.

An ideal database would allow data sets to be stored, accessed, and processed in an unresampled form, so as to preserve the data's integrity since much information can be lost in successive geographical, spectral, or temporal resampling. An operational solution to this problem is the global hierarchical indexing scheme (Shock et al. 1998), which can optimize access to and processing of remotely sensed data and higher level data sets. The procedure means that data sets can be fused in various ways without suffering from multiple reprojections and resamplings. The data are located through an advanced indexing system, all calculations and modeling can take place in the same form, and only for visualization need the product be projected to a conventional geographical map projection.

Another type of computer science technology that may have a significant impact on land-surface modeling concerns data organization and management, and user interfaces. Object-oriented, meta-databases can allow different types of data to be searched and cross related easily, using purpose-oriented criteria. One such user interface (Tanin et al. 1996) consist of two phases, query preview and query refinement. The interface design is based on the concepts of dynamic queries and query previews, which rapidly guide users to eliminate undesired datasets, reduces the volume to a manageable size, and refines queries locally before submission over a network. In a prototype system for EOSDIS, users select approximate ranges of three attributes: geographical location, variables (vegetation, land class, precipitation, etc.), and temporal coverage. A query is formulated by selecting attribute values. As each value is selected, the preview bars in the other attribute groups adjust to reflect the number of datasets available. Although the aim of the user interface is simple, to guide a modeler rapidly to available data sets, its execution is not and advances made in this arena can be expected to have dramatic effects on the extent to which the accumulating archives of data are used and the quality of land-surface modeling in the future.

More land-surface variables can now be estimated using existing and developing remote-sensing techniques and larger areas of the Earth can be observed and at higher frequencies than ever before. Thus the possibility of routine monitoring of aspects of the land-surface at a global scale has become a practical possibility. Certain, limited monitoring activities already exist (e.g., FEWS, FAO ARTEMIS), but enormous potential exists for land-surface models to be applied to remotely sensed observations of land-surface processes. Not only do these have scientific interest, but they would find immediate use in a wide variety of practical applications. The stability of satellite platforms and sensors means they have special value for observing longer-term trends, e.g., Prince, Brown de Colstoun et al. (1998). An example is the integrated degradation early warning system (DEWS) currently under development, initially for southern central Africa, in which GLO-PEM, VIC-2, and other models will be used to provide assessments of unusual land-surface conditions that may signal longer-term degradation. Another example is the use of the 250 m MODIS bands that will be exploited to improve land-cover change detection, and to create novel forms of land-cover characterization (J.R.G.Townshend, pers. comm.). A 250 m land-cover change indicator product is being prepared, which, in a simplified form, will also be used as a trigger to prompt acquisition of fine resolution data from systems such as Landsat TM.

CONCLUSIONS

By exploiting the full spectral, spatial, and temporal dimensions of remotely sensed data, significantly increased information is available. Poised, as we are, at the threshold of the EOS-era with the promise of much enhanced remotely sensed observations and with new algorithms available, and with an archive of data from existing sensors continuing to accumulate, the prospects for productive partnerships between remote sensing and land-surface process specialists have never been better. Paradoxically, the gulf between remote-sensing science and applications seems not to have been bridged, but rather to have widened in recent years (Sellers et al. 1995), so that advanced capabilities are by no means fully exploited. This predicament is partly due to the enormous cost and hence lengthy preparations needed to deploy new sensors, commitments to which the applications communities are unaccustomed. The fact that remote sensing often provides nontraditional surface variables is also a barrier. While the stage is set for dramatic advances in our capability of studying the hydrology, primary production, and biogeochemistry of the Earth, these developments will only be realized with closer cooperation between process and measurement specialists.

ACRONYMS AND ABBREVIATIONS

AIRSAR	Airborne SAR (q.v.)
AMSR	Advanced Microwave Scanning Radiometer
ASTER	Advanced Spaceborne Thermal Emission and Reflectance Radiometer (EOS-AM-1)
AATSR	Advanced Along-Track Scanning Radiometer
AVHRR	Advanced Very High Resolution Radiometer, carried on NOAA series of satellites
AVIRIS	Airborne Visible-infrared Imaging Spectrometer
BOREAS	Boreal Forest Ecosystem–Atmosphere Study
BRDF	Bidirectional Reflectance Distribution Function
CZCS	Coastal Zone Color Scanner
DEWS	Degradation Early Warning System
ENSO	El Niño Southern Oscillation
EOS	Earth Observation System
EOSDIS	EOS (q.v.) Data Information System
ERS	European Remote Sensing Satellite
ETM+	Enhanced Thematic Mapper
FAPAR	Fraction of incident PAR (q.v.) Absorbed by a vegetation canopy
FAO ARTEMIS	Food and Agriculture Organization Africa Real-Time Environmental Monitoring by Imaging Satellites
FEWS	Famine Early Warning System
FIFE	First ISLSCP Field Experiment
GLO-PEM	Global Production Efficiency Model
GMS	Geostationary Meteorological Satellite (Japanese)
GOES	Geostationary Operational Environmental Satellite
HAPEX-Sahel	Hydrology Atmosphere Pilot Experiment in the Sahel
INSAT	INdian geostationary meteorological SATellite

JERS	Japanese Environmental Research Satellite
LAI	Leaf Area Index
LAPA	Landsat Acquisition, Processing, and Analysis
LVIS	Laser Vegetation Imaging Sensor
METEOSAT	Meteorological SATellite
MISR	Multi-angle Imaging SpectroRadiometer (EOS-AM-1)
MODIS	MODerate-resolution Imaging Spectrometer (EOS-AM-1)
MSS	Multispectral Scanner
NDVI	Normalized Difference Vegetation Index
NWI	National Wetlands Inventory
PAR	Photosynthetically Active Radiation
POLDER	POLarization and Directionality of the Earth's Reflectance
RADARSAT	Canadian remote sensing satellite
RMSE	Root Mean Square Error
RUE	Rain Use Efficiency
SeaWiFS	Sea viewing Wide-Field Sensor
SLICER	Scanning Lidar Imager of Canopies by Echo Recovery
SPOT-HRV	Système Pour l'Observation de la Terre – High Resolution Visible
SVI	Spectral Vegetation Index
TM	Thematic Mapper
TOVS-HIRS/MSU	TIROS Operational Vertical Sounder – High Resolution Infrared Sounder/Microwave Sounding Unit
TRMM	Tropical Rainfall Mapping Mission
TVX	Thermal and Vegetation Index algorithm
VCL	Vegetation Canopy Lidar

REFERENCES

CEOS. 1997. Towards an Integrated Global Observing Strategy: 1997 CEOS Yearbook. Guildford, U.K.: Smith System Engineering Limited, U.K., for European Space Agency.

Coughlan, J.C., and J.L. Dungan. 1996. Combining remote sensing and forest ecosystem modeling: An example using the Regional HydroEcological Simulation System (RHESSys). In: The Use of Remote Sensing in the Modeling of Forest Productivity, ed. H.L. Gholz, K. Nakane, and H. Shimoda, pp. 135–158. Dordrecht, The Netherlands: Kluwer Academic.

Defries, R.S., M. Hansen, M. Steininger, R. Dubayah, R. Sohlberg, and J.R.G. Townshend. 1997. Subpixel forest cover in central Africa from multisensor, multitemporal data. *Remote Sens. Env.* **60**:228–246.

Defries, R., M. Hansen, and J.R.G. Townshend. 1995. Global discrimination of land cover types from metrics derived from AVHRR Pathfinder data. *Remote Sens. Env.* **54**:209–222.

Defries, R.S., and J.R.G. Townshend. 1994. NDVI-derived land cover classifications at a global scale. *Intl. J. Remote Sens.* **15**:3567–3586.

Dubayah, R. 1992. Estimating net solar radiation using Landsat Thematic Mapper and digital elevation data. *Water Res.* **28**:2469–2484.

Dubayah, R., J.B. Blair, J.L. Bufton, D.B. Clarke, J. Já Já, R. Knox, S.B. Luthcke, S. Prince, and J. Weishampel. 1998. The Vegetation Canopy Lidar Mission. Land Satellite Information in the Next Decade II. Washington, D.C.: American Society for Photogrammetry and Remote Sensing. Published on CD ROM.

Dubayah, R., and S. Loechel. 1997. Modeling topographic solar radiation using GOES data. *J. Appl. Meteorol.* **36**:141–154.

Dubayah, R., and V. van Katwijk. 1992. The topographic distribution of annual incoming solar radiation in the Rio Grande river basin. *Geophys. Res. Lett.* **19**:2231–2234.

Dye, D. 1992. Satellite estimation of the global distribution and interannual variability of photosynthetically active radiation. Ph. D. diss., University of Maryland.

Fallah-Adl, H., J. JáJá, S. Liang, J.R.G. Townshend, and Y.J. Kaufman. 1996. Fast algorithms for removing atmospheric effects from satellite images. *IEEE Comp. Sci. Eng.* **3**:66–67.

Frouin, R., and R.T. Pinker. 1995. Estimating photosynthetically active radiation (PAR) at the earth's surface from satellite observations. *Remote Sens. Env.* **51**:98–107.

Gholz, H.L., K. Nakane, and H. Shimoda, eds. 1997. The Use of Remote Sensing in the Modeling of Forest Productivity. Dordrecht, The Netherlands: Kluwer Academic.

Goward, S.N., and D. Dye. 1996. Global biospheric monitoring with remote sensing. In: The Use of Remote Sensing in Modeling Forest Productivity, ed. H.L. Gholtz, K. Nakane, and H. Shimoda, pp. 241–272. Dordrecht, The Netherlands: Kluwer Academic.

Goward, S.N., and K.F. Huemmrich. 1992. Vegetation canopy PAR absorptance and the normalized difference vegetation index: An assessment using the SAIL model. *Remote Sens. Env.* **39**:119–140.

Goward, S.N., and S.D. Prince. 1995. Transient effects of climate on vegetation dynamics: Satellite observations. *J. Biogeogr.* **22**:549–563.

Gurney, R.J., J.L. Foster, and C.L. Parkinson, eds. 1993. Atlas of Satellite Observations Related to Global Change. Cambridge: Cambridge Univ. Press.

Hall, F.G., J.R. Townshend, and E.T. Engman. 1995. Status of remote sensing algorithms for estimation of land surface state parameters. *Remote Sens. Env.* **51**:138–156.

Hanan, N.P., A. Bégué, and S.D. Prince. 1997. Errors in remote sensing of intercepted photosynthetically active radiation: An example from HAPEX-Sahel. *J. Hydrol.* **188–189**:676–696.

Hanan, N.P., S.D. Prince, and P.H.Y. Hiernaux. 1991. Spectral modelling of multicomponent landscapes in the semi-arid Sahel. *Intl. J. Remote Sens.* **12**:1243–1258.

Hansen, M., R. Dubayah, and R. DeFries. 1996. Classification trees: An alternative to traditional land cover classifiers. *Intl. J. Remote Sens.* **17**:1075–1081.

Harding, L.W., E.C. Itsweire, and W.E. Esaias. 1994. Estimates of phytoplankton biomass in the Chesapeake Bay from aircraft remote sensing of chlorophyll concentrations,1989–92. *Remote Sens. Env.* **49**:41–56.

Kalluri, S.N.V., Z. Zhang, S. Liang, J. JáJá, and J.R.G. Townshend. 1997. Retrieval of bidirectional distribution function (BRDF) at continental scales from AVHRR data using high performance computing. *IGARSS*, Singapore, pp. 174–176.

Kearney, M.S., A.S. Rogers, J.R.G. Townshend, W.L. Lawrence, K. Dorn, K. Eldred, D. Stutzer, F.E. Lindsay, and E. Rizzo. 1995. Developing a Model for Determining Coastal Marsh "Health." Third Thematic Conference on Remote Sensing for Marine and Coastal Environments. Seattle, WA.

Lakshmi, V., J. Susskind, and B.J. Choudhury. 1998. Determination of land surface skin temperatures and surface air temperature and humidity from TOVS HIRS2/MSU data. *Adv. Space Res.* **22**:629–636.

Liang, S., and J.R.G. Townshend. 1997. Angular signatures of NASA/NOAA Pathfinder AVHRR land data and applications to land cover identification. In: Proc. of IGARSS, Singapore, vol. 4, pp. 1781–1783.

Malingreau, J.P., F.A. Albini, M.O. Andreae, S. Brown, J.S. Levine, J.M. Lobert, T.A. Kuhlbusch, L. Radke, A. Setzer, P.M. Vitousek, D.E. Ward, and J. Warnatz. 1993. Quantification of fire characteristics from local to global scales. In: Fire in the Environment: The Ecological, Atmospheric, and Climatic Importance of Vegetation Fires, ed. P.J. Crutzen and J. G. Goldammer, pp. 329–343. Dahlem Workshop Report ES 13. Chichester:Wiley.

Martin, M.E., and J.D. Aber. 1997. High spectral resolution remote sensing of forest canopy lignin, nitrogen, and ecosystem processes. *Ecol. Appl.* **7**:431–443.

Means, J.E., S.A. Acker, D.J. Harding, J.B. Blair, M.A. Lefsky, W.B. Cohen, M.E. Harmon, and W.A. McKee. 1998. Use of large-footprint scanning airborne lidar to estimate forest stand characteristics in the western cascades of Oregon. *Remote Sens. Env.*, in press.

Meeson, B.W., F.E. Corprew, J.M.P. McManus, J.W. Closs, K.-J. Sun, D.J. Sunday, and P.J. Sellers. 1995. ISLSCP Initiative I. Global data sets for land-atmosphere models, 1987–1988. Greenbelt, MD: Published on CD by NASA.

Menenti, M., S. Azzali, A. de Vries, D. Fuller, and S.D. Prince. 1993. Vegetation monitoring in Southern Africa using temporal Fourier Analysis of AVHRR/NDVI observations. In: Intl. Symp. on Remote Sensing in Arid and Semi-Arid Regions (ISRSASR), Aug. 25–28, 1993, pp. 287–294. Lanzhou, China: ISRSASR.

Menenti, M., and W.G.M. Bastiaanssen. 1995. Mesoscale Climate Hydrology: Earth Observing System — Definition Phase. Wageningen, The Netherlands: Winand Staring Centre (SC-DLO).

Myneni, R., C.D. Keeling, C.J. Tucker, G. Asrar, and R.R. Nemani. 1997. Increased plant growth in northern high latitudes from 1981–1991. *Nature* **386**:698–702.

Myneni, R.B., S. Maggion, J. Iaquinta, J.L. Privette, N. Gobron, B. Pinty, D.S. Kimes, M.M. Verstraete, and D.L. Williams. 1995. Optical remote sensing of vegetation: Modeling, caveats, and algorithms. *Remote Sens. Env.* **51**:169–188.

Nemani, R., L. Pierce, S. Running, and S.N. Goward. 1993. Developing satellite-derived estimates of suface moisture status. *J. Appl. Meteorol.* **32**:548–557.

Nilson, T., and U. Peterson. 1994. Age dependence of forest reflectance: Analysis of main driving factors. *Remote Sens. Env.* **48**:319–331.

NOAA (National Oceanic and Atmospheric Administration). 1990–92. Federal Meteorological Handbook. No. 11. Asheville, NC: NOAA National Climate Data Center.

Petty, G.W. 1995. The status of satellite-based rainfall estimation over land. *Remote Sens. Env.* **51**:125–137.

Prince, S.D. 1991a. A model of regional primary production for use with coarse-resolution satellite data. *Intl. J. Remote Sens.* **12**:1313–1330.

Prince, S.D. 1991b. Satellite remote sensing of primary production: Comparison of results for Sahelian grasslands 1981–1988. *Intl. J. Remote Sens.* **12**:1301–1311.

Prince, S.D., and W.L. Astle. 1986. Satellite remote sensing of rangelands in Botswana. I. Landsat MSS and herbaceous vegetation. *Intl. J. Remote Sens.* **7**:1533–1553.

Prince, S.D., E. Brown de Colstoun, and L. Kravitz. 1998. Evidence from rain use efficiencies does not support extensive Sahelian desertification. *Glob. Change Biol.* **4**:359–374.

Prince, S.D., E. Brown de Colstoun, and H. Strand. 1997. Land cover characterization of the HAPEX-Sahel West Central Super Site with satellite data. In: Hapex-Sahel West Central Super Site, Methods, Measurements and Selected Results, ed. P. Kabat, S.D. Prince, and L. Prihodko, pp. 241–258. Wageningen, The Netherlands: Winand Staring Centre (SC-DLO).

Prince, S.D., S.J. Goetz, R.O. Dubayah, K.P. Czajkowski, and M. Thawley. 1998. Inference of surface and air temperature, atmospheric precipitable water and vapor pressure deficit using AVHRR satellite observations: Validation of algorithms. *J. Hydrol.* **212–213**:320–250.

Prince, S.D., and S.N. Goward. 1995. Global primary production: A remote sensing approach. *J. Biogeogr.* **22**:815–835.

Prince, S.D., and S.N. Goward. 1996. Evaluation of the NOAA/NASA Pathfinder AVHRR Land Data Set for global primary production modeling. *Intl. J. Remote Sens.* **17**:217–221.

Prince, S.D., and M. Steininger. 1999. Biophysical stratification of the Amazon basin. *Glob. Change Biol.* **9**:1–22.

Ranson, K.J., G. Sun, J.F. Weishampel, and R.G. Knox. 1997. Forest biomass from combined ecosystem and radar backscatter modeling. *Remote Sens. Env.* **59**:118–133.

Ritchie, J.C., and M. Menenti. 1997. Laser altimetry of vegetation canopies. In: Hydrologic Atmospheric Pilot Experiment in the Sahel (HAPEX-Sahel): Methods, measurements and selected results from the West Central Supersite, ed. P. Kabat, S.D. Prince, and L. Prihodko, pp. 117–125. Report 130. Wageningen, The Netherlands: Winand Staring Centre: SC-DLO.

Rizzo, E., A.S. Rogers, M.S. Kearney, J.R.G. Townshend, and W.T. Lawrence. 1996. Changes in Blackwater Marsh, Maryland, 1983–1993, determined by aerial photography and thematic mapper data. In: Technical Papers, Second Annual ASPRS/ACSM Meeting, Baltimore, MD, pp. 200–229. Baltimore, MD: ACSM.

Sellers, P.J., ed. 1995. Remote Sensing of Land Surface for Studies of Global Change. Remote Sensing of Environment 51. New York: Elsevier.

Sellers, P.J., B.W. Meeson, F.G. Hall, G. Asrar, R.E. Murphy, R.A. Schiffer, F.P. Bretherton, R.E. Dickinson, R.G. Ellingson, C.B. Field, K.F. Huemmrich, C.O. Justice, J.M. Melak, N.T. Roulet, D.S. Schimel, and P.D. Try. 1995. Remote sensing of the land surface for studies of global change: Models-algorithms-experiments. *Remote Sens. Env.* **51**:3–26.

Shock, C.T., C. Chang, and A. Sussman. 1998. The design and evaluation of a high-performance earth science database. *Parallel Computing* **24**:65.

Skole, D., and C.J. Tucker. 1993. Tropical deforestation and habitat fragmentation in the Amazon: Satellite data from 1978 to 1988. *Science* **260**:1905–1910.

Smith, P.M., S.N.V. Kalluri, S.D. Prince, and R. DeFries. 1997. The NOAA/NASA Pathfinder AVHRR 8-km land data set. *Photogram. Remote Sens.* **63**:12–32.

Tanin, E., R. Beigel, and B. Shneiderman. 1996. Incremental data structures and algorithms for dynamic query interfaces. *SIGMOD Record* **25**:21–24.

Townshend, J.G.R., ed. 1992. Improved global data for land applications: A proposal for a new high resolution data set. Report of the Land-Cover Working Group of IGBP-DIS. Stockholm: Intl. Geosphere-Biosphere Programme.

Townshend, J.R.G., V. Bell, A.C. Desch, C. Havlicek, C.O. Justice, W.E. Lawrence, D. Skole, W.W. Chomentowski, B. Moore III, W. Salas, and C.J. Tucker. 1995. The NASA Landsat Pathfinder Humid Tropical Deforestation Project. Land Satellite Information in the Next Decade. Tysons Corner, VA: American Society of Photogrammetry and Remote Sensing.

Tucker, C.J., H.E. Dregne, and W.W. Newcomb. 1991. Expansion and contraction of the Sahara desert from 1980 to 1990. *Science* **253**:299–301.

Tucker, C.J., and P.J. Sellers. 1986. Satellite remote sensing of primary production. *Intl. J. Remote Sens.* **7**:1395–1416.

Walthall, C.L. 1997. A study of reflectance anisotropy and canopy structure using a simple empirical model. *Remote Sens. Env.* **61**:118–128.

Wang, J. 1995. Some features observed by the L-band push-broom microwave radiometer over the Konza Prairie during 1985–1989. *J. Geophys. Res.* **100**:25,469–25,479.

Waring, R.H., J. Way, E.R. Hunt, L. Morrissey, K.J. Ranson, J.F. Weishampel, R. Oren, and S.E. Franklin. 1995. Imaging radar for ecosystem studies. *Bioscience* **45**:715–723.

Whitlock, C.H., T.P. Charlock, W.F. Staylor, R.T. Pinker, I. Laszlo, A. Ohmura, H. Gilgen, T. Konzelma, R.C. DiPasquale, C.D. Moats, S.R. LeCroy, and N.A. Ritchey. 1995. First Global WCRP Shortwave Surface Radiation Budget Dataset. *Bull. Am. Meteorol. Soc.* **76**:905–922.

Yang, J., and S.D. Prince. 1997. A theoretical assessment of the relation between woody canopy cover and red reflectance. *Remote Sens. Env.* **59**:428–439.

4

Remote Sensing – What Will We Get?

W. MAUSER[1], M. RAST[2], and H. BACH[3]

[1]Institute of Geography, Dept. of Geographical Remote Sensing, University of Munich, 80333 Munich, Germany
[2]Land Surface Processes Division, ESA-ESTEC, 2200 AG Noordwijk ZH, The Netherlands
[3]Vista – Remote Sensing Applications in Geosciences, 80333 Munich, Germany

ABSTRACT

An overview of the current development of remote sensing of the Earth's land surface is given and the implications of the ability of new sensors, namely to measure temporal series of land-surface parameter fields on the way landscapes and their biological and hydrological processes will be modeled is investigated. Based on the parameters that are needed in landscape models and on the remote sensing data we can expect in the future, the list of parameters is analyzed towards the potential of being remotely sensed. This is performed with special focus on the new features of both optical spectroradiometers and synthetic aperture radar (SAR) sensors.

To ensure the constant flow of the necessary input parameters and variables, parameter models operate at the center of the data fusion process to convert remote sensing measurements into a set of model input parameters and variables. Different strategies to use remote sensing derived parameters in models are demonstrated. They span from simple derivation of input parameters through regression models over model forcing and internal recalibration to inversion and spatial validation of landscape models. It is demonstrated how optical spectroradiometers can contribute land-surface information on vegetation type, leaf area index (LAI), biomass, plant water, fractional vegetation cover, surface albedo, and surface temperature. In addition and to overcome the weather dependency of optical systems, data from active microwave sensors on future missions will improve the parameter environment, in which the models operate by adding soil moisture, surface roughness, and topography.

INTRODUCTION

The scope of this chapter is to give an overview of the development of remote sensing of the Earth's land surface as can be foreseen within the next ten years. It investigates the implications this will have on the way landscapes, and the biological and

Integrating Hydrology, Ecosystem Dynamics, and Biogeochemistry in Complex Landscapes
Edited by J.D. Tenhunen and P. Kabat

hydrological processes within them, will then be modeled. This task is in itself not simple, because the right blend of facts and already initiated developments must be combined with a vision of the future.

Progress has been made in the modeling of landscapes on a wide range of scales (Running 1994; Ford et al. 1994; Feddes and Koopmans 1995; Schädlich and Mauser 1996; Famiglietti and Wood 1994; Myneni et al. 1995). Simple land-surface models were initiated in the 1980s due to the need of meteorologists for a better representation of land-surface to atmosphere interactions in global and mesoscale circulation models. Today it has become clear that modeling of the spatial interactions of hydrological and biological processes on the landscape level is itself a worthy scientific task. It is a means by which we can check the depth of our understanding of these processes and interactions. Models are developing to a stage of reliability to be valuable tools for decision making within a practical framework. Landscape models and their predictive abilities will eventually form the basis to balance the conflicts intelligently between nature and anthropogenic influences in a future world that is marked by the need for sustainable landscape management within the framework of growth in human populations and, hopefully, prosperity.

Models require data on several levels of their execution cycle. Data are necessary to estimate parameters describing the state of the system under consideration, as dynamic state variables expressing changes within the system, and for the validation of the model results. Since landscape models describe a process that takes place in space and time, the critical data must be spatially distributed. This is where remote sensing as the science of the interaction between natural objects and electromagnetic waves arrives on stage. Remote sensing allows a shift from point measurements to areal measurements required for modeling landscapes. The basic underlying assumption, which makes this science interesting for landscape modeling, is that model parameters can be derived quantitatively through measurements of the interactions of the Earth's surface with electromagnetic waves.

Remote sensing has come a long way since the first photographs were taken of the Earth's surface using a balloon in the last century. Twenty-five years have now passed since the launch of the first Landsat satellite. Since then, Earth observation through remote sensing has developed and diversified. Today, more than 45 earth observation satellites are providing data about the Earth's environment. Those dedicated to land observations include a wide range of geophysical parameters which are in the optical domain related to albedo, reflectance, and emission properties of the target observed and in the active microwave domain to the roughness and dielectricity constant of the target influencing the radar signal backscatter behavior. Still the level of analysis and interpretation of remote sensing data rarely exceeds simple land-use classifications to achieve quantitative estimates of ecosystem parameters. The main reason for this is that the presently available instruments to study land-surface functioning on a regional scale do not yet have suitable spatial, spectral, and temporal resolution to observe relevant processes. This is documented in Figure 4.1, where spatial and temporal observation requirements to measure land-surface dynamics adequately are compared

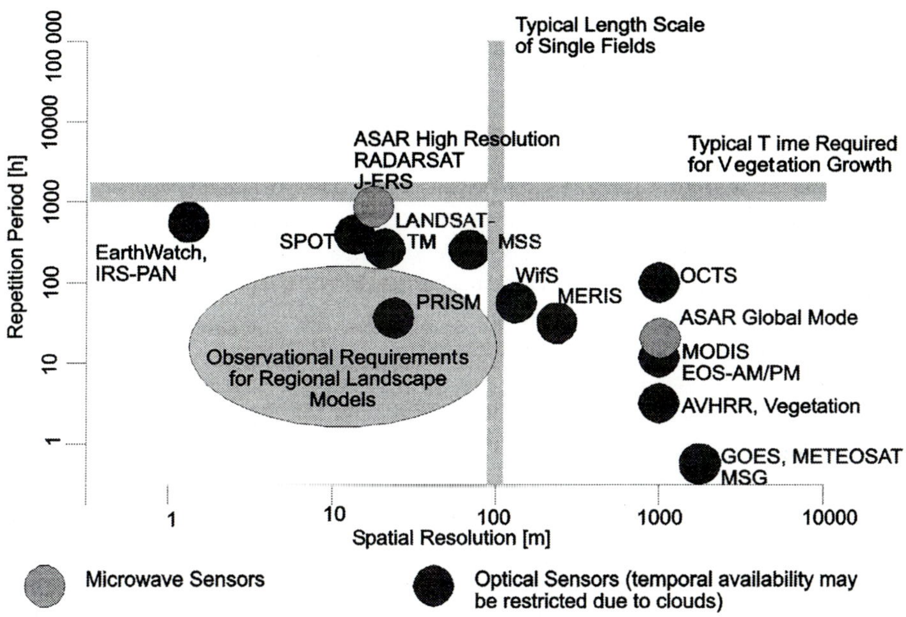

Figure 4.1 Comparison of spatial and temporal resolution available and upcoming Earth observation sensors with the requirements of regional landscape models.

with existing and upcoming satellites. In Figure 4.1, it is assumed that single fields or patches of different vegetation communities should be identifiable with respect to their spatial extent. The temporal resolution should be adequate to follow the course of vegetation development, which is assumed to last approximately 45 days. One can easily see that either timing or spatial resolution does not fit observational needs in the different cases.

Although numerous Earth observation satellites are currently in use, the next ten years will bring a dynamic development of remote sensing towards better spectral, better temporal, and better spatial coverage of the Earth's surface. It will most likely bring the commercial exploitation of Earth observation through remote sensing. This means that the technology and instrument part of remote sensing is maturing.

To utilize fully the potential of the upcoming instruments to determine fields of model parameters quantitatively, the full synergy between the complete palette of available sensors will have to be exploited together with all available information from ground-based networks. The focus of this chapter is on the combined analysis of existing soil–vegetation–atmosphere transfer (SVAT) and biogeochemical (BGC) models, the parameter needs of these landscape models, and the capabilities of future Earth observation sensors to deliver these parameters. Case studies, using already available space sensors or aircraft models of future spaceborne sensors, are used to support our concepts.

LANDSCAPE MODELS: WHICH PARAMETER DO THEY NEED?

Landscape models of surface processes fill the gap between locally verifiable point models and global land-surface process models working in conjunction with general circulation models (GCMs). Their general purpose is the spatially distributed description of the coupling of energy, water, and carbon fluxes at the land surface on a regional scale. A region in terms of scale can be defined as a part of the Earth's surface, where the composition of the landscape is assumed to remain constant. Landscape models are aggregated models. They need spatially distributed input parameters, which can be gathered both with optical and microwave sensors. As will be seen, landscape models will be the main drivers for four-dimensional data assimilation (4DDA) in the future. These are based on a synergistic utilization of remote sensing and ancillary data.

One approach, which is common to all distributed models of land-surface processes, is the combination of a set of submodels, each fulfilling a certain task within the overall model. These submodels can usually be divided into two different types:

- *process models*, which describe land-surface processes with algorithms requiring a given set of environmental parameters and variables (e.g., soil water model, plant growth model, atmospheric boundary layer model, erosion model);
- *parameter models*, which retrieve and supply spatially distributed fields of model input parameters. They apply methods to convert remote sensing measurements into input parameters, which process models can understand (e.g., models for leaf area index [LAI] from normalized difference vegetation index [NDVI] [Myneni et al. 1997] or sigma-naught into soil moisture [Rombach and Mauser 1997], interpolation models for meteorological station data, irradiance model from a digital terrain model).

Both types of submodels are equally important to a successful landscape model. The parameter models are considered the most promising field for application of remote sensing data, since most inadequacies in the description of the land-surface processes on the landscape level are due to inadequate input data. Models of the energy and water budget for land-surface units are called soil–vegetation–atmosphere transfer (SVAT) models. To also describe the biochemical reactions involved in vegetation growth, biomass production, and development influenced by geochemistry (minerals, fertilizer), SVAT models are often combined with BGC cycle models.

To review the current use and future capacities of remote sensing, existing landscape models were examined with respect to two criteria:

- Parameters and variables, which are derived from remote sensing data (optical and/or SAR), are currently used in the model, or the structure of the model allows the use of parameters and variables derived from remote sensing data.

- Statistical models based on a direct regression between remote sensing measurements and ecosystem model outputs (e.g., direct determination of evapotranspiration from thermal data) are not considered.

From the large number of published models of land-surface processes, six were selected that fulfill these criteria. These models and their main characteristics are summarized in Table 4.1. The six models are subdivided into three groups.

1. The main modeling task of PROMET (Mauser and Schädlich 1998) and SVATS (Famiglietti and Wood 1994a, b) is the hydrological cycle. PROMET was run and tested on various land cover types in Central Europe. The process models and formalism of SVATS are similar to PROMET. SVATS does not use remote sensing data yet and was applied on rangeland in the Northern U.S. within the FIFE (first ISLSCP field experiment) test site.
2. SWATRE and SWACROP (Feddes and Koopmans 1995; Kabat et al. 1992) were both developments of the Agricultural University of Wageningen. They concentrate on the soil water balance and crop production in agricultural areas. They do not use remote sensing data as input yet.
3. FOREST-BGC and RHESSys were developed at the University of Montana (Running 1994; Ford et al. 1994). They model biogeochemical cycles and biospheric productivity for forest, grass, shrubs, and crops. The list of these models also provides an idea about the direction in which landscape models will develop in the near future. They will most likely be raster-based (to link up easily with mesoscale meteorological models) and spatially distributed. Coupling between the processes will be intensified (e.g., lateral flows and plant–water interactions).

From these models a list of necessary spatial input parameters as well as model parameters was identified. Parameters that can only be determined from space with a resolution of several kilometers (like snow water equivalent using passive microwave) have not been included because there is no feasible technology to achieve the spatial resolution needed in regional studies. The list is shown in Table 4.2 together with the minimum required temporal frequency of observation. The temporal requirements for the observations have been derived from the development cycle of vegetation and from sampling theory. This roughly means that a gradient or a transition in a nonlinear response should be represented by at least five samples.

The temporal observation requirements vary from once ever for static soil parameters, over 1–2 weeks for vegetation parameters, to 1–3 days for parameters that are very variable with time (soil moisture, emergence date). In the third column, the future potential for assessing these parameters with remote sensing is indicated based on the author's experience. This best guess also considers that the assessment of most parameters using conventional or ground-based methods is impossible or at least very inaccurate when applied regionally as opposed to point scale, where conventional methods usually are more accurate than remote sensing.

Table 4.1 Comparison of characteristics of different land-surface process and landscape models.

	PROMET	SVATS	SWATRE	SWACROP	FOREST-BGC	RHESSys
Main Modeling Task	Hydrologic cycle	Hydrologic cycle	Soil water balance	Crop production	Biogeochemical cycle	Biospheric production
Main Land-surface Category	Agriculture Water Forest Meadows Urban	Rangeland	Agriculture	Agriculture	Forest ecosystems	Forest ecosystems
Geographical Region	Central Europe	Northern U.S.	Netherlands, Spain	Netherlands	Northern U.S.	Northern U.S.
Main Features/ Advantages	Spatial modeling on different scales, use of remote sensing data	Spatial modeling, lateral flows on catchment scale	Spatial modeling, multi-layer soil moisture model	C-cycle model	Spatial modeling, C and N submodels, use of remote sensing	Spatial modeling, lateral flows, use of remote sensing data
Drawbacks/ Disadvantages	No lateral flows, only one soil layer	No remote sensing, only one soil layer	No remote sensing	No remote sensing, not spatially distributed	Restricted to forests, simple soil water representation	Simple soil water representation
References	Mauser (1991); Mauser and Schädlich (1998)	Famiglietti (1992); Famiglietti and Wood (1994 a, b)	Belmans et al. (1983); Feddes (1995); Feddes and Koopmans (1995)	Feddes et al. (1988); Kabat et al. (1992)	Nemani and Running (1989); Running and Coghlan (1988); Running (1991, 1994)	Ford et al. (1994); White and Running (1994)

Table 4.2 Biogeophysical parameters identified as input in aggregated models and the required temporal observation frequency: ✓ = possibility exists; (✓) = possible in principle but severe difficulties; – = not possible yet.

Parameters	Required Observation Frequency	Remotely Sensable	Necessary For
Cover type	1 year	✓	Parameterization of vegetation or surface
Leaf area index	7–14 days	✓	Transpiration, interception
Vegetation height	7–14 days	✓	Aerodynamic resistance
Biomass	7–14 days	✓	Production, transpiration, respiration
Fractional vegetation cover	7–14 days	✓	Evaporation, transpiration
Emergence date	1–3 days	–	Yield formation
Spectral surface albedo, fPAR	7–14 days	✓	Energy balance, photosynthesis
Surface roughness	7–14 days	✓	Aerodynamic resistance, structure
Soil moisture	1–3 days	✓	Movement of water/nutrient in soil, evaporation, transpiration, biologic activity, percolation, surface runoff
Static soil hydraulic properties	once	–	Water and nutrient movement in soil, evaporation, transpiration
Root zone depth	7–14 days	–	Transpiration, soil water balance
Plant chlorophyll content	7–14 days	(✓)	Plant growth, nutrient supply, production
Plant nitrogen and minerals	7–14 days	(✓)	C-fixation, plant fertilization, groundwater pollution
Soil minerals	monthly	(✓)	Soil fertility, erosion
Radiation balance and shortwave radiation	0.5 hours	(✓)	Available energy, aPAR
Topography	once	✓	Irradiance in reliefed terrain, spatial interpolation of meteorological data, lateral water flows
Meteorological variables	0.5 hours	(✓) T_{air} – humid. ✓clouds (✓) rain – wind	Drivers
Snow water equivalent	1–3 days	(✓)	Water storage in the snow cover
Surface temperature	0.5 hours	✓	Energy balance, validation

REMOTE SENSING: WHAT DATA CAN WE EXPECT IN THE FUTURE?

The next generation Earth observation sensors for terrestrial remote sensing can be divided into three major categories:

- *large-scale* (coarse spatial resolution) sensors with a high repeat rate (target revisit time) and global coverage in less than two days enabling more frequent observation and thus multitemporal monitoring with a spatial resolution of the order of 250 m and daily repetition.
- *medium-scale* (medium spatial resolution) sensors with a high repeat rate and regional coverage. These sensors allow a detailed analysis of land-surface parameters and their temporal changes on the regional scale not regularly covering the whole globe. The spatial resolution will be of the order of several tens of meters and the revisit time will be 2–3 days.
- *small-scale* (high spatial resolution) sensors with a low repeat rate. These sensors allow a detailed analysis of land-surface parameters; only a multitemporal process analysis is difficult due to the low frequency of observation. In this category of sensors the capabilities to retrieve land-surface parameters depends strongly on the spectral information covered by the system.

Typical sensors representing the large-scale sensor category that were recently launched or will be available in the near future are:

- *AVHRR (on NOAA – K,L,M,N)*: This radiometer provides images in five broad spectral bands between 0.5 and 12.4 micron over a 3000 km swathwidth. Initially intended to monitor cloud cover and temperature distribution, the AVHRR has become a very valuable tool for ecosystem processes and vegetation monitoring due to its twice daily global coverage. This will even be enhanced through an additional band in the SWIR region of the spectrum.
- *MERIS (on the ESA Envisat)*: This 15 narrow band imaging spectrometer will also enable large-scale monitoring of vegetation state, biomass distribution and condition, agriculture, and forestry in the VIS/NIR. With the capability to achieve global coverage within three days and acquire data at dual resolution (300 m and 1200 m), it will contribute to large-scale ecosystems models.
- *VEGETATION (on SPOT 4 and above)*: SPOT 4 supplies daily data with the three spectral bands in the RED, NIR, and SWIR and an experimental band in the blue region of the spectrum. VEGETATION with its resolution of 1 km is optimized for studies of global vegetation dynamics.
- *ASAR – GLOBAL MODE (on the ESA Envisat)*: The primary objective of ASAR in its global mode of operation is to supply a global coverage of C-band microwave backscatter data at a spatial resolution of 1 km every 2–3 days in dual polarization. Since the dielectric constant together with surface roughness closely determine the backscatter properties of the surface, it will be useful for the observation of global and regional soil moisture dynamics.

- *AATSR (on the ESA Envisat)*: The primary objectives of the AATSR is to establish continuity of the ATSR–1 and ATSR–2 observations of precise sea surface temperature (SST), but also to exploit the science of quantitative remote sensing of land surfaces, particularly vegetation, through improved atmospheric correction by means of its two-angle view. This seven band radiometer will provide radiances over land and sea. With three spectral bands in the VNIR and four in the SWIR/TIR, AATSR will, in addition to temperature monitoring, contribute to experimental vegetation indices monitoring with a 1 km resolution over a 500 km swath.
- *MODIS (on the US EOS AM–1)*: MODIS's objective is to provide a comprehensive series of global observations of the Earth's land, oceans, and atmosphere in the visible and infrared regions of the spectrum covering the Earth every two days in 36 spectral bands ranging in wavelength from 0.4 μm to 14 μm. Two spectral bands are imaged at a nominal resolution of 250 m at nadir, with five bands at 500 m and the remaining 29 bands at 1,000 m. A ± 55° scanning pattern at an orbit of 705 km achieves a 2,330 km swath and provides global coverage every one to two days.
- *MISR (on EOS AM–1)*: In four VIS/NIR spectral bands, the MISR instrument measures global surface albedo, aerosol, and vegetation properties in multi-angle views analyzing the bidirectional reflectance distribution function in four different spatial resolutions between 240 m and 1.92 km and thus provides the needed reflectance anisotropy for quantitative reflectance measurements.

A typical sensor of the medium scale/regional coverage sensor category is:

- *PRISM (Candidate Land-Surface Processes and Interactions Mission of the European Space Agency [ESA])*: The Land-Surface Processes and Interactions Mission is one of ESA's candidates for a set of science-oriented Earth Explorer missions for the beginning of the next decade. The aim of the mission is to use a highly maneuverable satellite to provide a broad set of land-surface parameters, which can be verified in the field, at high temporal resolution. This is achieved by covering the complete optical spectral range and by measuring the bidirectional reflectance distribution function (BRDF). The underlying sensor concept is called PRISM (processes research by an imaging space mission). It is currently studied by ESA at the level of Phase A (ESA 1998). In Table 4.3 the properties of this mission are described.

  If implemented, this mission will for the first time close the gap, shown in Figure 4.1, between temporal and spatial resolution by sacrificing global coverage. It is therefore well suited for the study of land-surface processes and their interactions with the atmosphere on the regional scale, also enhancing their realistic representation in large-scale models. It will also:

  (a) help bridge the gap between knowledge of local- and global-scale processes (complementing the missions of large scale/coarse resolution

Table 4.3 Properties of PRISM.

SPECTRAL PROPERTIES	
Spectral range region 1	0.45–1.11 μm 1.16–1.40 μm 1.49–1.79 μm 2.02–2.35 μm
Spectral sampling interval	< 15 nm (Goal: 12 nm) < 10 nm in range 680 ÷ 769 nm
Spectral range region 2	8.2–8.7 μm 8.7–9.2 μm
SPATIAL PROPERTIES	
Swath width	50 km
Spatial sampling interval	50 m
RADIOMETRIC PROPERTIES	
Signal resolution in region 1	12 bit
Absolute radiometric accuracy in region 1	Goal: < $\sqrt{(\mathrm{NEdL}^2 + 0.02\ \mathrm{L}^2)}$
Radiometric dynamic range in region 2	212–345 K
Radiometric resolution in region 2	5 K (Tmin–240 K), 2 K (240–20 K), 1 K (270–345 K)
POINTING PROPERTIES	
Across-track range	± 475 km ± 30 degrees
Across-track step	5 km 0.36 degrees
BRDF PROPERTIES	
Range observation angle (along-track)	±60, ±45, ±30, 0 degrees
Spectral bands	40–60 selectable bands in region 1 and 2 bands in region 2
Spectral sampling	10–50 nm

imaging sensors such as AATSR, AVHRR, MERIS, VEGETATION, MODIS);

(b) provide the necessary inputs to land-surface processes models;

(c) provide the means to develop methodologies to quantify biospheric and hydrospheric change processes on the regional scale;

(d) support the determination of energy, water, and biochemical fluxes at the land surface.

Typical sensors of the small-scale observation sensor category are:

- *LANDSAT 7*: The earth-observing instrument, the enhanced thematic mapper plus (ETM+) on Landsat 7, replicates the capabilities of the thematic mapper

(TM) instruments on Landsat 4 and 5. The ETM+ also includes new features that make it more versatile for global change studies, land cover monitoring and assessment, and large area mapping: In addition to the existing TM series, the primary new features on Landsat 7 are a panchromatic band with 15 m spatial resolution and a TIR channel with 60 m spatial resolution.

- *SPOT*: The high-resolution visible (HRV) imaging system of the SPOT satellites delivers data in two spectral modes: (1) panchromatic with a spatial resolution of 10 meters and (2) multispectral, with three spectral bands, from 500–590, 610–680 and 790–890 nanometers and a spatial resolution of 20 meters. With future SPOT satellites the ground resolution will be increased to 5 m and 3 m (instead of 10 m) in panchromatic mode, and 10 m (instead of 20 m) in the multispectral mode.
- *ASTER (on EOS AM–1)*: ASTER is an imaging instrument that will be used to obtain detailed maps of surface temperature, emissivity, reflectance, and elevation. The primary objective for the ASTER mission is to obtain high spatial resolution global, regional, and local targeted data in 14 channels from the visible through the thermal infrared wavelength regions in 14 spectral bands at three different spatial resolutions: 4 VNIR bands at 15 m, 5 SWIR bands at 30 m, and 5 TIR bands at 90 m. The swathwidth of ASTER is 60 km.

LAND-SURFACE PARAMETERS: WHICH PARAMETERS CAN BE SENSED REMOTELY?

To define land-surface parameters that can be acquired using remote sensing methods, the investigations of this chapter cannot restricted to the present capabilities and availability of sensors, but should consider future spaceborne sensors. Planned optical spectroradiometers and SAR systems were compiled by Prince (Chapter 3, this volume). Their expected potential to assess the land-surface parameters, which are needed for the models, are now analyzed.

Optical Spectroradiometers

As an example of a future optical remote sensing sensor, PRISM is analyzed as a high to medium spatial resolution regional sensor for its potential to determine model parameters for integrated landscape models. For this, one is currently limited to results based on airborne sensors or which have evolved (with certain limitations) from spectral modeling. The spectral specifications of the airborne visible/infrared imaging sensor (AVIRIS) is very close to PRISM, although AVIRIS has no pointing capabilities and is therefore not able to measure the BRDF. Looking at results already achieved with AVIRIS shows us what kind of parameters and variables can be retrieved with PRISM in the future. Further, directional reflectance modeling of different surfaces and field measurements indicate the large potential to determine structural information of the surface (Goel and Thompson 1984) from BRDF measurements.

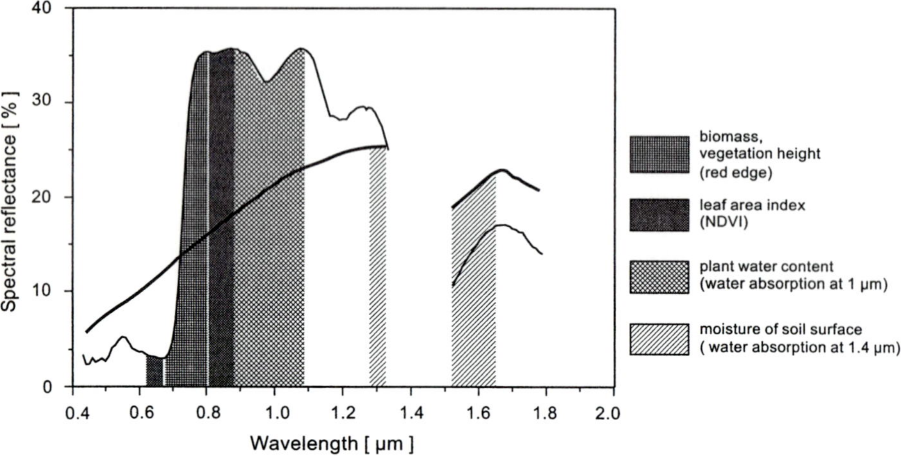

Figure 4.2 Information content of different spectral ranges derived from AVIRIS data (Bach et al. 1995).

A summary graph showing the information content of different spectral ranges in the optical range is given in Figure 4.2. Using the spectral reflectance of the red and near infrared to calculate the NDVI, the LAI of canopies can be determined. A high spectral resolution is necessary to observe the wavelength shift of the red edge of a vegetation spectrum, which can be correlated with the biomass and vegetation height of plant canopies. Through modeling the water absorption feature at 1 μm, the plant water content can be obtained. The soil surface moisture can be determined by modeling the water absorption at 1.4 μm (Bach et al. 1995).

On the basis of these spectral features and using models as well as semi-empirical approaches, the spatial distribution of plant parameters can be determined. Examples of this processing are shown in Plate 4.1.

The spatial distribution of LAI, biomass, vegetation height, and plant water content of the corn (maize) fields in a test area in the Upper Rhine Valley are quantified in Plate 4.1. Because the underlying semi-empirical relations and absorption models are only validated for corn at present, only the corn fields are colored in this figure according to their canopy properties. Plate 4.1 gives a quantitative impression of the expected spatial variability of land-surface parameters on the local and regional scale. More research is necessary on this topic to be able to obtain similar relations for various plant types. This is necessary to cover the processes at the landscape level, which differ according to different types of vegetation. It can be expected and has already been partly shown that similar results may be obtained in other land-use domains.

The improvements in parameter determination that can be achieved by using hyperspectral instead of multispectral optical data are illustrated in Figure 4.3 for the vegetation height of corn plants. The parameter vegetation height was chosen here, because it can be measured without destruction and correlates strongly with biomass during the investigated phenological phase of corn development. The height interval

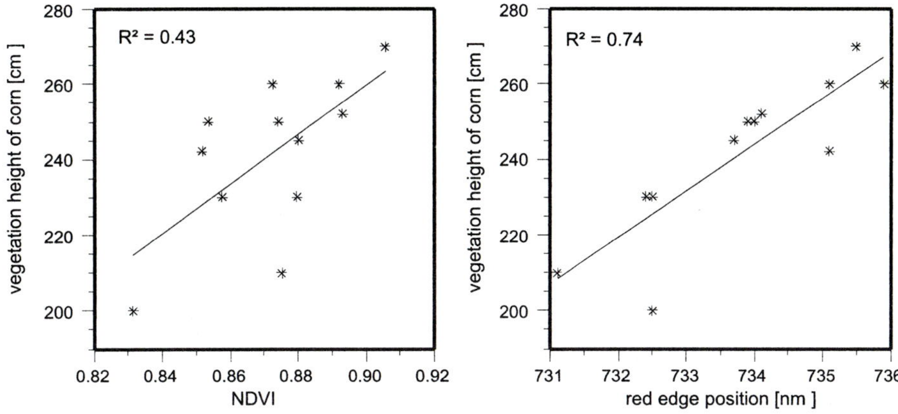

Figure 4.3 Comparison of the retrieval capacity for vegetation height of corn between multispectral data (represented through the parameter "NDVI") and hyperspectral data (represented through the parameter "red edge position").

between two and three meters is especially relevant because during this development stage, grain yield is determined. On the other hand, this height interval is difficult to differentiate with conventional multispectral remote sensing indices such as NDVI, because the corn plant has already reached canopy closure and the soil surface is completely covered. The left side of Figure 4.3 shows that, for multispectral data from which the NDVI has been calculated, the correlation with height of plants between two and three meters is not significant. This directly results from signal saturation caused by canopy closure. However, if one calculates the red edge position of the spectra, the correlation becomes highly significant.

Besides this direct parameterization and modeling of reflectance spectra, information at the subpixel scale can be obtained using spectral unmixing approaches. The idea of spectral unmixing is that the signature of a mixed pixel is the linear combination of the reflectance spectra of the contained endmembers. The weighting factors of this linear combination are the fractions of the endmembers of the mixed pixel. By knowing or assuming the spectral signatures of the endmembers, it is possible to retrieve information on the fractions of the endmembers in the mixed pixel. This approach allows the determination of the fractional vegetation cover of a pixel that is not fully covered by plants.

An example of this is given in Figure 4.4 for an AVIRIS image and the detected endmembers vegetation and bright soil. The unmixing algorithm calculates the images that are shown in the center of Figure 4.4. For each pixel, the percentage of coverage by vegetation is obtained. This fractional vegetation cover is not only qualitative in nature. The amount can be verified by comparing the unmixing result with measured fractional vegetation cover of corn fields, which were determined using ground photography and standard image processing routines. The good coincidence between unmixing results and measurements can be seen in the lower part of Figure 4.4.

In the case of corn it was shown that all vegetation parameters of Table 4.2, which are required as input for landscape models, can be obtained from optical spectroradiometer data. The examples show that the high spectral resolution of a sensor, like PRISM, will significantly improve the determination of model input parameters compared to existing spaceborne instruments. Beyond the given examples, recent literature indicates that biochemical constituents of plants such as chlorophyll and cellulose content can be derived from hyperspectral remote sensing data (Baret and Jacquemoud 1994; Wessman 1994). Further research has to be conducted in this field to stabilize the relationships that have already been found and to formulate a more theoretically based understanding of the relation between spectral reflectance and model parameters. This will enable the extension of the method to a large variety of plant species.

Synthetic Aperture RADAR (SAR) Sensors

In comparison with optical systems, SAR sensors work in the microwave region of the spectrum. To a large degree, this makes them independent of the weather situation in terms of clouds, rain, and snow. The basic interaction of microwaves with the land surface is different from the optical region. Whereas absorption of radiation by molecules plays the most important role in the optical region surface, volume scattering by dipols becomes more important in the microwave region. This enables the determination of parameters such as surface roughness, surface dielectric constant (which strongly relates to soil moisture), and topography using these sensors. Several SAR systems already work in orbit, among which ERS-1/2 and RADARSAT and J-ERS-1 are the most prominent. These systems record with a single frequency/single polarization in the C- or L-band. Therefore their current information content can be compared to a black and white image in comparison with a color photograph.

The future development of SAR systems is characterized by:

- More flexibility as with ESA's ENVISAT-ASAR system, which is currently being built and will be launched in 2000. It will give two polarizations and a range of spatial resolutions/revisit time combinations ranging from 1 km and 1 day repeat to 12.5 m and 35 days repeat. With this sensor, the right repetition time for surface dielectric constant observation on the regional to global scale (1 km resolution) is already achieved.
- More frequencies and polarizations. Aircraft campaigns using the AIRSAR system and the SIR-C/X-SAR shuttle mission, which measured the fully polarimetric signal C- and L-band have proven the usefulness of multifrequency/multipolarization SAR data.

With the current ERS sensors it can be shown that parameters like soil moisture and topography can be determined from SAR data. Figure 4.5 gives a comparison of field

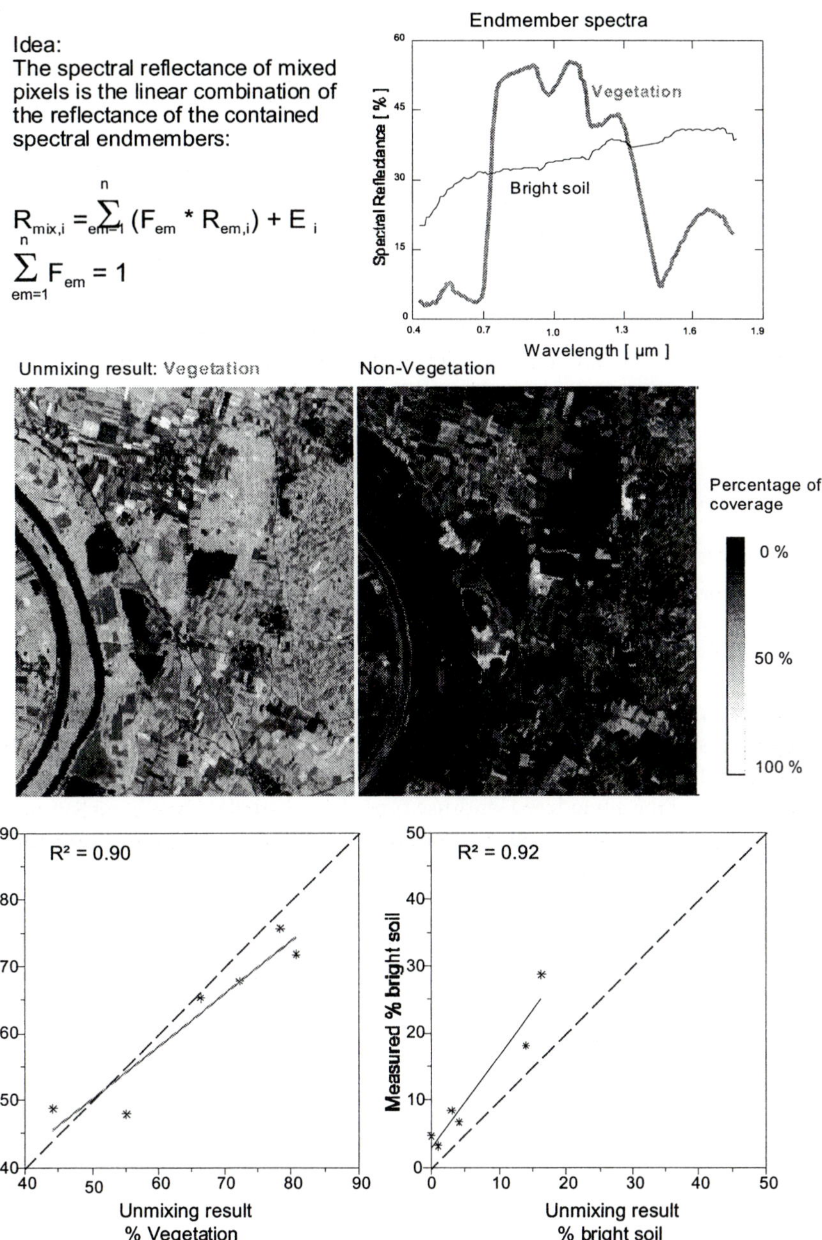

Figure 4.4 Spectral unmixing of hyperspectral data for the determination of the fractional vegetation cover (Bach 1995).

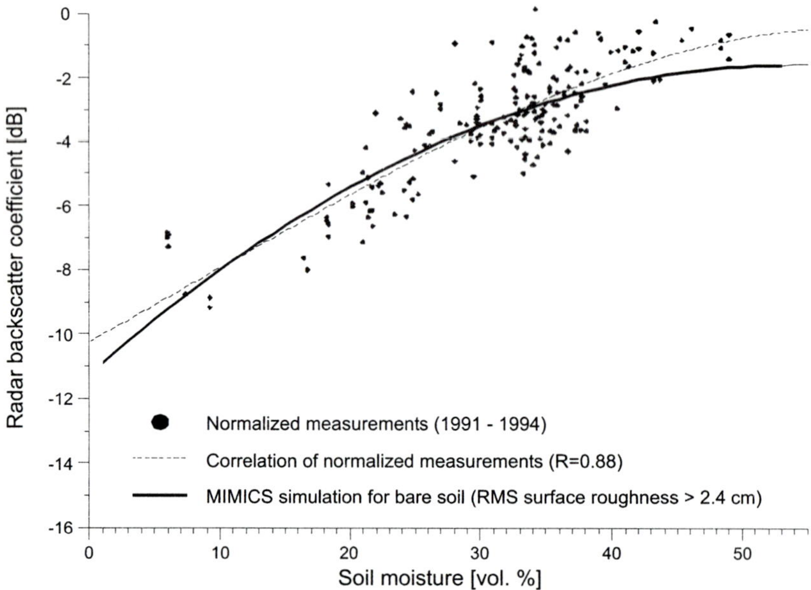

Figure 4.5 Correlation between backscatter coefficient and surface soil moisture derived from ERS-1/2 data (Rombach and Mauser 1997).

measurements of soil moisture for different annual land surfaces with the adjusted backscattering signal measured with ERS-1/2. The backscattering signals, which were measured over five growing seasons, were corrected for the influence of different levels of roughness of the considered land-use types based on plant height measurements in the field. As can be seen, the roughness corrected relation between soil moisture and backscatter signal compares well with modeled backscatter using MIMICS (Ulaby et al. 1990). As a first-order assumption, the moisture of the land surface can be determined down to a depth of approximately ½ wavelength, which means a depth of 2.5 cm for C-band, 10 cm for L-band, and 30 cm for P-band.

First results have already been achieved in comparing soil moisture measurements from microwave sensors with model results. Plate 4.2 shows the comparison of root zone soil moisture as determined with PROMET with surface soil moisture measured with the ERS-1 sensor using the relation shown in Figure 4.5. On May 29, soil moisture was low because of a prolonged dry period. After some heavy rains in June, the soil is very wet in the surface and the soil moisture in the root zone also increased.

Summary of Remote Sensing Capabilities

Table 4.4 summarizes which model parameters can be retrieved from remote sensing and also tries to give a best guess estimate of the suitability of different sensor configurations for the retrieval of input parameters for integrated landscape models. The

Table 4.4 Applicability of remote sensing data for the determination of landscape model parameters.

Input and internal parameter	Required temporal observation frequency	Applicability of remote sensing data for the present status of process models			
		Optic spectrometer	SAR C-band 35 days	SAR C-band 3 days	SAR X-, C-, L-band 3 days
Vegetation type	1 year	++	±	+	++
Leaf area index	7–14 days	++	–	–	+
Vegetation height	7–14 days	±	–	+	++
Biomass	7–14 days	++	±	±	++
Fractional vegetation cover	7–14 days	++	–	±	±
Emergence date	1–3 days	±	–	+	++
Spectral surface albedo	7–14 days	++	––	––	––
Surface roughness	7–14 days	–	+	+	++
Soil moisture	1–3 days	–	±	+	++
Static soil hydraulic properties	once	––	––	––	±
Root zone depth	7–14 days	––	––	––	––
Chlorophyll content	7–14 days	++	––	––	––
Plant nitrogen + minerals	7–14 days	+	––	––	––
Soil nitrogen + minerals	monthly	±	––	––	––
Topography	once	+	±	+	++
Snow water equivalent	daily	+	±	+	+
Surface temperature	–	++	––	––	––

++ = very good, + = good, ± = fair , – = poor, –– = not possible

potential of remote sensing is classified in *not possible*, *poor*, *fair*, *good*, and *very good*. In the following considerations, foreseeable future developments in data analysis (synergistic analysis, combination of remote sensing and models, inclusion of a priori knowledge from conventional sources, etc.) are taken account of as far as possible. The determination of plant parameters such as biomass and vegetation height, which depends on knowledge of individual species, serve as an example for this approach.

The same holds for soil moisture, which can only be obtained with C-band SAR for certain land cover types but not under forest. Information about land use is essential for the assessment of almost all other model parameters. It must not necessarily be determined with the same sensor used for parameter retrieval. Instead, complementary sensors with higher spatial and lower temporal resolution could be used to do the land-use classification. Not only the synergistic use of optical and SAR data, but also additional a priori knowledge and information from GIS systems are essential to improve parameter retrieval. Such data are helpful in reducing the dimensionality of the parameter space for the model. A good, yet simple example for this approach is to use topography to exclude certain species from further analysis beyond a certain elevation because of their known sensitivity to cold temperatures.

In Table 4.4, the potential of microwave sensors is subdivided in three different categories. The first one is a C-band SAR with a 35-day revisit cycle. This ERS/ASAR-type of sensor can be expected to still be available throughout the next years. The improvements that would be possible if the revisit time would be increased to three days is shown in the second category. User requirements for the next generation SAR systems ask for a polarimetric, multifrequency system. Therefore, a full polarimetric X-, C-, and L-band is demonstrated as a third group. For many model parameters, it can be seen that the retrievable information from SAR measurements increases from group 1 to 3.

As has been shown above, plant parameters that normally should be observed with a temporal frequency of one or two weeks can in principle be assessed with an optical spectroradiometer. Due to the weather dependency of optical observations, the revisit frequency of the sensor must be about an order of magnitude higher than the necessary model update frequency of the parameter. This eventually ensures a reliable supply to the model of vegetation parameters using optical data.

SAR data of any group will help to make the retrieval of vegetation parameters with optical sensors more reliable by delivering a temporally constant frame of observations, in which optical observations are embedded. In addition there are parameters that are required in aggregated models with a higher observation frequency of once per day to once every 2–3 days. At present, these parameters are generated by the models internally, which can lead to considerable instabilities in the modeling process. The two most prominent parameters are soil moisture and plant emergence date. For these parameters there is a good chance of retrieval using microwave systems of group 2 and 3. The next section will look into the mechanisms of incorporating remotely sensed parameters into aggregated models, their importance, and their influence on the model results in more detail.

REMOTE SENSING DATA: HOW TO UTILIZE THEM IN LANDSCAPE MODELS

Remote sensing data can be utilized and incorporated into landscape models in different ways, some of which are described below.

Determination of Model Input Parameters

The simplest way of using remote sensing data is to provide model input parameters that are static and do not change with time. These parameters can be obtained by specific retrieval models, which convert remote sensing measurements into parameter values. They have to be assessed once before the application of the model. One example for such an approach is the determination of the land-use type using remote sensing data. With both optical and SAR data, a land-use classification can be performed to spatially determine the land cover type. This information is essential for any process model of evapotranspiration and biospheric productivity. The accuracy of land-use classifications can be greatly enhanced if fuzzy theory and ancillary information on limitations to the growth of different species in terms of altitude, illumination, and temperature is known (Stolz and Mauser 1996).

Updating of Model Input Parameters through Forcing

If a model input parameter is needed more frequently because it changes with time, multitemporal remote sensing data can be used to update the parameters through model forcing. How this method can work is illustrated in Figure 4.6 for the LAI, which is a central parameter to most SVAT-BGC models. Optical remote sensing data sources deliver data in the form of radiances in irregular time intervals (dotted line in Figure 4.6). A parameter model is needed to derive values of model input parameters (rectangles) from these measurements. In a second step the discretely available measurements from remote sensing sources are converted into a continuous stream of values of model parameters through intelligent interpolation (temporal development of LAI in Figure 4.6). This information can then be used directly in the calculations of the process model, which results in transpiration values.

Recalibration of Internal Model Parameters

Remotely sensed observations can not only be used to provide surrogate values for one or more parameters that drive the model, they can also be used to adjust or correct the model during execution. This is illustrated in Figure 4.7, where soil moisture is provided for the modeling of the soil water balance. It is also a required input for the calculation of transpiration and evaporation, and at the same time, an output of the calculations of the soil water balance.

In Figure 4.7, a SAR sensor delivers data in the form of backscatter values in regular intervals (dotted line). A parameter model is applied to convert the backscatter values to soil moisture values of the soil surface (rectangles). Soil moisture is dependent on water balance inputs and outputs (precipitation, evapotranspiration, percolation, capillary rise) in a strongly nonlinear manner. Since precipitation of a few hours duration leads to a sudden increase of soil moisture, the discrete and regular time steps of the SAR measurements cannot simply be interpolated to a temporal course of soil

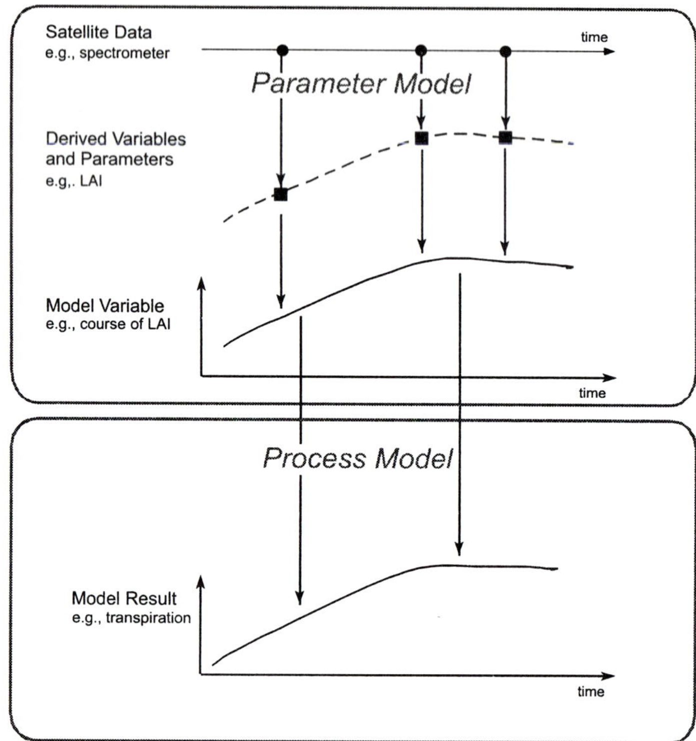

Figure 4.6 Updating of model input parameters through model forcing.

moisture. The temporal resolution of existing and planned SAR sensors is obviously too coarse for this. Therefore, soil moisture observations cannot be directly used as model input. The observations, however, can be compared at certain points in time with the soil moisture that results from the continuous modeling of the soil water balance (lower part of Figure 4.7). The differences in the spatial patterns of modeled and observed soil moisture can then be minimized through recalibration of the landscape model. The result is an adjusted and stable time course for soil moisture, which is externally controlled through measured values and therefore also close to the observations.

Parameter Determination through Model Inversion

A further step can be conducted, if one not only recalibrates the process model but inverts it on the basis of the observations to determine land-surface parameters. This approach has proven valuable in many cases where the model structure is relatively simple and relationships between parameters are unique. As an example, through parameter optimization using inverse modeling of the model SWATRE, scale-dependent

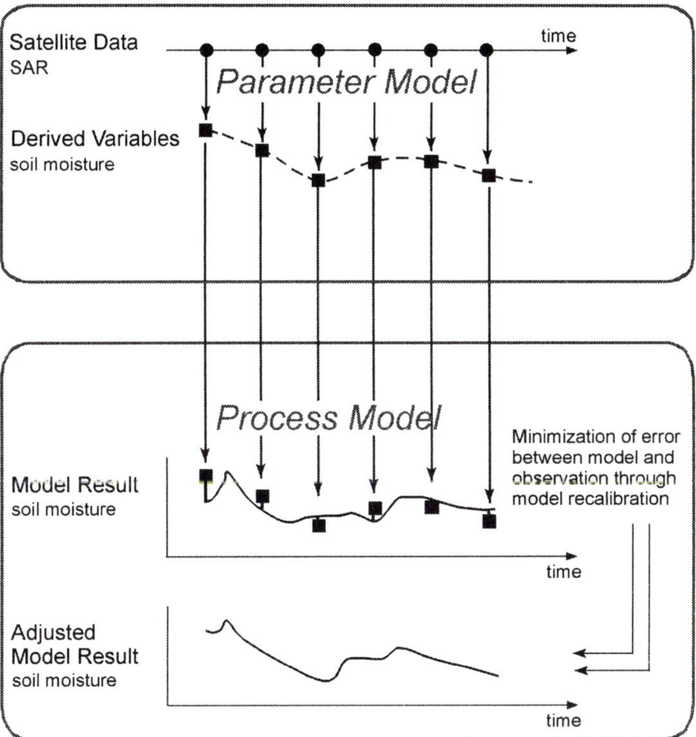

Figure 4.7 Recalibration of internal model parameters.

effective soil hydraulic functions could be inferred (Feddes 1995). On the basis of measurements of evapotranspiration and soil moisture in different depths, the hydraulic properties of the soils were determined. Feddes showed that his approach worked on 32 hypothetical soil samples, but has not yet tested it with remote sensing data due to the missing detailed soil moisture information. This is expected to be obtained in the future, when multifrequency and multipolarization SAR data is available.

A simplified illustration of this type of model inversion is given in Figure 4.8. A multilayer model of the soil water balance is run, assuming three different soil types (sand, sandy loam, and clay). The model result of surface soil moisture is then compared to soil moisture measurements conducted with microwave sensors. The model is inverted by determining the soil type, for which the temporal patterns of measured and calculated surface–soil moisture fit best. Weighting functions for the relative importance of the retrieved surface soil moisture using different SAR-frequencies, which correspond to different penetration depths, must be taken into account in this approach. In Figure 4.8, the SAR measurements show that the soil in this example is a sandy loam.

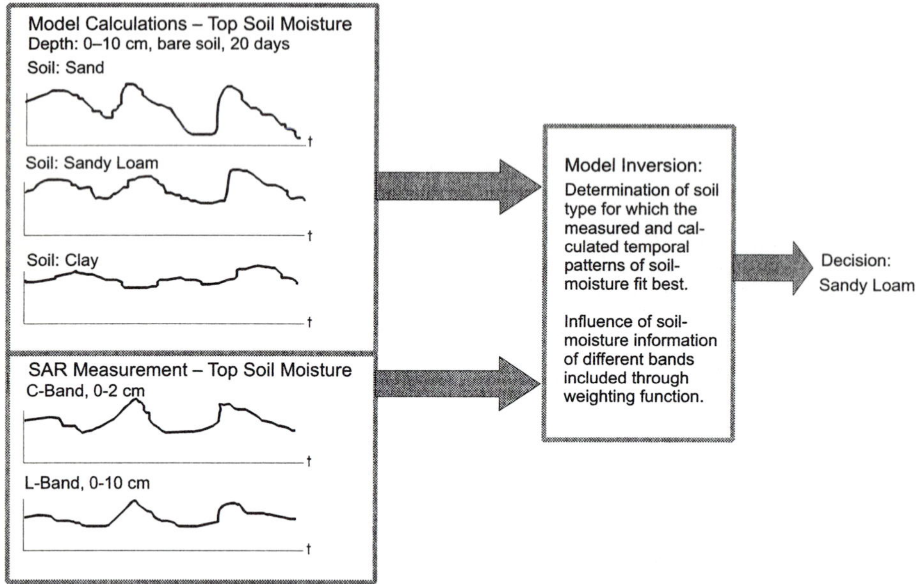

Figure 4.8 Soil-physical parameter determination through inversion of SVAT-models.

Validation of Landscape Models

The last but not less-important possibility for using remote sensing data in landscape models lies in the verification of the model results. The verification of the SVAT-BGC model output on a local scale for a point or a single field is possible. Several measurement techniques have been developed, such as Bowen ratio, eddy correlation, sap flow, and gas exchange.

But how can a model run, which is performed on a 10,000 km^2 mesoscale region, be verified or at least validated? In the upper part of Figure 4.9 the pattern of evapotranspiration is shown calculated using PROMET. It is calculated for May 15, 1992, 2:00 p.m. for an area of ca. 150 × 100 km in Upper Bavaria with a spatial resolution of 1 km^2. Direct measurements of transpiration are not possible on this scale. One possibility for the validation of this spatial pattern, however, lies in the fact that evapotranspiration alters surface temperature and thermal emission through cooling. Therefore, a comparison of the modeled evapotranspiration at a certain time and the surface temperature as measured using NOAA-AVHRR at this particular time is an independent path for model validation during cloud-free conditions. This at least holds true as long as surface temperature is not used as an input parameter in the model, which is true for the considered case.

Even though the times of the two data takes do not exactly correspond, the two images in Figure 4.9 are very similar in their spatial patterns, with lakes being the coldest and urban areas being the warmest regions. The forests (i.e., light areas in the upper left

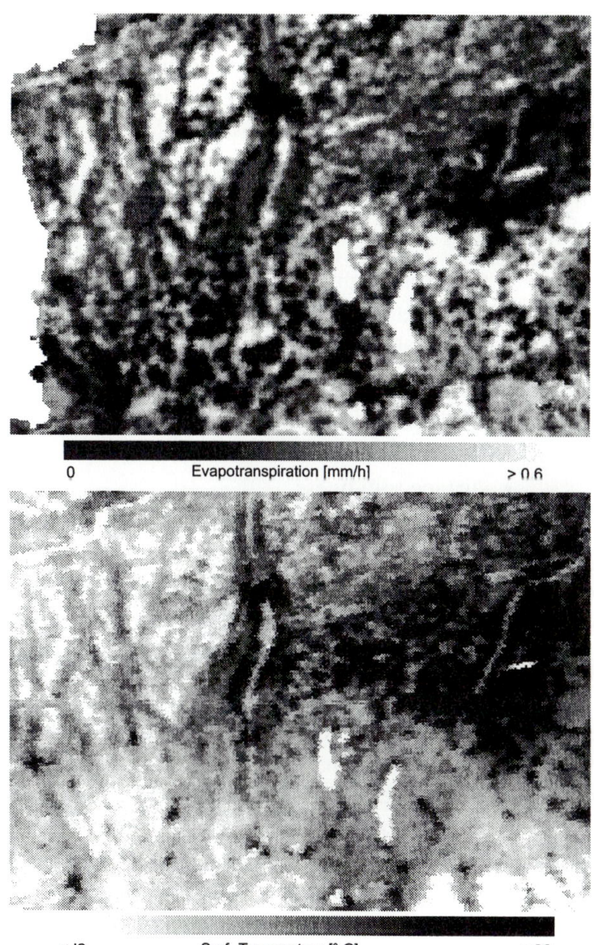

Figure 4.9 Model verification on the regional scale; top: spatial distribution of modeled hourly evapotranspiration in Upper Bavaria on June 1, 1992 14:00; bottom: Thermal image of Upper Bavaria of NOAA-AVHRR on May 15, 1992, 14:12.

part of the evapotranspiration image) are the coldest vegetated areas in the test region, indicating the large evapotranspiration of forests. After some modifications and further development, this approach would allow not only a qualitative validation but also a quantitative verification of mesoscale modeling.

SUMMARY

The primary goal of future applications of available remote sensing data of the Earth's surface is to use remote sensing to determine time series of parameter fields describing the dynamics of the land surfaces with an adequate quality to be used in regional landscape models. Missions of this quality are already on their way (ENVISAT, EOS) or planned (Earth Explorer PRISM). Existing SVAT and BGC models already describe spatially the flux of water and carbon at the boundary between soil, plants, and atmosphere. Our capabilities to fuse both parameter fields from remote sensing and models are still marginal. Therefore, the development of adequate data-assimilation and data fusion techniques is of great importance.

From the analysis of existing point and landscape models it was shown that the following parameter fields have to be supplied:

- net-radiation components,
- vegetation type,
- plant chlorophyll, nitrogen,
- emergence date,
- biomass,
- plant water content,
- soil hydraulic properties,
- ground water level.
- air-temperature, humidity, wind velocity,
- leaf area index, aPAR,
- plant cellulose, lignin,
- canopy height,
- fractional soil cover,
- soil moisture,
- surface roughness,

With the assumed development of available sensors within the next ten years the potential of remote sensing in landscape modeling can be summarized as follows:

- The majority of the parameters and variables required by landscape models can in principle be retrieved using advanced optical spectroradiometers with high repetition rates (1–3 days) and medium spatial resolution. This definition of the sensor category ensures that the derived parameters are locally verifiable. In this context, it is essential for these future systems to provide high spectral resolution, enabling not only the diagnosis of the absorbing/reflecting feature, but due to the absorption behavior and the shape of the absorption curve, provide access to more quantitative measurements. For this, knowledge of BRDF taking into account surface anisotropy is also needed. Using the optically derived parameters, it is likely that the landscape models will improve in their performance although important parameters like surface soil moisture have to be generated internally. This may lead to instabilities and inaccuracies of the models.
- Active microwave sensors of the ENVISAT/ASAR type with a short revisit time of 1–3 days are qualified to deliver unique model parameters, such as topsoil moisture and surface roughness. To improve the level of accuracy, additional data sources for rainfall, plant development, soil cover, etc. are necessary. This information is available within the landscape model structure. This data will have a spatial resolution of 1 km, which is not sufficient for regional studies. It would have to be improved to 50 m to be compatible with PRISM-type optical instruments.
- For the determination of meteorological parameters with high temporal resolution of one hour or less the geosynchronuous satellites of the next

generation (GOES, MSG) can be utilized. With the help of this data the estimation of the temporal and spatial distribution of radiation and (at least for the tropics) the estimation of rainfall will considerably improve. Since meteorological fields are very dynamic the spatial resolution of 1–4 km is adequate even for regional studies.

To ensure the constant flow of all necessary input parameters and variables, data from all available data sources (both remote sensing and ancillary) have to be fused to one data stream. Parameter models operate at the center of this data fusion process to convert remote sensing measurements into a set of model input parameters and variables. Several strategies to use remote sensing derived parameters in models were shown. They span from simple delivery of input parameters to the internal re-calibration of the model. Optical spectroradiometers can contribute land-surface information on vegetation type, LAI, biomass, plant water, fractional vegetation cover, surface albedo, and surface temperature. The drawbacks of optical systems, e.g., weather dependency and lack of penetration, must however also be considered. In addition to data of the optical spectrometer and thermal bands, data from active microwave systems on future missions will improve the parameter environment, in which the models operate by adding soil moisture, surface roughness, and topography. These systems have the advantage of reliable data availability and sensitivity to the moisture and roughness status of the land surface.

Nevertheless it is necessary to achieve success in spatial landscape modeling. Additionally, beyond basic proofs of the capabilities of remote sensing data for a few plants, many relationships between remote sensing measurements and model parameters have to be established empirically. This means that major emphasis should be given on ground measurements in conjunction with remote sensing measurements for detailed correlation analysis and with the aim of a more general view on the information content of the upcoming remote sensing instruments in terms of ecosystem functions. This will help to shift more consequently from a species oriented towards a function oriented parameterization of the land surface, and may be the only way to provide spatially distributed model parameters for complex plant communities or land-surface cover.

Overall, the improvement in understanding of the fluxes and processes on the landscape level, which can be expected when moving from point measurements to spatial measurements, is well worth the effort.

REFERENCES

Bach, H. 1995. Die Bestimmung hydrologischer und landwirtschaftlicher Oberflächenparameter aus hyperspektralen Fernerkundungsdaten. *Münch. Geogr. Abh.* **B21**:1–175.

Bach, H., A. Demircan, and W. Mauser. 1995. The use of AVIRIS data for the determination of agricultural plant developing and water content. In: MAC Europe'91 Final Results Workshop Proc., ed. M. Wooding, F. Lodge, and E. Attema, pp. 91–101. ESA WPP–88. Paris: European Space Agency.

Baret, F., and S. Jaquemoud. 1994. Modeling canopy spectral properties to retrieve biophysical and biochemical characteristics. In: Imaging Spectrometry as a Tool for Environmental Observations, ed. J. Hill and J. Megier, pp. 145–168. Euro Courses, Remote Sensing, vol. 4, EUR 15679. Brussels, Luxemburg: Kluwer Academic.

Belmans, C., J.G. Wesseling, and R.A. Feddes. 1983. Simulation model of the water balance of a cropped soil: SWATRE. *J. Hydrol.* **63**:271–286.

ESA (European Space Agency). 1998. The Land Surface Processes and Interactions Mission — Mission Requirements Document. Noordwijk: European Space Agency.

Famiglietti, J.S. 1992. Aggregation and scaling of spatially-variable hydrological processes: Local, catchment-scale and macroscale models of water and energy balance. Ph.D. diss. Princeton Univ., Princeton, NJ.

Famiglietti, J.S., and E.F. Wood. 1994a. Application of multiscale water and energy balance models on a tallgrass prairie. *Water Res.* **30(11)**:3079–3093.

Famiglietti, J.S., and E.F. Wood. 1994b. Multiscale modeling of spatially variable water and energy processes. *Water Res.* **30(11)**:3061–3078.

Feddes, R.A. 1995. Remote sensing — Inverse modelling approach to determine large scale effective soil hydraulic properties in soil-vegetation-atmosphere systems. In: Space and Time Scale Variability and Interdependencies in Hydrological Processes, ed. R.A. Feddes, International Hydrology Series, pp. 33–42. Wageningen: Cambridge Univ. Press.

Feddes, R.A., P. Kabat, P.J.T. van Bakel, J.J.B. Bronswijk, and J. Halbertsma. 1988. Modeling soil water dynamics in the unsaturated zone — State of the art. *J. Hydrol.* **100**:69–111.

Feddes, R.A., and R.W.R. Koopmans. 1995. Agrohydrology. Dept. of Water Resources, Publication KI50-305. Wageningen: Wageningen Agricultural University.

Ford, R., S.W. Running, and R. Nemani. 1994. A modular system for scalable ecological modeling. *IEEE Comp. Sci. Eng.* **1070**:32–44.

Goel, N.S., and R.L. Thompson. 1984. Inversion of vegetation canopy reflectance models for estimating agronomic variables. V. Estimation of Leaf Area Index and Average Leaf Angle using measured canopy reflectances. *Remote Sens. Env.* **23**:453–477.

Kabat, P., B.J. Broek, and R.A. Feddes. 1992. SWACROP: A water management and crop production simulation model. *ICID Bull.* **41(2)**:61–84.

Mauser, W. 1991. Modeling the spatial variability of soil-moisture and evapotranspiration with remote sensing data. In: Proc. Intl. Symp. Remote Sensing and Water Resources, Enschede, Aug. 20–24, 1990, pp. 174–182. Lingen: R. van Ackern Pub.

Mauser, W., and S. Schädlich. 1998. Modeling the spatial distribution of evapotranspiration on different scales using remote sensing data. *J. Hydrol.* **212/213**:251–267.

Myneni, R.B., S.O. Los, and G. Asrar. 1995. Potential gross primary productivity of terrestrial vegetation from 1982–1990. *Geophys. Res. Lett.* **22**:2617–2620.

Myneni, R.B., R.R. Nemani, and S.W. Running. 1997. Estimation of Global Leaf Area Index and Absorbed PAR using radiative transfer models. *IEEE Trans. Geosci. Remote Sens.* **35(6)**:1380–1393.

Nemani, R.R., and S.W. Running. 1989. Testing a theoretical climate-soil-leaf area hydrological equilibrium of forests using satellite data and ecosystem simulation. *Agr. For. Meteorol.* **44**:245–260.

Rombach, M., and W. Mauser. 1997. Multiannual analysis of ERS surface soil moisture measurements of different land uses. In: Space at the Service of Our Environment, pp. 27–34. Proc. Third ERS Symp., Florence. ESA SP–414, vol. 1. Noordwijk: European Space Agency.

Running, S.W., and J. Coughlan. 1988. A general model of forest ecosystem processes for regional applications. I. Hydrologic balance, canopy gas exchange and primary production processes. *Ecol. Mod.* **42**:125–154.

Running, S.W. 1991. Computer simulation of regional evapotranspiration by integration landscape biophysical attributes with satellite data. In: Land Surface Evaporation — Measurement and Parameterization, ed. T.J. Schmugge and J.C. André, pp. 559–370. Heidelberg: Springer.

Running, S.W. 1994. Testing FOREST-BGC ecosystem process simulations across a climatic gradient in oregon. *Ecol. Appl.* **4(2)**:238–247.

Schädlich, S., and W. Mauser. 1996. Spatial evapotranspiration calculation on a microscale test site using the GIS-based PROMET model. *IAHS Publ.* **235**:649–657.

Stolz, R., and W. Mauser. 1996. A fuzzy approach for improving landcover classifications by integrating remote sensing and GIS data. In: Progress in Environmental Remote Sensing Research and Applications, ed. E. Parlow, pp. 33–41. Rotterdam: Balkema.

Ulaby, F.T., K. Sarabandi, K. McDonald, M. Whitt, and M.C. Dobson. 1990. Michigan microwave canopy scattering model. *Intl. J. Remote Sens.* **11(7)**:1223–1253.

Wessman, C.A. 1994. Estimating canopy biochemistry through imaging spectrometry. In: Imaging Spectrometry as a Tool for Environmental Observations, ed. J. Hill and J. Megier, pp. 57–70. Euro Courses, Remote Sensing, vol. 4, EUR 15679. Brussels and Luxemburg: Kluwer Academic.

White, J.D., and S.W. Running. 1994. Testing scale dependent assumptions in regional ecoysystem simulations. *J. Veg. Sci.* **5**:687–702.

Standing, left to right:
Josep Peñuelas, Phillipe Martin, Wolfram Mauser, Ramakrishna Nemani, Robert Gurney, Carol Wessman
Seated, left to right:

5

Group Report: Remote Sensing Perspectives and Insights for Study of Complex Landscapes

C.A. WESSMAN, Rapporteur

W. CRAMER, R.J. GURNEY, P.H. MARTIN, W. MAUSER, R. NEMANI, J.M. PARUELO, J. PEÑUELAS, S.D. PRINCE, S.W. RUNNING, and R.H. WARING

INTRODUCTION

Opportunities to apply remote sensing techniques to landscapes have increased significantly in the last decade as a result of advances in technology, quantitative techniques, and the combination of biophysical models with remotely sensed data. These advances allow us to monitor and model the implications of rapidly changing terrestrial surfaces more realistically and accurately than ever before. As a result, we are on the verge of being able to determine, as well as predict with models, responses to changes in land use, vegetation cover, and interactions with climatic factors that together affect the state of the global environment.

Improvements in remote sensing technology make analyses more quantitative than previously possible. The first generation of satellite sensors was uncalibrated and thus unable to provide an accurate measure of reflectance in any spectral band. In addition, bandwidths were broad, limiting spectral discrimination between, for example, land surface and atmospheric properties. As a result, until recently, the primary use of satellite-derived data was in classifying landscape units into broad categories that differed significantly in the relative reflectance in one spectral band to another. In contrast, the next generation of satellite sensors, soon to be launched, will be calibrated and thus provide more quantitative measures of landscape features (Kaufman et al. 1998; Justice et al. 1998). In addition, the data sets derived from the new satellite sensors will be more useful to the general scientific community because the information will be widely distributed as a suite of carefully chosen biophysical variables. These

Integrating Hydrology, Ecosystem Dynamics, and Biogeochemistry in Complex Landscapes
Edited by J.D. Tenhunen and P. Kabat

advances should dramatically accelerate the use of satellite data by nonremote sensing specialists for regional-scale science and resource management questions.

In this chapter we stress advances in remote sensing that offer new insights in studies of landscape and regional complexity. Our two major objectives are to provide a framework for appraising variation across landscapes, and to illustrate how remote sensing can better be integrated into modeling the response of complex landscapes.

WHEN AND WHERE DOES LANDSCAPE COMPLEXITY MATTER?

Landscape Complexity and Remote Sensing

How much detail is required in a landscape analysis depends on the question asked. Complexity of the landscape will vary with respect to the spatial distribution of its components such as vegetation, topography, water, and soil, and surface physical properties such as reflectance, temperature, and dielectric constant. These variables clearly have functional significance as well, being associated, as they are, with significant differences in the rates that waters flow, carbon is stored, and gases are exchanged with the atmosphere. As the structure and pattern of landscape variables alter the magnitude and direction of these processes, the processes will, in turn, modify structure and pattern (Shugart 1996). The degree to which this link between structure and function is expressed is the key to the use of remote sensing in landscape studies which we seek (Wessman and Asner 1998).

Over time, changes in the type, age, and density of vegetation are among the most dynamic features of a landscape. Because satellites provide consistent and frequent coverage of the Earth's surface, data derived from satellite-borne sensors provide the only widely available source of information to quantify spatial and temporal changes in vegetation and land use. Where large areas are represented by vegetation with a similar composition and activity, the analysis is greatly simplified. On the other hand, if the landscape pattern is heterogeneous, more complexity is introduced, which in turn requires increased resolution in measurement accuracy and spatial registration of data.

Some variables can be assessed directly through remote sensing, such as the height of vegetation or the fraction of light intercepted by leafy canopies (e.g., Sellers et al. 1992; Lefsky et al. 1998). Other variables can only be inferred, such as the populations of animals associated with different types of vegetation, the seasonal rates of canopy photosynthesis, and litter decomposition rates. The inferences may be based on fundamental biophysical relationships and therefore be quite accurate, or be derived from widely scattered ground surveys, which may no longer be representative. The reliability of any inference clearly depends on the consistency between measured and inferred relationships.

Detection of Spatial and Temporal Scales of Complexity

No landscape is perfectly homogeneous spatially or temporarily. Across relatively small landscapes, differences in soil properties or patterns of land use contribute much to the observed complexity. With increasing areal extent, local influences have less effect; variation in climate and landform become the dominant contributors to spatial variation (Meentemeyer and Box 1987; Wiens 1989). Similarly, the critical factors associated with temporal variation across landscapes change at the time dimension increases. Seasonally, most of the variation observed is associated with the amount of leaf area displayed. Over decades, disturbance from fire, floods, windstorms, harvesting activities and land conversion account for most of the variation. With consistent coverage now available through remote sensing, changes at the landscape or regional scale can be detected and an appropriate scale selected to answer the questions posed.

Spatially, the lower limit of discrimination ultimately is set by the spectral, spatial, and temporal resolution of the original remotely sensed data. The actual limit of discrimination is generally less than that prescribed by the satellite-borne instruments alone. Resolution is reduced below the theoretical maximum because of unfavorable viewing angles, atmospheric conditions, and errors introduced during the processing of data through imprecise registration and contamination from adjacent cells.

The time series of scenes acquired through remote sensing also can serve as a basis for identifying general relationships to aid in scaling to broader levels of integration. One means of discovering relationships that scale is to apply simulation models to predict changes in vegetation structure and function with progressively larger aggregation of cells (*pixels*). The difference between model runs with increasing size aggregations indicates whether functional relationships are linear and stable or nonlinear and unstable. Relationships often change over time. For example, shortly after a major storm, runoff is a simple function of soil permeability at saturation and may be similar across a wide area. During a drought, however, few soils remain saturated and the spatial variation in permeability increases. Likewise, processes such as photosynthesis and transpiration can be easily estimated across large areas when it is not necessary to delineate where soil drought is a major constraint. "Continuity in coverage" is thus a key requirement to define an appropriate spatial resolution necessary to integrate functional responses across landscapes.

Defining Landscape Complexity on the Basis of Controlling Factors

In addition to providing the power to detect change across landscapes, remote sensing can help stratify landscapes into categories that reflect the dominant agent creating the observed patterns. In this section we present examples where remote sensing can help distinguish whether observed landscape variation is inherent or externally imposed. We include examples where inherent differences in life forms of vegetation explain variation (savannas), and where topography creates climatic gradients, along with situations where external forces dominate, such as in the conversion of forests to other vegetation, and where urban development is spreading across a landscape.

Savanna: An Example of Inherently Complex Vegetation

Tropical and subtropical mixed woodland and grassland systems, often known as savannas, are an example of a complex system where the different processes that operate in the grassland and woodland components interact with each other to produce significant patchiness in the ecosystem (Scholes and Archer 1997). Savannas have distinctive hydrological, vegetation, and biogeochemical responses, which, to a large extent, can be assessed with remote sensing techniques. The range of spatial scales currently recognized in savanna ecosystem dynamics spans the ranges of a number of existing sensor systems. Beyond the measurement of the spatial extent of the system as a whole, remote sensing can contribute information on the temporal and spatial values and variation of some functional properties of the system.

An important element in the maintenance of pattern in savanna ecosystems is the spatial extent and temporal frequency of fires. These are now detected from satellites with increasing precision (e.g., Belward et al. 1994; Pereira and Setzer 1996). Analyses of subpixel fractional cover (i.e., spectral mixture analysis) provide spatial information on the patterns of tree and grass cover and their potential change over time with management pressures (e.g., Roberts et al. 1993; Wessman et al. 1997; Asner et al. 1997). Spectral mixture analysis coupled with radiative transfer models that simulate the interaction of solar energy with plant and soil surfaces on a physical basis (Myneni et al. 1989, Jacquemoud et al. 1995, Asner and Wessman 1997) provide a means to quantify leaf and plant area available for vegetation–atmospheric exchanges (Asner, Wessman, and Archer 1998; Asner, Wessman, and Schimel 1998). The amount and seasonality of the fraction of absorbed photosynthetically active radiation (fPAR), monitored with spectral vegetation indices, indicates carbon allocation patterns (e.g., Prince and Goward 1995). Passive sensors have generally not been able to provide a measure of standing biomass, but with the launch of satellites carrying active radar or lidar the potential for obtaining better quantitative estimates of aboveground structure will greatly improve (Lefsky et al. 1997, 1998). When active sensors are in orbit, quantitative three-dimensional data sets will be available to verify predicted growth, harvest, and conversion rates of major types of vegetation to another life form or land use. Other remotely sensable measurements important to savanna ecology include, in decreasing order of certainty, surface temperature, aboveground biomass, air temperature, vapor pressure deficit, soil moisture, and rainfall.

Topographically Variable Regions

Large variations in topographic features generate dramatic gradients in climatic conditions. In general, 1000 m of elevation is equivalent to a shift of 600 km in latitude. Even a hill of a few hundred meters can alter the precipitation by more than 50%, while at the same time affecting temperature and wind profiles. The direction and steepness of the slope can also create significant variation in the received solar radiation. Topographic variation is particularly critical to consider when predicting hydrologic responses because topography largely controls the direction that water flows, the way that snow is deposited, and the amount of water available to vegetation and streams throughout the year.

As a matter of stratification, topographic features are among the most stable and, thus, the first layer of detail that should be imposed in making a landscape analysis. Inherent variation in topography can be related to the kind of vegetation present, its seasonal activity, and its likelihood of experiencing natural disturbances. In addition, the location and intensity of human activities can often be predicted on the basis of topographic relationships. Any models proposed to test the kind and frequency of landscape disturbance clearly will benefit from the availability of comprehensive data sets being accumulated from a variety of remote sensing platforms. Consistency in coverage, preciseness in registration, and availability of data sets are challenging prerequisites, but ones that the remote sensing community feels strongly should be provided to meet the widespread need for more comprehensive landscape analyses. At present, use of remote sensing technology is inhibited because of deficiencies in one or more of these three prerequisites.

Conversion of Natural Vegetation

The most rapid change in the present terrestrial environment involves the conversion of natural vegetation to agriculture or urban landscapes (Turner et al. 1990). The extent that these changes alter the flow and storage of water, carbon, and minerals depends on the specific location and composite activities occurring across a region. Generally, the rates of water movement and sediment and chemical transfers increase through fragmentation of contiguous patches of vegetation, the intensity of historic land use, and the current frequency and intensity of fire and other disturbances. Ideally, the frequency of coverage should be sufficient to define the rates that all types of disturbance occur. With appropriate data, landscape process models can improve estimates of net changes in standing biomass, and the rates that water, carbon, and nutrients cycle through diverse landscapes.

Urban Environments

Urbanization is a major manifestation of human activity, and in some areas is expanding rapidly. Human settlements create patterns that affect lateral flows of water, nutrients, and pollutants into and across landscapes. The resulting interactions are difficult to predict but urbanization, which appears to simplify the natural environment, can also introduce considerable complexity by altering natural gradients and by creating new sources of chemical inputs to the environment. The emissivity and spectral absorption and reflectance properties of urban areas differ substantially from natural surfaces and thus can be easily perceived from satellites. There is an opportunity to improve estimates on the rates of settlement development, and to quantify changes in landscape properties that affect the mesoclimate and hydrology of whole regions.

WHAT ARE THE MAIN MODES OF APPLICATION OF REMOTE SENSING?

Remote sensing data fall naturally into three broad, spatially-based categories that represent patches, landscapes, and regions. There is a link between the size of area and importance of various ecological processes. According to hierarchy theory, within the patch lies an explanation for variation observed in landscapes, and within the landscape lies an explanation for regional changes (Allen and Hoekstra 1992). Thus an increase in foliage mass in patches of vegetation might explain higher rates of evaporation. Whether precipitation had increased significantly would require analysis at the landscape level to follow changes in stream and reservoir dimensions and the duration that soil surfaces remain wet. If much of the runoff is stored and used to irrigate arid areas, this could explain increases in humidity and precipitation over a region. Alternatively, a region-wide drought could be imposed that would bring about opposite trends at all levels and lead to increased outbreaks of fire and defoliating insects. To unravel these kinds of hierarchical interactions, different types of remote sensing are obviously required.

In this hierarchical context, pattern and periodicity are two land-surface properties that can be most reliably sensed remotely. Pattern is descriptive of the spatial distribution of any variable. Satellite-based sensors acquire information on landscape units as small as 10 m on a side and as large as the entire globe. The periodicity of the growing season can thus be defined at a broad array of scales, but data from satellites are only available for the last 25 years. Fluctuations in the periodicity of the growing season, periods of drought, or floods can also be obtained through an adequate sampling of satellite-derived data. To quantify functional relationships at the landscape level requires a coupling of pattern and periodicity with models that usually require additional information on topography, soils, and climate. Regardless of how many different types of sensors are placed on satellites, there will always be need for some direct measurements to confirm underlying assumptions and to verify model predictions. As models and measurement techniques continue to improve, we will better be able to anticipate and to document the consequences of expanding human activities.

INTEGRATION OF MODELING AND REMOTE SENSING

The Evolving Landscape Perspective

Classically, ecological analyses identified variation in landscape patterns on the basis of "plant communities" that differed in their taxonomic composition. Communities were assumed to represent a predictable series that followed disturbance, in particular physiographic units. Ecologists recognized that groups of species occupied similar environments, but the environment was usually left quantitatively undefined in such aspects as water availability, fertility, or susceptibility to various types of disturbance. Geographers broadened the classification of vegetation by recognizing different life cycles (annuals and perennials) and life forms (evergreen needle-leaf and broad-leaf trees, sprouting and nonsprouting shrubs, etc.). Physiologists brought a functional perspective to community analysis by recognizing differences in photosynthetic pathway, abilities of some plants to fix nitrogen from the atmosphere, and to tolerate specific environmental stresses (e.g., frost, drought, salt, infertile soils). Ecosystem studies emphasized classifications based on key properties (e.g., rooting depth, leaf area, surface litter accumulation, and turnover) that result in different rates of energy and material capture and flow through systems distributed across landscapes. At the landscape level, ecosystem functions and interactions were difficult to predict until modeling procedures were developed to extrapolate climatic and soil information geographically. Remote sensing contributes further to extrapolation of ecosystem processes by providing estimates of canopy structure as it changes over time and space. This latter contribution provides a basis for distinguishing responses to climatic variation from those associated with disturbance.

We are now at the threshold where dynamic models exist to predict ecosystem and community response at finer spatial and temporal resolutions, and where different remotely sensed variables are increasingly available to verify initial and changing states of systems across landscapes. Nevertheless, the integration of remote sensing into landscape process models requires a significant adjustment of the basic model formulation and the field sampling strategies for model parameterization and validation. Modelers must keep in mind the physical nature of remote sensing. Whatever properties have an influence on the reflection, absorption, and transmission of photons will influence the measure made by the sensor. Models attempting to address landscape level processes need to be deliberately designed to use remotely sensed variables or produce estimates of variables that can be remotely sensed (model what can be measured, not measure what can be modeled.)

The capability of remote sensing to detect spatial and temporal patterns provides a strong reason for its use in the design of field measurement campaigns. If the local measurements can be inferred either directly or indirectly from variables that are available from remote sensing, then the spatially comprehensive coverage provided by remote sensing instruments can be exploited to estimate variables at landscape to regional scales. The inference may be from a mechanistic model or from an empirical correlation, with appropriate cautions in both cases. Remote sensing simplifies field-sampling strategies for model parameterization/validation because it, as the scaling tool, can delineate landscape structural or functional units for optimal sampling design (Prince and Steininger 1999). There are well-proven scientific approaches that, whenever possible, first characterize the slow-moving parts of a system (and also always compare model output with observations). These principles have been applied during the development of retrieval theory in remote sensing, but not always in problems in ecology and hydrology using remote sensing. A field approach to characterize aspects of the landscape through an understanding of the factors which influence, if not control, reflectance is the one most likely to yield maximum benefit from the observations. Strategic sampling of relevant landscape units provides the range of variation within the landscape with which to bound model parameters.

In the past, the use of remote sensing has been most important for the identification of land-surface types which have provided the first level of stratification in ecosystem analysis. Once major structural ecosystem types could be identified from space, the corresponding biogeochemical models could be parameterized differently for the different classes. Recently, the temporal dimension of satellite measurements has been exploited to improve the classification, mainly by better separation of deciduous from evergreen canopies. This has also resulted in improved flux estimates. A second role of remote sensing in ecosystem modeling has been to validate estimates of dynamic ecosystem state variables, such as leaf area index (LAI) or soil moisture. Here, the ecosystem model has been used (either with or without using a satellite-based land cover classification) to estimate prognostically ecosystem characteristics as a function of the environment (the state of which may be derived from ground measurements or from satellites). LAI, for example, could then be compared between the satellite-derived estimate and the state variable of the ecosystem model, and deviations between the two could be analyzed and hopefully be used to improve the model formulation. It must be noted that this test does not represent model validation in a strict sense, because the satellite-LAI is also a model-based estimate, and because no systematic factorial experiment with different states of the ecosystem is possible. Nevertheless, such tests are generally helpful in adding understanding to a prognostic simulation of ecosystem processes. If a model indeed is designed to run with climate and site information only, it may then also be used to extrapolate into the future.

More recent developments approach the problem from a different angle. Fundamentally, the most important input to ecosystem processes is solar radiation, and its reflection is one of the fundamental quantities that can be sensed by satellites. Based on this consideration, it is more meaningful to treat the ecosystem model as a model simulating ecosystem processes in terms of their reflectance, rather than LAI or soil moisture. Following this logic, satellite observations can be used either to detect ecosystem processes (most importantly, APAR) directly, or to validate the capacity of the ecosystem model to estimate the fraction of incoming solar radiation that is absorbed by photosynthesis (fPAR). In both cases, the problem of estimating LAI, which is always problematic, since most satellite-derived vegetation indices saturate at LAI values higher than about four, is avoided altogether (Fischer et al. 1996). A recent application of this logic was given by Myneni et al. (1997) who showed that broad-scale trends in changing ecosystem function might well be detectable from presently available time series of satellite measurements.

Remote Sensing and Applications to Functional Groups of Plants

Canopy structure and dynamics play a significant role in energy and mass exchange processes of vegetation, providing one link between ecology and remote sensing. Using simple canopy structure variables such as the aboveground biomass, leaf longevity, and leaf size/shape that are inferred remotely, vegetation has been grouped into simple classes or functional groups (such as deciduous broad-leaf, deciduous needle-leaf, evergreen broad-leaf, evergreen needle-leaf, grass and broad-leaf annuals; Running et al. 1995). While such a simple classification scheme may not satisfy a wide variety of applications at regional scales, it appears to be adequate for representing global variations in biogeochemical cycling. Canopy structural characteristics used to define functional groups also influence energy absorption and emission. We can conclude, therefore, that frequent observations of surface reflectance and surface temperature over time may allow us to classify vegetation into the above functional groups or related groups. Seasonal trajectories of surface reflectance and temperature in various landscapes also allows us to monitor landscape dynamics in terms of changes in function (e.g., phenology, growth) and structure (e.g., land cover changes).

Characteristics of other remotely sensed variables such as fPAR which capture basic aspects of mass and energy exchange can help define spatial distribution of functional groups. For example, seasonal curves of Normalized Difference Vegetation Index (NDVI) values from the Advanced Very High Resolution Radiometer (AVHRR) can provide estimates of changes in fPAR throughout the year and help define timing of the start and end of the growing season (Goward et al. 1985; Soriano and Paruelo 1992; Paruelo and Lauenroth 1995; Paruelo et al. 1997; Running et al. 1995). Multivariate techniques (i.e., principal components analysis, tree regression analysis, cluster analysis, etc.) may provide insights on the main directions of variability. A description of the gradients of attributes listed above or a hierarchical description of functional, remote-sensing-based units would provide a complementary approach to the definition of land area functional units based on descriptions of carbon, water, and N fluxes (Valentini et al., this volume).

FUTURE CONTRIBUTIONS OF REMOTE SENSING TO LANDSCAPE STUDIES

Currently available satellite remote sensing instruments measure radiation in very broad bands in the visible (VIS), short-wave infrared (SWIR) and thermal wavelengths (7 for LANDSAT TM, 2 [VIS/NIR] for AVHRR). Successful and robust applications using these sensors have primarily included broad landscape classification, vegetation index measures, and surface temperature. However, the limited number of wavelengths of these sensors restricts both the total information content of the signal received, and the methods which can be used to extract information from that signal.

New developments in imaging detector technology now allows high spectral resolution remote sensing which typically involves the collection of reflected radiation in over 200 wavebands at about 10 nm intervals through the visible and shortwave infrared region (400–2400 nm). There is not yet a space-borne instrument of this type. Data collected to date have come from helicopter and aircraft instruments. One major application of high spectral resolution data of relevance to landscape studies is their use in determining the biochemistry of forest canopies, allowing the inference of ecosystem process rates. For instance, it appears that mean concentrations of foliar nitrogen and lignin can be estimated for whole forest canopies, at least from aircraft (Wessman et al. 1988; Martin and Aber 1997). Physiological status and stress conditions may be assessed through absorption characteristics of plant pigments and water content (Gao and Goetz 1995; Peñuelas et al. 1995; Gamon et al. 1998). These variables are important in several landscape-level ecosystem models. A large number of narrow wavebands also allows a wider range of analysis techniques to be used, with a stronger physical basis. These techniques are critical in separation of individual variables across a heterogeneous landscape, and become important in landscapes such as savannas in which mixing of functional groups becomes prevalent. In soil and geology mapping, high spectral resolution data allow the compilation of libraries of a large number of unique endmembers for more accurate separation into larger numbers of endmember classes. However, there are currently limitations to high spectral resolution remote sensing. They include the absence of a satellite-borne system, limited swath width from airborne systems, and little testing of algorithms has been carried out due to limited data availability. Still, the value of fine spectral resolution measurements, along with planned new instrument deployment, suggests that this technique will increase in importance over the next decade.

Most operational remote sensing instruments provide information that can only be interpreted in two dimensions, usually as a map. Stereo imagery and shadowing models enable some limited information about the vertical dimension to be obtained. Remote sensing that can be used for ranging involves active instruments such as radar and, more recently, lidar. Retrieval of topography or vertical structure using radar has severe limitations because of the complicated way that the land surface reflects microwave radiation. Lidar, on the other hand, is simpler to interpret. Early use of lidar was limited to narrow-beam devices that only record the first return but new laser systems are being developed that enable the top, bottom, and internal structure of the vegetation to be measured.

Because vegetation reflectance varies with changing sun and sensor-viewing geometry, measurements of the land surface made from multiple angles provide information on canopy- and landscape-level structural characteristics (Goel 1988; Myneni et al. 1989). Multi-view angle measurements of vegetation reflectance will allow improved access to canopy structural characteristics (e.g., LAI) and retrieval of important biophysical variables such as fPAR (Braswell et al. 1996; Privette et al. 1996). A lack of instrumentation has restricted development of these techniques for ecological research, but planned instrumentation will allow these methods to achieve an operational level of use (Barnsley et al. 1994; Asner, Braswell et al. 1998).

The joint use of satellite remote sensing and models of surface processes has advanced to the point at which useful landscape to global scale observations can be carried out routinely. Monitoring systems are in use in Europe for policing the Common Agricultural Policy and in several parts of the world for commercial and government intelligence gathering. Our capabilities, however, extend beyond mapping and include using observations to constrain models of many environmental processes. For example, remote sounding of atmospheric temperature and moisture is used in numerical weather prediction to give estimates of precipitation and evaporation over land surfaces. Research is in progress to investigate the possibility of establishing a degradation early warning system in southern-central Africa in which multiple land-surface properties will be used to diagnose the incidence of medium-term land degradation.

OUTLOOK

Improved Implementation

There has been a significant maturing in remote sensing to provide well-calibrated instruments for specific scientific applications. A new generation of sensors is being designed by funding agencies to answer specific scientific questions as part of the entire project. This approach makes remote sensing data more available to the general scientific community than has been done in the past. In fact, related implementation processes need to be applied to current and archived data as well, in order to improve the use of an historically important database. Certainly, the lack of ready and easy access to data has been a serious hindrance to the general use of remote sensing in, for example, ecological and hydrological sciences and natural resource management.

New Integrated Modeling

As discussed earlier, integration of remotely sensed data is critical to landscape scale process modeling. Several models currently exist which utilize remote sensing data as input or constraints (Field et al. 1995; Hunt et al. 1996). New models are designed to make full use of our biophysical understanding, and include models of the interaction of radiation with the land surfaces, often using radiative transfer theory and assimilation techniques. However, greater interaction is needed between ecologists and physicists to broaden the use of the observations, and graduate education to promote this interdisciplinary approach needs to be encouraged.

New Techniques and Technologies

New technologies such as radar, lidar, multiangular sensors, and imaging spectrometry are needed for increased ability to track a number of variables in time and space, with reduced interference from the atmosphere and other confounding factors. Measurements in higher spectral and spatial resolutions, at multiple angles, and with combined spectral capabilities (e.g., optical and radar) will facilitate our ability to address the multivariate nature of landscapes. Assessing complex landscapes with single-value indices, as is often done with limited spectral or angular information, leaves ecological problems significantly underdetermined.

New technologies are becoming increasingly available through international small-satellite programs and growing interest from the commercial sectors. Satellites can now be made smaller and cheaper (in construction and launch), leading to more rapid deployment and lower capital risk. As a consequence, delays in observational programs are significantly reduced and broader capabilities in spectral and temporal coverage are more likely. In other words, new technology may yield far more new thinking, leading the way for remote sensing systems to be designed for the scientific problem and not the reverse.

New Collaborative Experiments

The application of remote sensing to the understanding of landscapes necessarily involves the joint measurement of surface processes by direct methods and the collection of the relevant remotely sensed data. New experiments are needed where we can work on clarifying where these remote sensing technologies best combine with landscape process models. The recent large, interdisciplinary field measurement campaigns such as FIFE, HAPEX, EFEDA, ABRACOS, and BOREAS (Sellers et al. 1988; Bolle 1995; Hall and Sellers 1995; Sellers et al. 1995; Gash and Nobre 1997;Goutorbe et al. 1997; Menenti 1998) have not only provided these data but also encouraged the investigation of new techniques for the use of remotely sensed data in landscape models. New observational field campaigns should also be considered in areas where concern already exists for regional change and where remote sensing can contribute now to the understanding of landscape processes and their potential management. The continued investigation of existing data sets, continued field studies in established sites, and new campaigns in new areas will all play a part in the future development of remote sensing science applications to landscape process studies.

RECOMMENDATIONS

1. Implementation of new technologies in order to ensure that remote sensing is driven by scientific hypothesis and not the other way around.
2. New integrated modeling for merging remote sensing and landscape process models, emphasizing interdisciplinary collaboration between physicists, hydrologists, ecologists, and atmospheric scientists.
3. New technologies, including multiangular sensors, imaging spectrometry, lidar, and radar, at a range of temporal and spatial scales in order to tackle the complex of patterns and processes at landscape and regional scales.
4. New collaborative experiments to test new sensor/model syntheses, clarify the integration of remote sensing techniques in landscape process studies, and implement studies in targeted areas of rapid environmental change.

REFERENCES

Allen, T.F.H., and T.W. Hoekstra. 1992. Toward a Unified Ecology. New York: Columbia Univ. Press.

Asner, G.P., B.H. Braswell, D.S. Schimel, and C.A. Wessman. 1998. Ecological research needs from multiangle remote sensing data. *Remote Sens. Env.* **63**:155–165.

Asner, G.P., and C.A. Wessman. 1997. Scaling PAR absorption from the leaf to landscape level in spatially heterogeneous ecosystems. *Ecol. Mod.* **103**:81–97.

Asner, G.P., C.A. Wessman, and S. Archer. 1998. Scale dependence of absorption of photosynthetically active radiation in terrestrial ecosystems. *Ecol. Appl.* **8(4)**:1003–1021.

Asner, G.P., C.A. Wessman, and J.L. Privette. 1997. Unmixing the directional reflectances of AVHRR sub-pixel landcovers. *IEEE Trans. Geosci. Remote Sens.* **35(4)**:868–878.

Asner, G.P., C.A. Wessman, and D.S. Schimel. 1998. Spatial heterogeneity of plant canopy structure and function in a sub-tropical savanna from imaging spectrometry. *Ecol. Appl.* **8(4)**:1022–1036.

Barnsley, M.J., A.H. Strahler, K.P. Morris, and J.-P. Muller. 1994. Sampling the surface bidirectional reflectance distribution function. 1. Evaluation of current and future satellite sensors. *Remote Sens. Rev.* **8**:271–311.

Belward, A.S., P.J. Kennedy, and J.M. Gregoire. 1994. The limitations and potential of AVHRR GAC data for continental-scale fire studies. *Intl. J. Remote Sens.* **15(11)**:2215–2234.

Bolle, H.J. 1995. Identification and observation of desertification processes with the aid of measurements from space. Results from the European field experiment in desertification-threatened areas (EFEDA). *Env. Mon. Asses.* **37(1–3)**:93–101.

Braswell, B.H., D.S. Schimel, J.L. Privette, et al. 1996. Extracting ecological and biophysical information from AVHRR optical data: An integrated algorithm based on inverse modeling. *J. Geophys. Res.* **101**:23,335–23,345.

Field, C.B., J.T. Randerson, and C.M. Malmstrom. 1995. Global net primary production: Combining ecology and remote sensing. *Remote Sens. Env.* **51**:74–88.

Fischer, A., S. Louahala, P. Maisongrande, L. Kergoat, and G. Dedieu. 1996. Satellite data for monitoring, understanding and modelling of ecosystem functioning. In: Global Change and Terrestrial Ecosystems, ed. B. Walker and W. Steffen, vol. 2, pp. 566–591. Cambridge: Cambridge Univ. Press.

Gamon, J.A., L. Serrano, and J. Surtus. 1998. The photochemical reflectance index: An optical indicator of photosynthetic radiation-use efficiency across species, functional types, and nutrient levels. *Oecologia* **112**:492–501.

Gao, B.-C., and A.F.H. Goetz. 1995. Retrieval of equivalent water thickness and information related to biochemical components of vegetation canopies from AVIRIS data. *Remote Sens. Env.* **52**:155–162.

Gash, J.H.C., and C.A. Nobre. 1997. Climatic effects of Amazonian deforestation: Some results from ABRACOS. *Bull. Am. Meteorol. Soc.* **78(5)**:823–830.

Goel, N.S. 1988. Models of vegetation canopy reflectance and their use in estimation of biophysical parameters from reflectance data. *Remote Sens. Rev.* **4**:1–212.

Goutorbe, J.P., T. Lebel, A.J. Dolman, J.H.C. Gash, P. Kabat, Y.H. Kerr, B. Monteny, S.D. Prince, J.N.M. Sticker, A. Tinga, and J.S. Wallace. 1997. An overview of HAPEX-Sahel: A study in climate and desertification. *J. Hydrol.* **189**:1–4.

Goward, S.N., C.J. Tucker, and D.G. Dye. 1985. North American vegetation patterns observed with the NOAA–7 advanced very high resolution radiometer. *Vegetatio* **64**:3–14.

Hall, F.G., and P.J. Sellers. 1995. First International Satellite Land Surface Climatology Project (ISLSCP) Field Experiment (FIFE) in 1995. *J. Geophys. Res.* **100(D12)**:25,383–25,395.
Hunt, E.R., S.C. Piper, R. Nemani, C.D. Keeling, R.D. Otto, and S.W. Running. 1996. Global net carbon exchange and intra-annual atmospheric CO_2 concentrations predicted by an ecosystem process model and three-dimensional atmospheric transport model. *Glob. Biogeochem. Cyc.* **10**:431–456.
Jacquemoud, S., F. Baret, B. Andrieu, F.M. Danson, and K. Jaggard. 1995. Extraction of vegetation biophysical parameters by inversion of the PROSPECT+SAIL models on sugar beet canopy reflectance data. Application to TM and AVIRIS sensors. *Remote Sens. Env.* **52**:163–172.
Justice, C.O., E. Vermote, J.R.G. Townshend, R. Defries, D.P. Roy, et al. 1998. The Moderate Resolution Imaging Spectroradiometer (MODIS): Land remote sensing for global change research. *IEEE Trans. Geosci. Remote Sens.* **36(4)**:1228–1249.
Kaufman, Y.J., D.D. Herring, K.J. Ranson, and G.J. Collatz. 1998. Earth Observing System AM1 Mission to Earth. *IEEE Trans. Geosci. Remote Sens.* **36(4)**:1045–1055.
Lefsky, M.A., W.B. Cohen, S.A. Acker, T.A. Spies, G.G. Parker, and D. Harding. 1997. Lidar remote sensing of forest canopy structure and related biophysical parameters at the H.J. Andrews Experimental Forest, Oregon, USA. In: Natural Resource Management using Remote Sensing and GIS, ed. J.D. Greer, pp. 79–91. Washington, D.C.: American Society for Photogrammetry and Remote Sensing.
Lefsky, M.A., D. Harding, W.B. Cohen, G. Parker, and H.H. Shugart. 1998. Surface Lidar remote sensing of basal area and biomass in deciduous forests of eastern Maryland, USA. *Remote Sens. Env.,* in press.
Martin, M.E., and J.D. Aber. 1997. High spectral resolution remote sensing of forest canopy lignin, nitrogen, and ecosystem processes. *Ecol. Appl.* **7(2)**:431–443.
Meentemeyer, V., and E.O. Box. 1987. Scale effects in landscape studies. In: Landscape Heterogeneity and Disturbance, ed. M.G. Turner, pp. 15–34. New York: Springer.
Menenti, M. 1998. Assimilation of multispectral measurements in interactive land surface models. *Synerg. Use Multis. Data Land Proc.* **22(5)**:611–624.
Myneni, R.B., C.D. Keeling, C.J. Tucker, G. Asrar, and R.R. Nemani. 1997. Increased plant growth in the northern high latitudes from 1981 to 1991. *Nature* **386**:698–702.
Myneni, R.B., J. Ross, and G. Asrar. 1989. A review on the theory of photon transport in leaf canopies. *Agr. For. Meteorol.* **45**:1–153.
Paruelo, J.M., H.E. Epstein, W.K. Lauenroth, and I.C. Burke. 1997. ANPP estimates from NDVI for the Central Grasslands Region of the United States. *Ecology* **78(3)**:953–958.
Paruelo, J.M., and W.K. Lauenroth. 1995. Regional paterns of normalized Difference Vegetation Index in North American Shrublands and Grasslands. *Ecology* **76(6)**: 1888–1898.
Peñuelas, J., I. Filella and F. Baret. 1995. Semiempirical indices to assess carotenoids/ chlorophyll a ratio from leaf spectral reflectance. *Photosynthetica* **31**:221–230.
Pereira, A.C., and A.W. Setzer. 1996.Comparison of fire detection in savannas using AVHRR's channel 3 and TM images. *Intl. J. Remote Sens.* **17(10)**:1925–1937.
Prince, S.D., and S.N. Goward. 1995. Global primary production: A remote sensing approach. *J. Biogeogr.* **22(4–5)**:815–835.
Prince, S.D., and M. Steininger. 1999. Biophysical stratification of the Amazon basin. *Glob. Change Biol.* **9**:1–22.
Privette, J.L., W.J. Emery, and D.S. Schimel. 1996. Inversion of a vegetation reflectance model with NOAA AVHRR data. *Remote Sens. Env.* **58**:187–200.

Roberts, D.A., M.O. Smith, and J.B. Adams. 1993. Green vegetation, non-photosynthetic vegetation, and soils in AVIRIS data. *Remote Sens. Env.* **44**:255–269.

Running, S.W., T.R. Loveland, L.L. Pierce, R.R. Nemani, and J.E.R. Hunt. 1995. A remote sensing based vegetation classification logic for global land cover analysis. *Remote Sens. Env.* **51**:39–48.

Scholes, R.J., and S.R. Archer. 1997. Tree-grass interactions in savannas. *Ann. Rev. Ecol. Syst.* **28**:517–544.

Sellers, P.J., J.A. Berry, G.J. Collatz, C.B. Field, and F.G. Hall. 1992. Canopy reflectance, photosynthesis and transpiration. III. A reanalysis using improved leaf models and a new canopy integration scheme. *Remote Sens. Env.* **42**:187–216.

Sellers, P.J., F.G. Hall, G. Asrar, D.E. Strebel, and R.E. Murphy. 1988. The First ISLSCP Field Experiment (FIFE). *Bull. Am. Meteorol. Soc.* **69(1)**:22–27.

Sellers, P., et al. 1995. The Boreal Ecosystem-Atmosphere Study (BOREAS): An overview and early results from the 1994 field year. *Bull. Am. Meteorol. Soc.* **76(9)**:1549–1577.

Shugart, H.H. 1996. The importance of structure in understanding global change. In: Global Change and Terrestrial Ecosystems, ed. B. Walker and W. Steffen, pp. 117–126. Cambridge: Cambridge Univ. Press.

Soriano, A., and J.M. Paruelo. 1992. Biozones: Vegetation units defined by functional characters identifiable with the aid of satellite sensor images. *Glob. Ecol. Biogeogr. Lett.* **2**:82–89.

Turner, B.L., II, W.C. Clark, R.W. Kates, J.F. Richards, J.T. Mathews, and W.B. Meyer. 1990. Earth as Transformed by Human Action. Cambridge: Cambridge Univ. Press.

Wessman, C.A., J.D. Aber, D.L. Peterson, and J.M. Melillo. 1988. Remote sensing of canopy chemistry and nitrogen cycling in temperate forest ecosystems. *Nature* **335**:154–156.

Wessman, C.A., and G.P. Asner. 1998. Ecosystems and problems of measurement at large spatial scales. In: Successes, Limitations, and Frontiers in Ecosystem Science, ed. M. Pace and P. Groffman, pp. 346–371. New York: Springer.

Wessman, C.A., C.A. Bateson, and T. Benning. 1997. Detecting fire and grazing patterns in the Konza tallgrass prairie using spectral mixture analysis. *Ecol. Appl.* **7**:493–511.

Wiens, J.A. 1989. Spatial scaling in ecology. *Fun. Ecol.* **3**:385–397.

6

Land-surface Influences on Atmospheric Dynamics and Precipitation

R.A. PIELKE, SR.,[1] G.E. LISTON,[1] L. LU,[1] and R. AVISSAR[2]

[1]Dept. of Atmospheric Science, Colorado State University,
Fort Collins, CO 80523, U.S.A.
[2]Dept. of Environmental Sciences, Cook College, Rutgers University,
New Brunswick, NJ 08903, U.S.A.

ABSTRACT

Land-use change strongly influences climate on the local, regional, and global scales. Vegetation dynamics interact with climate in a coupled nonlinear manner. The temporal limits on climate prediction are controlled by (a) our understanding of the interactions between each aspect of the Earth's environmental system, and (b) their degree of nonlinearity. These limits have not been determined. Global-, regional-, and local-scale environmental effects cannot, in general, be considered independently from interactions across scales. The use of land-surface temperature records to detail climate change/climate variability must include the influence of local and temporal land-use change over the period of record.

INTRODUCTION

There is general agreement on the importance of land-surface characteristics on microclimate (e.g., Rosenberg et al. 1983; Cotton and Pielke 1995; Valentini et al., this volume). When a forest patch is clear-cut, for example, daytime temperature near the surface of the clear-cut increases relative to the same height within the forest canopy (e.g., Bastale et al. 1993).

The importance of landscape on mesoscale and regional-scale weather and climate is also seldom questioned. In a recent contribution, O'Brien (1995), for instance, has documented how deforestation in part of Chiapas State in Mexico has resulted in an altered climate from that which occurs in the undisturbed forested region. Pielke et al. (1997), as illustrated in Plate 6.1, demonstrates the very significant role that land use

Integrating Hydrology, Ecosystem Dynamics, and Biogeochemistry in Complex Landscapes
Edited by J.D. Tenhunen and P. Kabat

has in generating thunderstorms. In this figure, identical initial and lateral boundary condition meteorology were used; the only difference was that in the bottom simulation, a short grass prairie was assumed, while for the top simulation, the current heterogeneous landscape was prescribed. As shown in Grasso (1996), Ziegler et al. (1997), and Shaw et al. (1997), the use of the current landscape in the model is a necessary condition for a realistic simulation of thunderstorm activity in this region.

Even on a global scale, as shown in Chase et al. (1996), regional landscape changes in the tropics, in particular, alter climate thousands of miles away — in the mid- and high-latitude polar jet flow. Figure 6.1 shows how the long-wave tropospheric wind flow was substantially altered when current, as contrasted with potential, leaf area index (LAI) was specified as a lower boundary condition in the NCAR CCM2 GCM. Figure 6.2, also from Chase et al. (1996) shows how regional precipitation patterns were changed in the model over southeast Asia as a result of the model simulated change in LAI. It was the change in the GCM-modeled thunderstorm activity that resulted from the LAI change, that teleconnected to the higher latitudes, and changed the polar jet flow.

There are additional feedbacks between precipitation, transpiration, evaporation, percolation, soil moisture storage, and runoff. Changes in monthly precipitation also significantly affect vegetation growth. These changes in vegetation growth alter precipitation through changes in the partitioning of latent and sensible heat fluxes, and different surface albedo, and the resulting influence on cumulus cloud precipitation (e.g., Pielke et al. 1998).

The existence of such significant nonlinear feedbacks suggests that the development of coupled hydrologic–atmospheric models as operational predictive tools will be difficult and limited in the time of forecast skill. The use of such coupled models, however, will be valuable for scenario construction and process-oriented studies.

It is well known that the contrast between land and water generates breezes (i.e., sea, lake, and land breezes), which are mesoscale circulations. Several papers and textbooks describe in length the mechanism involved in the generation of these circulations, and their impact on the weather and the climate (e.g., Pielke 1984, among many others). More recent investigations have indicated that other landscape discontinuities, e.g. irrigated land in arid areas, deforestation, and afforestation, provide also an environment appropriate for the development of mesoscale circulations (e.g., Pielke and Avissar 1990; Pielke et al. 1991; Avissar and Chen 1993; Avissar and Liu 1996; Kabat, this volume; Meixner and Eugster, this volume).

For instance, Avissar and Liu (1996) studied the relative contribution of turbulence and mesoscale circulations on clouds and precipitation, in homogeneous and heterogeneous landscapes. Figure 6.3 illustrates the domains considered for their analysis, and Plate 6.2 provides the resulting horizontal distribution of accumulated precipitation after one day of simulation.

The domains shown in Figures 6.3d and 6.3e are spatially homogeneous. Thus, no mesoscale circulation is generated in these domains, and clouds and precipitation develop as a result of moist, turbulent eddies reaching their level of condensation. The

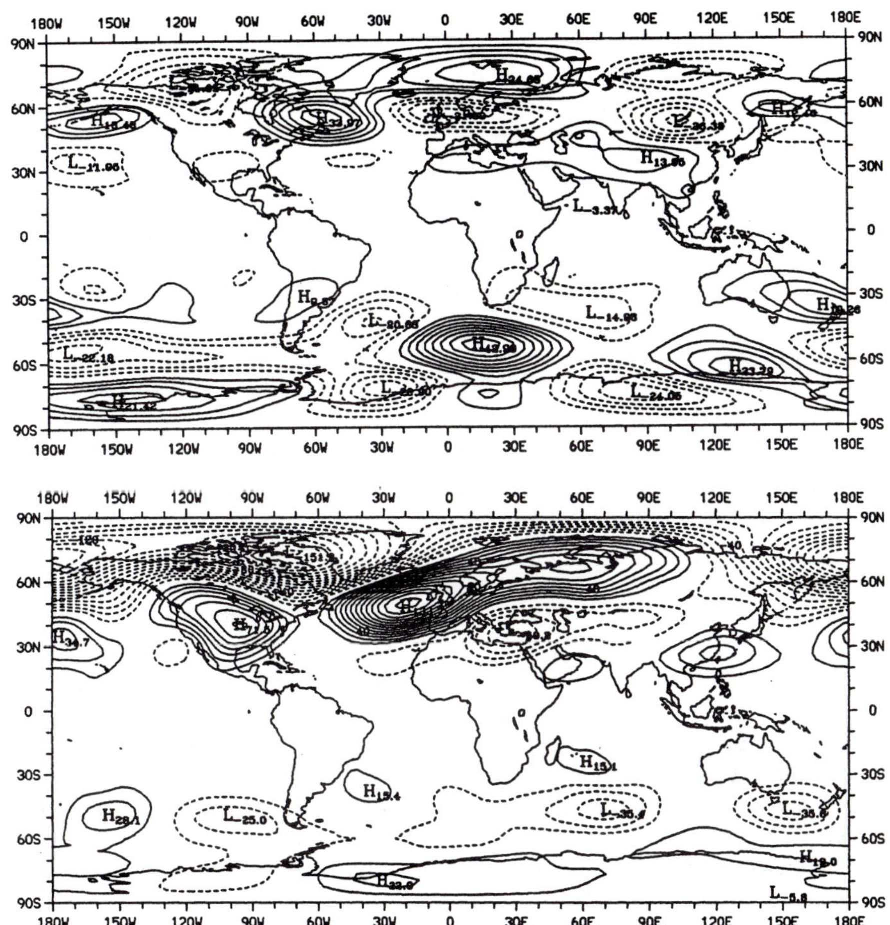

Figure 6.1 Ten-year averaged differences in 500 mbar heights zonal wavenumbers 1–6 only (actLAI–potLAI). (a) January, contour 10 m, and (b) July, contour 5 m (from Chase et al. 1996).

position of these turbulent eddies appears to be random and, not surprisingly, a random-like pattern of accumulated precipitation is obtained with these simulations. There are, however, significant differences between these two simulations. Thin clouds develop early in the day over wet land. During the morning hours, they are enhanced by the relatively large amount of water injected into the atmosphere from the evaporating land surface. The larger the relative part of the domain covered by wet land, the more water is available in the atmosphere to produce clouds. During the afternoon hours, relatively deep clouds develop as a result of this turbulent process. By contrast, when the domain consists of dry land, clouds are small in both depth and width, if created at all. In general, they do not produce noticeable precipitation. Since there is no

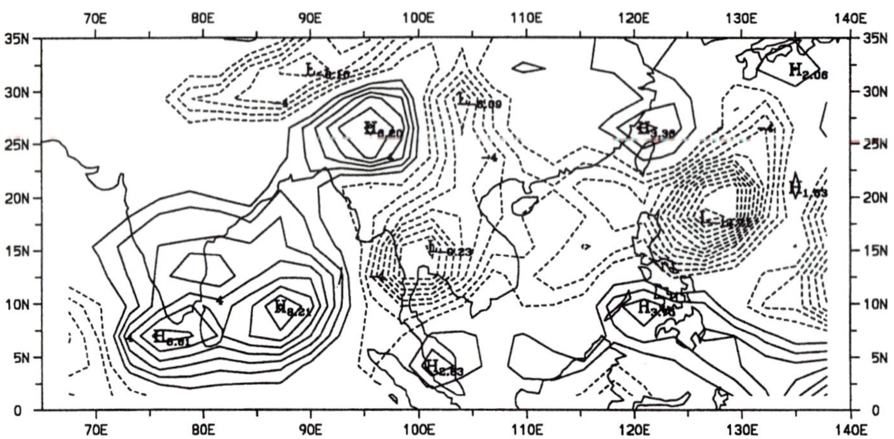

Figure 6.2 July total precipitation differences (actLAI–potLAI) 65–130 deg E, 10 deg S – 35 deg N in mm per day (from Chase et al. 1996).

supply of moisture from the dry land surface, clouds form only if the atmospheric background is relatively humid. These differences are also obviously reflected in the accumulated precipitation presented in Figures 6.3d, e.

Clearly, in Plate 6.2a, precipitation concentrates in the originally dry land, where the convergence of the mesoscale circulations generated by the forest–pasture contrast resulted in a strong, moist vertical motion. It is interesting to note that this is also the case in Plate 6.2b and c, in spite of the fact that the mesoscale circulations generated by these domains are relatively very weak. It should be emphasized that even though no mesoscale circulation developed in the domain shown in Figure 6.3d, the mean precipitation obtained in that domain is larger than in the domain illustrated in Figure 6.3c. However, when a strong mesoscale circulation develops within the domain, as in the domain shown in Figure 6.3a, the maximum accumulated and mean precipitation in the heterogeneous domain are much larger than in the homogeneous domain.

Note that because precipitation concentrates in originally dry land, a negative feedback between the landscape and the atmosphere tends to eliminate the landscape discontinuities, and spatially homogenize the land water content. This has interesting ecological implications which need to be investigated. Also, this implies that these types of clouds and precipitation are unlikely to be produced every day at the same location, unless the dry land consists of a permeable soil with a rapid drainage capacity in its upper layer. On a diurnal basis, this negative feedback has little impact, since precipitation occurs only during the afternoon hours, while the morning hours are mostly relevant for its development.

The conclusion from these, and other related studies, is that land use plays a significant role on local, regional, and global climate.

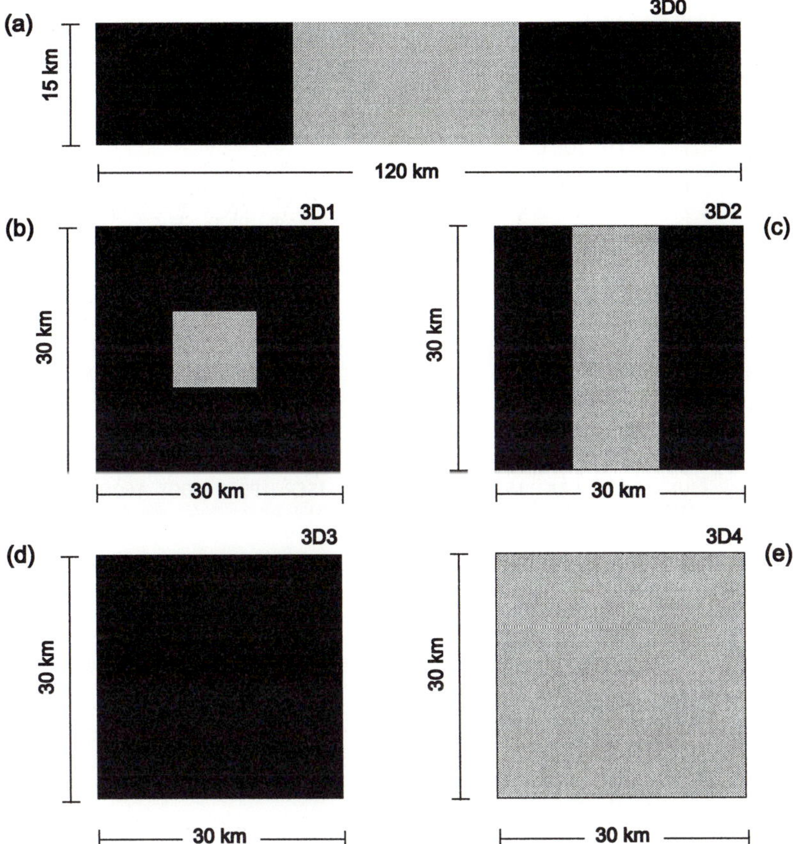

Figure 6.3 Schematic representation of the simulated domains. Green and yellow areas represent dense, unstressed forest and dry pasture, respectively. The numerical grid area used to represent these domains was 250 × 250 m^2 (adapted from Avissar and Liu 1996).

TWO-WAY INTERACTION

Two-way interactions between land-covered atmosphere clearly occur (Pielke et al. 1998; Raupach 1998). These interactions can be on the diurnal scale (biophysical), on the seasonal scale (biophysical, biogeochemical), and on the multi-year time scale (biogeochemical, biogeographical). On the seasonal time scale, prescribed LAI significantly affects the meteorological model simulation of temperature and precipitation (Figure 6.4). Correspondingly, the prescription of temperature, and particularly precipitation, dramatically affects a biogeochemical model simulation of LAI (Figure 6.5) and root density (not shown).

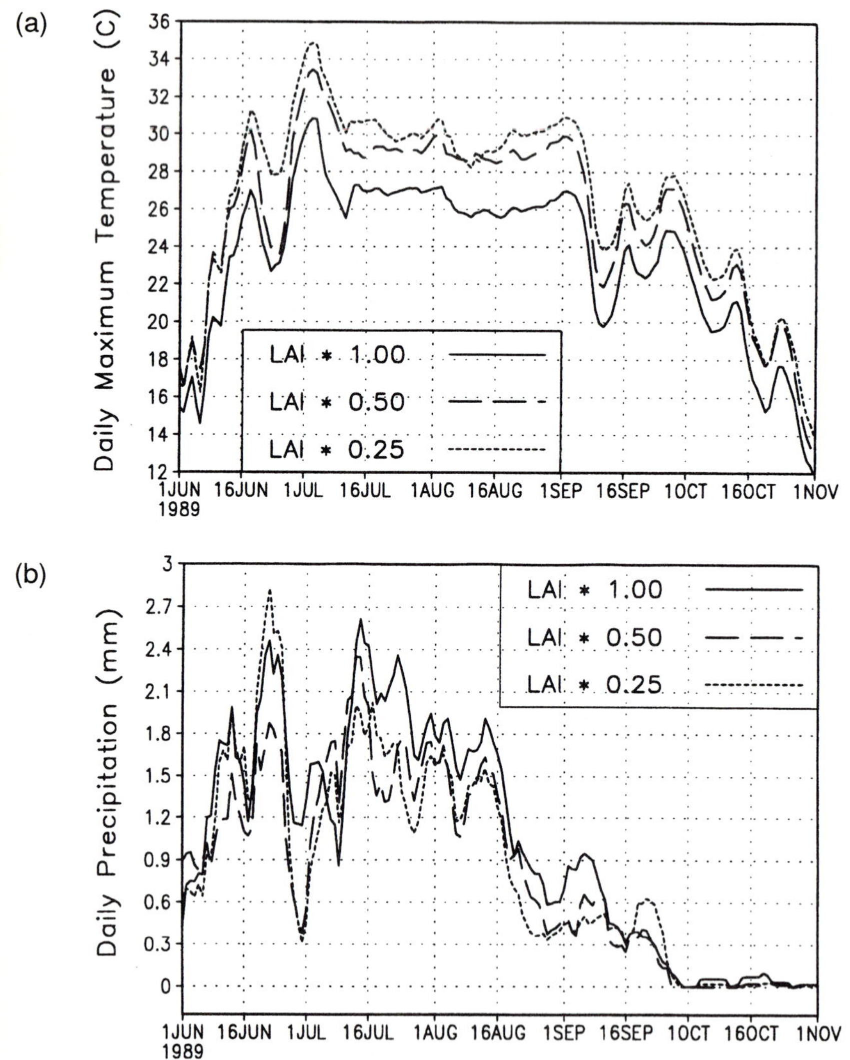

Figure 6.4 (a) The effect of changing leaf area index (LAI) on daily maximum temperature (degree C) over Colorado. (b) The effect of LAI on daily precipitation (mm) over Colorado.

Since both the meteorological and ecological models are strongly influenced by what are dependent variables in the other models, this feedback must be considered in any climate simulation. The conclusion, therefore, is that vegetation dynamics interact with climate and weather through a coupled nonlinear interaction.

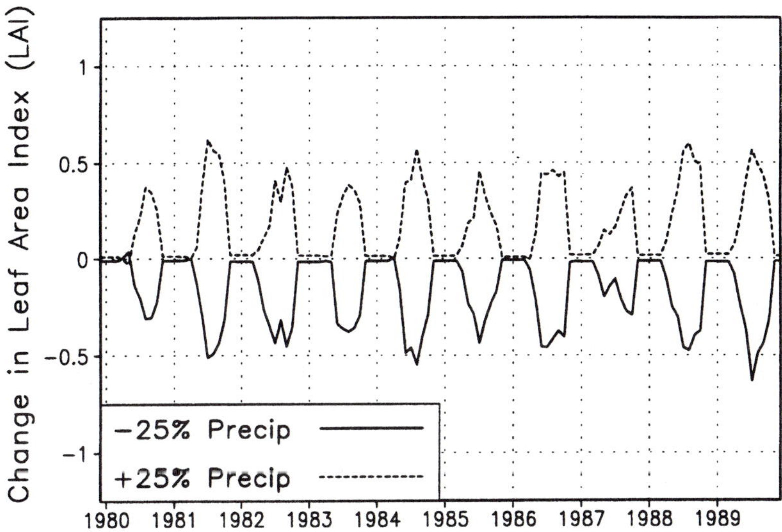

Figure 6.5 The effect of changing precipitation on leaf area index (LAI) for short grassland over eastern Colorado. Solid curves indicate the changes of LAI when precipitation decreased 25% and dashed curves indicate the LAI changes when precipitation increased 25%.

PREDICTABILITY

In any nonlinear system, the time period of predictability is dependent on the degree of nonlinearity and the level of accuracy that the feedbacks within the system can be represented (Drazin 1992; Gleick 1987; Lorenz 1987). In the context of the climate system, the temporal limits on climate prediction are determined by (a) our understanding and ability to represent quantitatively the interactions between each important aspect of the Earth's climate system, and (b) the degree of nonlinearity of these interactions. Figures 6.4 and 6.5 provide an example of one of these feedback mechanisms. In the context of regional and global climate prediction, these limits have not been determined. Moreover, as shown in the INTRODUCTION, global- and regional-scale climate effects cannot be considered independently, but interaction across scales must be considered.

CONCLUSION

The use of hydrologic models with atmospheric input has been very successful as operational predictive tools for flash flood forecasting (Foufoula-Georgiou and Krajewski 1995; Krzysztofowicz 1995). Similarly, in the context of drought prediction, forecasts (El Niño events), can provide valuable information (Betsill et al. 1997).

Hydrologic information (such as LAI, soil moisture, snow cover; LAI can be considered as a hydrologic parameter because the amount of vegetation directly influences transpiration of water to the atmosphere) has also shown value in daily weather forecasting (Ziegler et al. 1997).

Coupled hydrologic–atmospheric–ecological models are also currently being developed, although the extent, or even whether they can improve operational forecast skill, needs to be assessed. Process modeling will, of course, improve our understanding of how weather and climate operate, even if additional predictive skill is not achieved.

There are several provocative conclusions that follow as a result of these limits on predictive capability. First, since GCMs in the context of climate change are likely to provide only information in a scenario context (due to the incompleteness of representing all of the important external and internal forcings, and because of the significance of nonlinear feedbacks [Oreskes et al. 1994]), downscaling of GCM output (either using a regional atmospheric model as an interface or statistical downscaling) cannot achieve an accuracy superior to the GCM simulation except to the extent that surface forcings in the hydrologic study area are included (e.g., land-use change in an area of complex topographic relief). Also when GCMs achieve high enough spatial resolution globally (around 5–10 km horizontal grid intervals), there will be no need for atmospheric downscaling (at least in terms of land-surface forcing), since the atmosphere appears to homogenize effectively above the surface layer for spatial scales that are that size and smaller (e.g., Avissar and Schmidt 1998; Raupach and Finnigan 1995). However, increasing resolution is not necessarily the best way to improve the simulation. It is possible that an ensemble of GCM model runs with lower resolution will provide a more robust statistical representation. Second, GCM climate change simulations that have been performed up to the present are scenarios, not operational predictions. The IPCC "projections" need to be interpreted in this fashion and recognized as a subset of possible future climate states.

In the investigation of feedbacks, all important atmospheric, ecological (biophysical and biogeochemical), oceanographic, and societal (land-use agricultural practices, urbanization, deforestation, etc.) effects must be incorporated. These process simulations, as stated earlier, still cannot be interpreted as predictions if the nonlinear feedbacks are shown to be sufficiently important.

Finally, in the context of what is of use to policymakers with respect to environmental change, we suggest a focus on the hydrologic, biogeochemical, and ecosystem models themselves to assess what the modeled system is most sensitive to as a function of prescribed external and internal forcings. This focus should also include an assessment of the relative importance of these forcings. In the climate change arena, this recommendation would encourage hydrologic, biogeochemical, and ecosystem dynamics modelers to explore the full realistic range of conditions that potentially could affect their study area, rather than being constrained by the limited set of scenarios provided by GCM output. While GCMs cannot be used for operationally predicting future climate, they are nevertheless powerful research tools to assess atmospheric

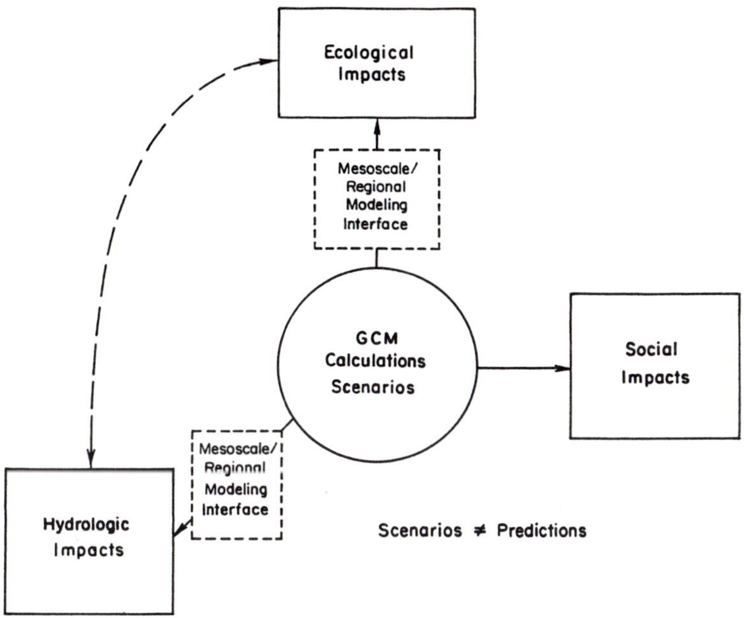

Figure 6.6 Current way to perform environmental assessments associated with climate change studies.

processes over long time scales in order to aid in the understanding of the nonlinear and linear feedback processes in the Earth's hydrologic system. In this framework, other human-caused effects, such as local land-use change, could be evaluated, and its importance assessed relative to changes in basin-averaged precipitation, temperature, etc. Figures 6.6 and 6.7 illustrate the two contrasting procedures used to assess hydrologic sensitivity to external and internal forcings. We recommend that the procedure of vulnerability risk assessment, illustrated simply in Figure 6.7, is of more value to policymakers. Indeed, there is a rich source of information on the assessment of risk and vulnerability in many disciplines. The procedure in Figure 6.7, of course, is already used by a range of decision makers, such as hydrologic engineers in the design of dams, where existing climate data (for past years) is used to estimate extreme event statistics, and to design dams accordingly.

Since both the meteorological and ecological models are strongly influenced by what are dependent variables in the other models, this feedback must be considered in any climate simulation. Vegetation dynamics interacts with climate and weather through a coupled nonlinear interaction.

Another conclusion is that the use of land-surface weather records to detect regional (and therefore, global) climate change/climate variability must include the influence of land-use change over time on the record. For example, urban/rural differences, forestation, afforestation, agricultural, and grazing changes, must be

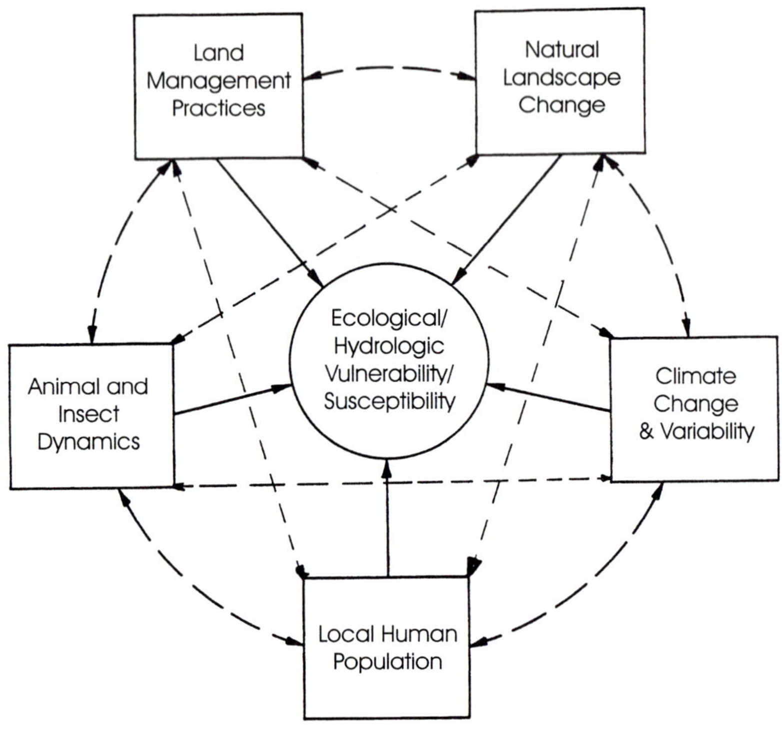

Figure 6.7 Use of ecological hydrologic vulnerability/susceptibility in environmental assessment.

included. The use of surface weather records in heterogeneous regions also requires that an assessment of the representative footprint of what the weather record is measuring be determined. This assessment must necessarily include microclimate (e.g., Rosenberg et al. 1983) and mesoclimate effects (e.g., O'Brien 1995).

ACKNOWLEDGMENTS

Work in this paper has been supported by the USGS Grant #1434–94–A–01275, NPS Contract #CA 1268–2–9004 CEGR–R92–0193, NSF Grant #ATM–9306754, NSF Grant #EAR–9415441, and NOAA Grant #NA56GP0200. This paper benefited from discussions with and suggestions from R.A. Pielke, Jr.

REFERENCES

Avissar, R., and F. Chen. 1993. Development and analysis of prognostic equations for mesoscale kinetic energy and mesoscale (subgrid-scale) fluxes for large scale atmospheric models. *J. Atmos. Sci.* **50**:3751–3774.

Avissar, R., and Y. Liu. 1996. A three-dimensional numerical study of shallow convective clouds and precipitation induced by land-surface forcings. *J. Geophys. Res.* **101**:7499–7518.

Avissar, R., and T. Schmidt. 1998. An evaluation of the scale at which ground-surface heat flux patchiness affects the convective boundary layer using a large-eddy simulation model. *J. Atmos. Sci.* **55**:2666–2689.

Bastable, H.G., W.J. Shuttleworth, R.L.G. Dallarosa, G. Fisch, and C.A. Nobre. 1993. Observations of climate, albedo and surface radiation over cleared and undisturbed Amazonian forest. *Intl. J. Climatol.* **13**:783–796.

Betsill, M.M., M.H. Glantz, and C. Crandall. 1997. Preparing for El Niño: What role for forecasts? *Environment* **39(10)**:6–13, 26–29.

Chase, T.N., R.A. Pielke, T.G.F. Kittel, R. Nemani, and S.W. Running. 1996. The sensitivity of a general circulation model to global changes in leaf area index. *J. Geophys. Res.* **101**:7393–7408.

Cotton, W.R., and R.A. Pielke. 1995. Human Impacts on Weather and Climate. New York: Cambridge Univ. Press.

Drazin, P.G. 1992. Nonlinear Systems. Cambridge: Cambridge Univ. Press.

Foufoula-Georgiou, E., and W. Krajewski. 1995. Recent advances in rainfall modeling, estimation, and forecasting. U.S. National Report to the IUGG 1991–1994. *Rev. Geophys.*, Suppl. 1125–1137.

Gleick, J. 1987. Chaos. New York: Viking Penguin.

Grasso, L. 1996. Numerical simulation of the May 15 and April 26, 1991 tornadic thunderstorms. Ph.D. Diss., Atmos. Sci. Paper #596, Colorado State Univ., Dept. of Atmospheric Science, Fort Collins, CO 80523.

Krzysztofowicz, R. 1995. Recent advances associated with flood forecast and warning systems. U.S. National Report to the IUGG 1991–1994. *Rev. Geophys.*, Suppl. 1139–1147.

Lorenz, E.N. 1987. Deterministic and stochastic aspects of atmospheric dynamics. In: Irreversible Phenomena and Dynamical Systems Analysis in Geosciences, ed. C. Nicolis and G. Nicolis, pp. 159–179. Dordrecht: Reidel.

O'Brien, K.L. 1995. Deforestation and climate change in the Selva Lacandona of Chiapas, Mexico. Ph.D. Diss., Pennsylvania State Univ.

Oreskes, N., K. Shrader-Frechette, and K. Belitz. 1994. Verification, validation, and confirmation of numerical models in the earth sciences. *Science* **263**:641–645.

Pielke, R.A. 1984. Mesoscale Meteorological Modeling. New York: Academic.

Pielke, R.A., and R. Avissar. 1990. Influence of landscape structure on local and regional climate. *Landsc. Ecol.* **4**:133–155.

Pielke, R.A., R. Avissar, M. Raupach, H. Dolman, X. Zeng, and S. Denning. 1998. Influence of short- and long-term ecosystem dynamics on weather and climate. *Glob. Change Biol.* **4**:461–475.

Pielke, R.A., G. Dalu, J.S. Snook, T.J. Lee, and T.G.F. Kittel. 1991. Nonlinear influence of mesoscale land use on weather and climate. *J. Clim.* **4**:1053–1069.

Pielke, R.A., T.J. Lee, J.H. Copeland, J.L. Eastman, C.L. Ziegler, and C.A. Finley. 1997. Use of USGS-provided data to improve weather and climate simulations. *Ecol. Appl.* **7**:3–21.

Raupach, M.R. 1998. Radiative, physiological, aerodynamic and boundary-layer feedbacks in the terrestrial surface energy balance. *Glob. Change Biol.* **49**:2–19.

Raupach, M.R., and J.J. Finnigan. 1995. Scale issues in boundary layer meteorology: Surface energy balances in heterogeneous terrain. *Hydrol. Proc.* **9**:589–612.

Rosenberg, N.J., B.L. Blad, and S.B. Verma. 1983. Microclimate, The Biological Environment. New York: Wiley.

Shaw, B.L., R.A. Pielke, and C.L. Ziegler, 1997. A three-dimensional numerical simulation of a Great Plains dryline. *Mon. Wea. Rev.* **125**:1489–1506.

Ziegler, C.L., T.J. Lee, and R.A. Pielke. 1997. Convective initiation at the dryline: A modeling study. *Mon. Wea. Rev.* **125**:1001–1026.

7

Ecological Controls on Land-surface Atmospheric Interactions

R. VALENTINI[1], D.D. BALDOCCHI[2], and J.D. TENHUNEN[3]

[1]University of Tuscia, Dept. of Forest Science and Environment, Via de Lellis, 01100 Viterbo, Italy
[2]Atmospheric Turbulence and Diffusion Division, NOAA/ARL, P.O. Box 2456, Oak Ridge, TN 37831–2456, U.S.A.
[3]Dept. of Plant Ecology, University of Bayreuth, 95440 Bayreuth, Germany

ABSTRACT

Quantification is needed of the land-surface atmospheric exchanges that are dependent on the type of vegetation cover in order to understand better the global carbon cycle, trace gas emissions and the nitrogen cycle, and the oxidative capacity of the atmosphere. The development of eddy covariance methods to measure exchanges with the atmosphere has allowed us for the first time to investigate directly heterogeneity in ecosystem function at landscape, regional, and continental scales as well as time-dependent changes with respect to seasonality and longer-term fluctuations in climate conditions, e.g., occurrences of cycles such as El Niño versus La Niña phenomena. Research networks for measuring net ecosystem exchange of CO_2 as an indicater of net ecosystem production have been highly successful and have provided new perspectives on carbon balance. Data bases of this type will add confidence to our estimates of process rates and regulation at global scales. In addition, these direct measurements of ecosystem function permit the study of vegetation cover effects on water use, nitrogen use, and light use efficiencies. From these, land–atmosphere functional units (LAFUs) may be recognized with similar properties. This suggests that, from the viewpoint of land-surface atmospheric exchange, derivation of landscape estimates of gas exchange may be easier than expected from the number of occurring plant communities. On the other hand, measurements also indicate that the properties of LAFUs may change in response to environmental stresses and with respect to seasonal time. Appreciation of the variation that occurs in patterns of exchange properties requires continued observation of a larger number of ecosystem types.

INTRODUCTION

The biosphere is that region of the Earth's surface where living organisms exist and are able to exchange mass and energy with the surrounding environment. This includes

Integrating Hydrology, Ecosystem Dynamics, and Biogeochemistry in Complex Landscapes
Edited by J.D. Tenhunen and P. Kabat

the land surface, where plants reside, the rhizosphere, where roots and microbes dominate, and water bodies. The cycling of material and the nature of ecosystems is so different over water bodies that we usually divide the biosphere into a geo-biosphere, comprising terrestrial ecosystems, and a hydro-biosphere, comprising acquatic ecosystems (Walter 1983). In this chapter we will focus on the geo-biosphere and fluxes associated with the major biogeochemical cycles that take place there, namely water, carbon, nitrogen, and the exchange of trace gases.

The hydrological cycle is highly influenced by terrestrial ecosystems both at global and regional scales. More than 20% of atmospheric water vapor is recycled through the terrestrial vegetation (Budyko 1982). Several observations and simulation experiments show the important role of terrestrial vegetation in the water cycle. In particular, for closed systems like the Amazon basin (Salati and Vose 1984), more than 70% of rainfall can be attributed to forest evapotranspiration. Other studies show that deforestation determines a decrease of the annual inputs of precipitation (Lean and Warrilow 1989), which suggests a strong feedback between surface characteristics (albedo, surface conductance, etc.) and mesoscale cloud formation.

Terrestrial vegetation also plays a fundamental role in the global carbon cycle. The net flux of carbon dioxide, between the biosphere and the atmosphere, consists of a balance between fluxes associated with photosynthetic assimilation by the foliage and respiratory effluxes from the foliage, roots, boles, and soil heterotrophs. In general, terrestrial ecosystems have higher rates of net primary production (NPP) compared to aquatic ecosystems. Some of this production accumulates as biomass and humus, which constitute net ecosystem production. The storage of carbon in forests, for example, accounts for about 60% of the organic C stored on the Earth's land surface. Gross primary production removes about 100 Gt C yr^{-1}, which in turn is released into the atmosphere. At this rate, the equivalent of the total atmospheric CO_2 passes through the terrestrial biota every seven years. The ability to measure carbon fluxes on the global scale is quite difficult, due to their size and relative errors. Nevertheless, an imbalance is estimated for the global carbon cycle on the order of 1.3 Gt C y^{-1}, which is thought should be accounted for by the net production of terrestrial ecosystems (Schimel 1995) (Table 7.1).

The nitrogen cycle, due to its strong dependence on biological fixation and trace gas emission, is closely related to terrestrial ecosystem function. Nitrogen availability is also a critical variable driving the vegetation–atmosphere water and carbon exchanges (Schulze et al. 1994). Compounds of nitrogen play an important role in many contemporary environmental issues, ranging from the perturbation of stratospheric ozone to the contamination of groundwater.

Important questions facing scientists and society today include: What factors are responsible for the striking increase of nitrous oxide in the atmosphere? What is the role of ammonia in atmospheric aereosol formation? What are the effects of nitrogen deposition on ecosystem function and what are the resulting feedbacks on the carbon and water cycles? Furthermore, the emission by plants of volatile organic compounds (monoterpenes, isoprene, etc.), which are linked to plant metabolism and

Table 7.1 CO_2 budget for 1980–1989 in GtC yr^{-1} (IPCC 1996).

CO_2 sources	
• Emissions from fossil fuel combustion and cement production	5.5 ± 0.5
• Net emissions from changes in tropical land use	1.6 ± 1.0
Partioning among reservoirs	
• Storage in the atmosphere	3.3 ± 0.2
• Ocean uptake	2.0 ± 0.8
• Uptake by northern hemisphere forest regrowth	0.5 ± 0.5
• Inferred terrestrial sink	1.3 ± 1.5

photosynthesis, potentially influence the tropospheric oxidative capacity. In recent years, increase in these substances together with other reduced species in air (CH_4, N_2O, CO, SO_2) suggests that the oxidative capacity of the atmosphere may be decreasing (Charlson 1992). The understanding of such processes, which are fundamental components of our climate system, depends on our ability to investigate processes occurring at the land–atmosphere interface and, thus, on our capacity to quantify exchange processes at the ecosystem scale.

TOWARD THE CONCEPT OF LAND–ATMOSPHERE FUNCTIONAL UNITS

The biospheric fluxes of carbon and other biogeochemical trace gases are the result of independent action by billions of individual organisms. These individual organisms interact with each other and with physical factors and anthropogenic impacts in ways that are often consistent across broad ranges in biological diversity. This consistency makes it possible to characterize the function of large collections of organisms (ecosystems or biomes) as aggregated entities for the study of the biosphere at the regional, continental, and global scale.

Traditionally, community ecologists have treated a community of plants as the sum of its individual components. From the time of Theophrastus until the middle of the nineteeenth century, little or no recognition was given to the important role of the whole system. Karl Möbius, a marine biologist who studied oyster communities in the late nineteenth century, was one of the first scientists to emphasize the importance of studying the entire biotic community.

Recognition of the integration of the components of a biotic community into a functional system led to the idea of the community as a *supraorganism*, which was the basis of the new concept of ecosystem introduced by Tansley in 1935. Further development of the analogy of the community acting as an individual organism is largely attributed to the American ecologist Clements. He compared the successional stages of a community as it developed from a pioneer stage, invading unoccupied space, to a mature community (or climax), with the developmental stages of an individual from birth to maturity. These ideas are to be considered in the light of a romantic view of nature and its processes. Actually there are fundamental differences between a succession and an individual life. Certain stages of development can be short-circuited in a succession, but not by a living individual organism. Above all, communities lack the strict genetic definition that is inherent in the individual. In spite of these reservations, it is generally recognized that communities are organic entities in which there is a considerable degree of internal interconnectivity and a coordinated response to external factors. Some ecologists (Kimmins 1987) reach the compromise of defining the community as a *quasiorganism,* stating that there are certain similarities and parallels between these two different levels of biological organization. It can be considered, in fact, that the ecosystem functions fundamentally like an organism through the cooperative activities of acquiring energy, water, and nutrients for reproduction, growth, and maintenance. More recent studies in the ecological disciplines of energetics and biogeochemistry have confirmed many of the early ideas concerning the holistic and integrated nature of the biotic community (Vitousek et al. 1997).

In view of the problem of a better definition for land-surface atmospheric exchange of trace gases and energy, there is a clear advantage in defining a *supra-* or *quasiorganism* characteristic for terrestrial ecosystems. Aggregation has clear advantages for tackling large-scale problems. If we look at the Earth's surface, it is evident that there is a complex mosaic of landscapes, reflecting the effects of a number of natural and socioeconomic forces, which interact with climate and regional-scale atmospheric processes. The effects of disturbances like forest fires, grazing, afforestation, and desertification make terrestrial ecosystems highly dynamic. Ecosystems are characterized by significant diversity in functional properties and are capable of differentially influencing atmospheric processes. In this respect, it is possible to introduce a new definition of the *supra-* or *quasiorganism* property with a specific focus on processes involved in atmospheric exchanges. We call such entities land–atmosphere functional units (LAFUs), which can be defined as *a patch of landscape which is able to exchange mass and energy with the atmosphere and show a coordinated and specific response to environmental factors*. In other chapters there is discussion of landscape functional units (LFUs). Our LAFU is a special case of these with a focus on the land surface to atmosphere interface.

The term "coordinated" means a response that is not only the sum of individual components but reflects the integration of complex and nonlinear interactions, leading to emerging properties not clearly apparent at the individual level. Some ecosystem biochemical and physiological processes scale almost seamlessly from individual leaves or organs to the whole plant canopy, while others show a different behavior when scaled from single organs to the community. An example is light absorption and utilization to drive carbon and water exchange of multi-storied and mixed-species canopies, whose behavior as a whole differs from the functional response of the single elements.

"Specific response" means that it is possible to classify LAFUs in terms of energy and mass exchange and derive whole system parameters that can represent recurrent properties in multiple landscape patches or ecosystem types.

In the following sections, a number of examples will be presented where this concept is examined with respect to selected, measured ecosystem responses.

WHAT TO MEASURE? WHY MEASURE? HOW TO MEASURE?

Mathematical models are a major tool for integrating, interrelating, and extending knowledge on the interactions between the biosphere and atmosphere. They provide a framework for scaling ecosystem processes from the single individual to the whole community, as well as enable the extrapolation to larger spatial case and longer period of times. They also simulate ecosystem responses, and their predictive capability may be used to examine relevant socioeconomic problems (food security, ecosystem productivity, and sustainability) and to provide a general framework for sensitivity analysis and studies of vulnerability. They are, therefore, the natural tool for defining and describing LAFUs. For instance, they reproduce ecosystem responses and can eventually detect emerging properties and functions, which can be scaled across biomes and land use/cover systems. However a biosphere–atmosphere gas exchange model should meet several design criteria. Foremost, it should consider the exchanges of energy and mass in concert. Flows of energy are extremely important, since the biosphere requires energy to perform work. Processes requiring energy and work include biosynthesis, evaporation, transport of nutrients, and carbon dioxide fixation. All of these processes are interrelated and models must consider feedbacks and interactions among these processes. Second, the rules for scaling from individual components of the community (i.e., leaves or trees) to the entire ecosystem must be clearly visualized and included in order to reproduce the behavior of the whole system on the basis of biophysical laws which also regulate exchanges of mass and energy at the individual and organic level. Often nonlinear phenomena occur that change the scaling rules when moving from single organs to the whole ecosystem, e.g., the change in light response of photosynthesis from saturation at leaf scale to a linearly increasing response at the community scale as in sparse canopies and grasslands (see below). The complexity of feedbacks and the uncertainties associated with scaling rules push for an urgent validation of models with experimental data.

At present, many modeling activities make use of ecophysiological, biophysical, and biogeochemical principles to calculate trace gases fluxes for several land use/cover types, going from the patch to regional and global scales. Examples include studies on carbon dioxide fluxes (Tenhunen et al. 1989; Falge et al. 1997; Denning et al. 1996; Williams et al. 1996), ozone deposition (Amthor et al. 1994), and energy exchange (Bonan 1995; Sellers et al. 1986). Nevertheless, since the use of these concepts is relatively new, they have not been tested widely against field measurements of mass and energy exchange, nor have these concepts been tested over an assortment of vegetation types. Experimental measurements of biosphere–atmosphere exchanges have the additional advantage of providing a new synthesis view of ecosystem processes, which allow us to improve our conventional classifications of vegetation types and forms, achieving a deeper insight into functional biodiversity at the landscape level.

Within this decade, developments in micrometeorological technology and theory made the study of the interactions between the vegetation and the atmosphere routine, particularly by recent improvement of the eddy covariance method, which provides a direct measure of biosphere–atmosphere mass and energy fluxes at the ecosystem level. In the past, most of our knowledge of carbon and water exchange of vegetation was based on measurements taken at the leaf level with enclosure chambers or portable cuvettes (Field et al. 1982). With the development of micrometeorological methods, in particular eddy covariance (Baldocchi et al. 1988), a large change in scale of focus was achieved, moving from a leaf level, species/individual-dependent approach, to an integration of overall ecosystem fluxes in a direct, nondestructive way. Historically, this technique has been used in intensive and short-term land-surface experiments; however, today a number of experiments are available at larger spatial and longer temporal scales, since it is now possible to make continuous measurements of carbon dioxide and water fluxes on a seasonal basis with hourly discrimination (Wofsy et al. 1993; Baldocchi et al. 1996). Equipment and methodologies give us a solid basis for site intercomparisons (Moncrieff et al. 1997).

Continuous eddy covariance measurements of CO_2 fluxes on an annual basis are the only direct tool for estimating net ecosystem production (NEP), which is a critical term for evaluating the carbon balance of an ecosystem and vitally important for assessments of the fate of excess CO_2 from combustion of fossil fuels (Wofsy et al. 1993; Goulden et al. 1996). Moreover, long-term flux measurements are needed to capture seasonal dynamics. Many of the processes driving water and carbon fluxes at the ecosystem level are strongly dependent on seasonal changes in climate. Seasonal changes of phenology, available energy, and biomass production significantly affect the rates and properties of water and carbon exchanges with the atmosphere. Furthermore, extreme events (extreme temperature, high wind velocity, drought conditions, floods) are not usually considered during short-term field campaigns, but they can have a strong impact on hydrological and carbon cycles. Finally, questions concerning the biospheric responses to interannual climate variability require continuous flux measurements. For example, how do terrestrial carbon fluxes differ during ENSO (El Niño Southern Oscillation) years, as compared to La Niña years in regions influenced by this event, or how do perturbations such as the Mt. Piñatubo eruption influence biospheric carbon exchange. These relevant questions can only be answered by long-term monitoring of biospheric fluxes.

According to the eddy covariance theory, the eddy flux of any scalar can be written as:

$$F_c = \overline{w\rho_c} \tag{7.1}$$

where F_c is the flux density of scalar c, w is the vertical wind speed, and ρ_c is the density (or concentration) of the transported species. The overbar represents the mean of the product over the sampling interval.

Records of wind speed, temperature, and concentration exhibit turbulent or an irregular form; it is convenient to regard these variables as the sum of the mean and fluctuating part. This process is known as Reynold's decomposition (Arya 1988). For wind speed and concentration, fluxes may be written as:

$$w = \overline{w} + w' \qquad \rho_c = \overline{\rho_c} + \rho_c' \tag{7.2}$$

where the primes represent the fluctuation about the mean. We rewrite Equation 7.1, making use of this decomposition, to produce:

$$F_c = \overline{w\rho_c} + \overline{w'\rho_c'} \tag{7.3}$$

which shows that the total vertical flux of any scalar is the sum of a mean vertical flux $\overline{w\rho_c}$ and an eddy flux $\overline{w'\rho_c'}$. It can be noted that in the full expansion of Equation 7.1, some terms involving the mean of a fluctuating component have been omitted from Equation 7.3 because, by definition, $\overline{w'}$ and $\overline{\rho_c'} = 0$. One assumption normally made is that over a suitable interval of time there is no mass movement of the air in the vertical, i.e., $\overline{w} = 0$. Thus, from Equation 7.3 we obtain the practical working equation for eddy covariance:

$$F_c = \overline{w'\rho_c'} + \text{correction terms.} \tag{7.4}$$

The assumption that $\overline{w} = 0$ is problematic, but sufficient experimental evidence suggests that it can be corrected for in subsequent analysis. The issue is that sensible and latent heat fluxes cause variations in air density and most nondispersive infrared gas analyzers measure the density of the trace gas species in air rather than the mixing ratio or mole fraction. The effect can be substantial for trace gases which have fluxes that are small in relation to their background concentration (Fowler and Duyzer 1989). The need for a correction is also dictated by the type of measuring system in use, i.e., measuring CO_2 by an open-path gas analyzer may require corrections of 50%–100% and opposite in sign simply due to simultaneous transfer of sensible heat in the middle of the day (Suyker and Verma 1993). By contrast, in closed-path systems, temperature fluctuations can be effectively eliminated by passage down a sampling tube and the correction term for sensible heat flux is eliminated.

Equation 7.4 shows that we must have (a) instrumentation that can sample vertical wind speed and scalar concentration and be able to perform either real-time analysis in which means are subtracted from raw data to yield the fluctuating components when cross-products formed or (b) the facility to store all raw data for later processing in the laboratory. A number of stations measuring carbon, water, and energy exchanges on a continuous long-term basis are currently operating in various parts of the world, in

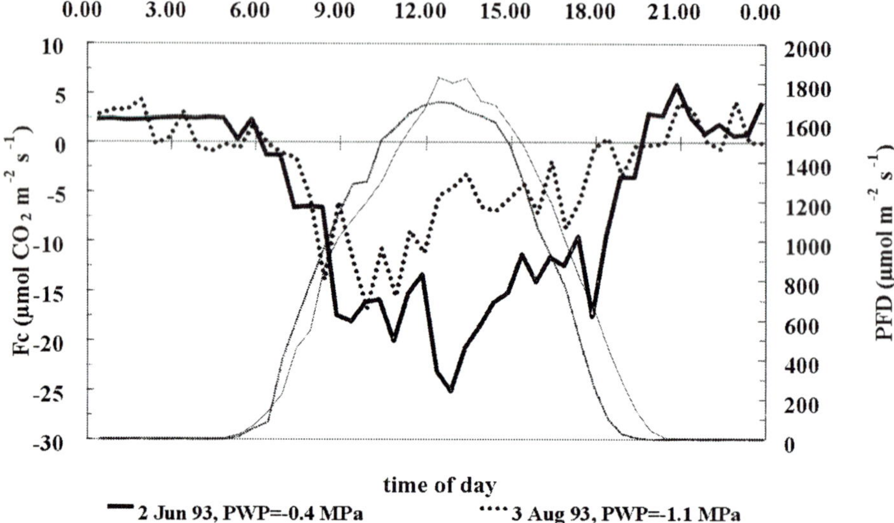

Figure 7.1 Carbon dioxide fluxes on two days (June 2 ideal conditions and August 3 water stress) for a beech canopy. PFD is the photosynthetic flux density on the two days.

different vegetation cover and biomes. A coordinated activity has been initiated (FLUXNET) and about 60 stations are listed around the world (Valentini et al. 1997).

An example of application of the eddy covariance technique is presented in Figure 7.1, where the diurnal pattern of carbon dioxide exchange for a beech forest in Central Italy during two contrasting periods (June and August) and reflecting two different water regimes (Valentini et al. 1996), is presented.

During the day, the canopy takes up carbon dioxide from the atmosphere due to the photosynthetic processes, which exceed respiratory losses, while during night, photosynthesis shuts down and respiration produces CO_2 which is released to the atmosphere. In August, CO_2 fluxes decrease as the result of a decline in predawn water potential (from –0.4 Mpa in June to –1.1 Mpa).

It is possible to extend in time the daily measurements and to provide budgets of fluxes. In Figure 7.2, an example of a seasonal budget of carbon for a temperate forest is presented, where NEE is the net ecosystem exchange of CO_2 directly measured by eddy covariance. The terms REE and GEE are respiration and gross ecosystem exchanges, respectively. They are inferred from nighttime fluxes as related to climatic factors, e.g., with a simple derived model. On an annual basis in this example, the forest is a sink of carbon with 472 g C m^{-2} y^{-1}, which was estimated directly by means of the net exchange measured continuously with eddy covariance (Valentini et al. 1996). Below, a number of examples indicate how these fluxes are controlled by biotic and abiotic factors.

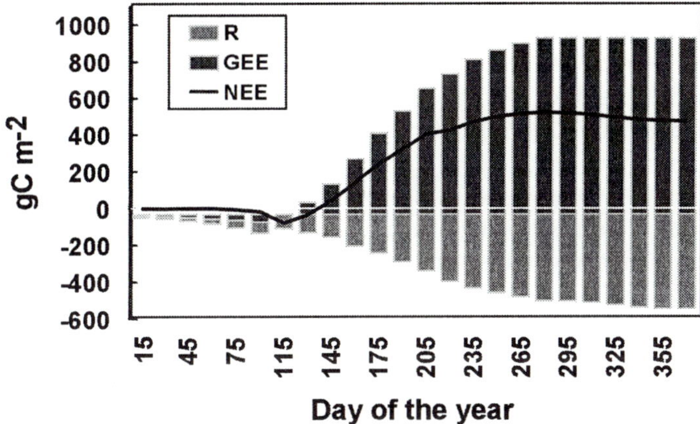

Figure 7.2 Seasonal cumulative net carbon exchange of a beech forest. R = respiration flux, GEE = gross ecosystem exchanges, and NEE = net ecosystem exchanges.

FACTORS CONTROLLING ECOSYSTEM FLUXES

Controls on Evapotranspiration

If we consider an "ideal" surface that is relatively smooth, horizontal, homogeneous, and extensive, with a very thin interface between the atmosphere above and the soil below, the energy balance of such a surface is relatively simple in that only the vertical fluxes of energy need to be considered. There are essentially four types of energy fluxes at an ideal surface: the net radiation (*Rn)* to or from the surface, the sensible (*H*) and latent heat (*LE*) fluxes to or from the atmosphere, and the heat flux (*G*) into or out of the soil.

An ideal, horizontally homogeneous plane surface is rarely encountered in practice. In many practical situations, such as over forest canopies, the surfaces are not adequately represented as thin interfaces, but more appropriately as thick layers, which are able to store energy and gaseous compounds. Thus, the surface energy balance includes the effect of a finite mass and heat capacity and the accumulation of gaseous compounds, which would allow energy to be stored in or released from the layer over a given time. Such changes must then be considered in the energy budget for the layer. An additional term contributing to the energy balance is the biochemical energy stored in the canopy layer (*P*). Photosynthetic processes utilize solar radiation for organic matter production, thus an energy flux is associated with these processes. On the other hand, respiratory fluxes are usually a source of energy. However, metabolic processes may be considered as negligible in relation to the magnitude of the other energy fluxes.

Thus it is possible, following the energy conservation law, to write the energy balance equation as:

$$Rn + H + LE + G + DS_T + DS_B + DS_E + P = 0 \tag{7.5}$$

where DS_T, DS_B, and DS_E are the storage fluxes for air temperature, biomass, and water vapor, respectively, while P is the energy stored in metabolic processes. Since these last terms are often negligible, we assume here and in the following that the available energy ($A = RN - G$) is partitioned into:

$$A = H + LE\,. \tag{7.6}$$

Traditionally, the partitioning of available energy into latent and sensible heat has been the subject of many studies focusing on crops as well as natural vegetation. Water balances, sap flow, lysimeters, and micrometeorological methods have been applied to evaluate water fluxes and energy partitioning in various ecosystems (Monteith 1975).

The partitioning of available energy at the surface can be considered "per se" a property of the vegetation type with a large variation among different terrestrial ecosystems.

Jarvis and McNaughton (1986) were among the first investigators to point out how energy partitioning can be affected by the biological and structural control of vegetation. They introduced the concept of atmospheric coupling, which they expressed by the Omega factor. One way to express the evapotranspiration flux for a given canopy is the Penmann–Monteith equation:

$$LE = \frac{sA + \dfrac{\rho c_p VPD}{Ra}}{s + \gamma\left(1 + \dfrac{Rs}{Ra}\right)} \tag{7.7}$$

where ρ and c_p are the density and specific heat of dry air, respectively; VPD is the water vapor pressure deficit at the reference height; s is the slope of saturation water vapor pressure versus temperature; γ the psychrometric constant; Rs and Ra the canopy stomatal and aereodynamic resistances, respectively. The aereodynamic resistance (Ra) is a function of both wind speed, atmospheric stability, and the structural properties of vegetation: height, leaf area, density, roughness of the canopy, etc. In particular for short, smooth canopies (like crops), Ra is very high indicating a less-efficient turbulent transport of mass and energy. On the contrary, rough, tall vegetation (like forests) shows smaller values of Ra, indicating a better aereodynamic efficiency.

There are two limits that reflect the effects of changes in the aereodynamic properties of vegetation, for which evapotranspiration can be affected:

$$\lim_{Ra \to \infty} LE = \frac{sA}{s + \gamma} \tag{7.8}$$

$$\lim_{Ra \to 0} LE = \frac{\rho c_p VPD}{\gamma Rs}\,. \tag{7.9}$$

The first (Equation 7.8) is the limit when the vegetation is short and poorly efficient from an aereodynamic point of view. Evapotranspiration depends largely on the available energy, irrespective of stomatal control. In the second (Equation 7.9), evapotranspiration is more directly controlled by *VPD* and the stomata. Broadly speaking, short vegetation, like crops, is more dependent on the available energy (they are decoupled from the atmosphere, according to terminology by Jarvis and McNaughton [1986]) while tall vegetation, like forests, is affected more by *VPD* and can control through stomata the water fluxes (they are coupled). If we want to investigate the effect of a small change of stomatal resistance on evapotranspiration (given constant all the other factors), i.e., to perform a sensitivity analysis of this process (see also Raupach and Finnigan [1988]), we notice that the fractional change of evapotranspiration with respect to *Rs* is:

$$\frac{Rs}{E}\frac{\partial E}{\partial Rs}=\frac{1}{\left(\frac{s}{\gamma}+1\right)\frac{Ra}{Rs}+1}. \tag{7.10}$$

The negative sign indicates a negative feedback, since any increase of *Rs* represents a decrease of evapotranspiration. However, increasing *Rs* increases the sensitivity of *LE* to changes in stomata opening. In Figure 7.3 we present a typical, theoretical behavior for three different types of vegetation cover (wheat, oak and pine forest), using reference values of *Ra* 80, 15, and 5 s m^{-1}, respectively. More negative values indicate a stronger control.

The sensitivity of *LE* to changes in *Rs* is much greater for the forests and in particular for conifers, since for any value of *Rs* it shows the most negative values. Thus both structure and function of vegetation, expressed by its aereodynamic properties and stomatal response, determine the extent of the control of evapotranspiration by the stomata.

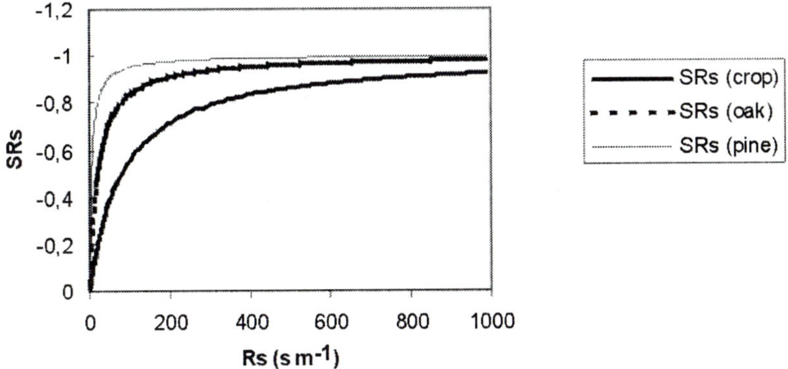

Figure 7.3 Sensitivity of evaporation (SRs) to fractional changes of canopy bulk stomatal resistance (Rs) for three vegetation types (crops, coniferous and deciduous forests).

It is interesting to note that a number of studies about vegetation–atmosphere energy exchange indicate that there is a functional convergence of vegetation types to enhance functional differences in the evapotranspiration control. It is evident that most of the aereodynamically rough canopies (forest, rough natural vegetation) with the lowest aereodynamic resistances show the highest canopy stomatal resistances, hence amplifying their coupling with the atmosphere and their strong control on evapotranspiration through *VPD* and stomata. On the other hand, most of the crops and grasslands show the opposite, with lower stomatal and higher aereodynamic resistances, enhancing their decoupling and hence their dependence on the available energy.

A survey of literature data concerning energy exchanges is presented in Figure 7.4, where the degree of control in evapotranspiration is expressed as the ratio of actual (*LE*) over equilibrium evapotranspiration (LE_{eq}):

$$LE_{eq} = \frac{sA}{s+\gamma}, \quad (7.11)$$

as a function of *Rs*.The data show a clear consistency among different vegetation types, indicating their functional properties with respect to evapotranspiration control. Crops are more dependent on the available energy; the opposite is generally true for forests.

Why do such properties (evapotranspiration control) change so much across different vegetation types? Why, for example, do forests need to control evapotranspiration by stomata more than grasslands? Finally, in particular within forest catogories, why are coniferous forests more sensitive than deciduous ones?

A possible explanation can be given if we consider that forests and shrublands usually tend to colonize environments with a less predictable and highly variable regime of precipitation, which increases the need for a better regulation of water fluxes. Furthermore, evergreen life forms, due to their longer vegetation activity, need to regulate evapotranspiration according to the climate conditions, while deciduous ecosystems control water fluxes only during their limited growing season (Valentini, Gamon et al. 1995; Valentini, Scarascia Mugnozza et al. 1995). These differences parallel common strategies observed in plant species referred to as "water spenders" and "water savers." It is interesting to note such control on evapotranspiration scale from the plant species level to ecosystems and, thus, it can be considered one of the emerging properties of the LAFU.

Carbon and Water Links

An example of linkages between carbon and water fluxes involves the conservative interdependence between NPP and evaporation. Many authors have shown how canopy photosynthesis or dry matter production is related to evaporation through the atmospheric humidity deficits (Cowan 1982; Tanner and Sinclair 1983). Other data show a positive correlation between leaf area index (LAI) and water balance of a canopy (precipitation minus evaporation) (Gholz 1982; Waring and Schlesinger 1985;

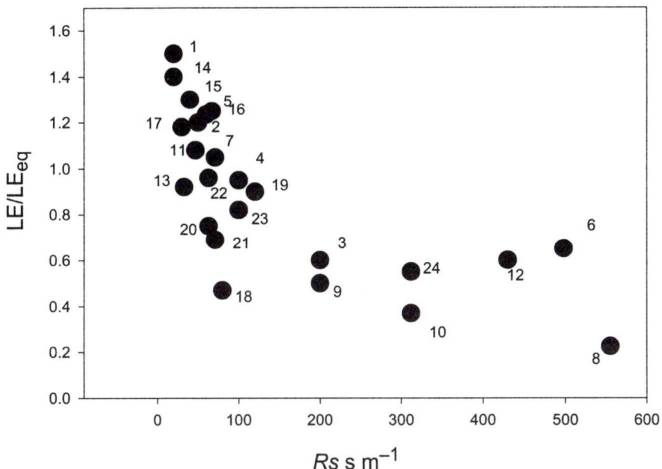

Figure 7.4 Relationship between canopy resistance (Rs s m^{-1}) and the ratio between latent heat and equilibrium evapotranspiration (LE/LE_{eq}) for various ecosystems.

1, 2	Baldocchi (1994)
3	Baldocchi and Vogel (1996)
4	Verma et al. (1986)
5, 6	Stewart and Verma (1992)
7, 8	R. Valentini, unpublished (1996 expedition in Central Siberia)
9	Valentini et al. (1996)
10, 11	Valentini et al. (1991)
11	Valentini et al. (1991)
12	F. Kelliher, unpublished
13	Anderson et al. (1984)
14, 15	McGinn and King (1990)
16	Baldocchi and Shankar (1995)
17	Monteney et al. (1985)
18	Jarvis (1994)
19	Stewart (1988)
20	Price and Black (1990)
21	Leuning and Attiwell (1978)
22	Grace et al. (1995)
23, 24	Miranda et al. (1996)

Woodward 1987; Bonan 1993). A conclusion drawn from these results is that the attainment of high rates of NPP comes at the expense of high evaporation rates.

In the late 1970s, Wong et al. (1979) reported that stomatal conductance was tightly coupled to leaf photosynthesis. They argued that stomata opened and closed to keep the ratio between intercellular and atmospheric CO_2 nearly constant (near 0.7 for C_3 plant). Ball et al. (1988), Leuning (1990), and Collatz et al. (1991) drew upon these observations their own laboratory experiments to publish a model that linked stomatal conductance (G_s) to leaf photosynthesis (P), relative humidity (h), and CO_2 concentration at the leaf surface (C_s):

$$G_s = \frac{1}{Rs} = m\frac{Ph}{C_s} + G_0\,, \tag{7.12}$$

where G_0 is a residual conductance needed to fit the linear regression and m is a scaling factor. The importance of this equation, even though it is an empirically derived equation, is that the coupling of carbon and water metabolism can be achieved through stomatal conductance. Despite the number of observations at the leaf level, which confirm such a relationship, few data are available at the canopy level. Is the linkage still valid at the ecosystem level? How does this relationship scale across different ecosystems?

Valentini et al. (Valentini, Gamon et al. 1995; Valentini, Scarascia Mugnozza et al. 1995) reported a first indication that for a grassland and a Mediterranean evergreen forest a relation of this form is still valid. Thus, also at the ecosystem level, the theory of water use optimization (Cowan 1986) is validated, since ecosystems with a limited productivity also tend to save water. This relationship also scales across different ecosystems. In Figure 7.5 maximum values of ecosystem CO_2 exchange versus canopy conductances are reported for a number of vegetation types. Here the general trend reported at the leaf scale also holds at the ecosystem level across a number of biomes. For instance, low productivity vegetation also shows lower stomatal conductances.

Further evidence of a strong coupling between carbon and water fluxes derives from the result of canopy model simulations, which show that theoretically evapotranspiration over crop and deciduous forest has a curvilinear dependence on the product of maximum carboxylation velocity (V_{Cmax}) and LAI (Figure 7.6). Normalized rates of evapotranspiration increased with any combination of photosynthetic capacity or canopy cover until the dependent variable (LAI × V_{Cmax}) approached a value equal to 300 (Baldocchi and Amthor 1999). For crops that are fertilized and attain V_{Cmax} values

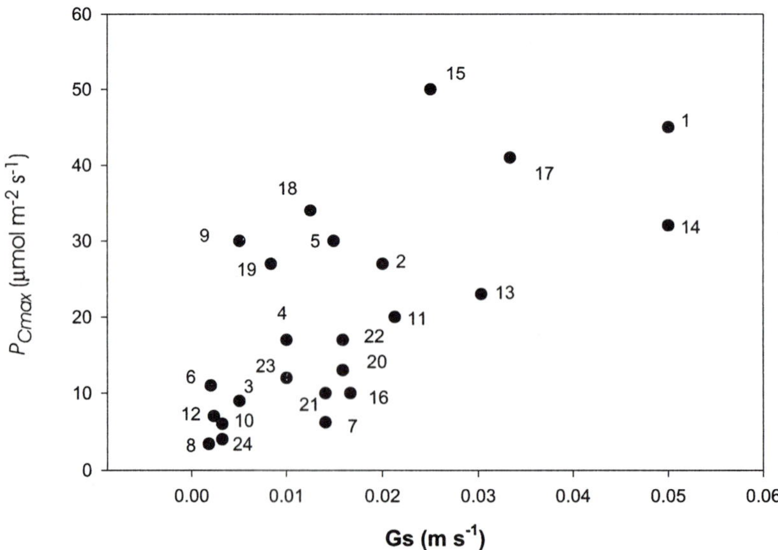

Figure 7.5 Relationship between maximum canopy conductance (Gs, m s^{-1}) and maximum canopy CO_2 fluxes (P_{Cmax} μmol CO_2 m^{-2} s^{-1}) measured by the eddy correlation technique (cf. legend in Figure 7.4).

approaching 100 μmol m^{-2} s^{-1} (Wullschleger 1993), this critical value would correspond with a LAI near 3.

The calculations shown in Figure 7.6 allow us to draw the conclusion that normalized rates of evapotranspiration are relatively low (below 1.0) over vegetation canopies that have low leaf area indices or low photosynthetic capacity.

Light Use Efficiency

The response of photosynthesis to light is a well-known process at leaf scale, both from a biochemical point of view (Farquhar et al. 1980) and an ecophysiological perspective (Björkman 1981). At leaf level, the typical response of net photosynthesis to light is a saturation curve above a certain photosynthetic photon flux and a linear response at low light intensity. This typical behavior reflects the properties of the photosynthetic system, which is light limited at low light and CO_2 limited at higher light levels. What is the response at the ecosystem level?

Since the first measurements and computations of canopy photosynthesis, it has been recognized that canopy CO_2 flux densities over closed crops are a quasi-linear function of available photon flux density (Q_p) (Ruimy et al. 1995). This is in strong contrast to the light response curve of a leaf, which can be approximated with a rectangular hyperbola (Thornley 1976). The "linearity" of this response has received much attention and criticism. A number of micrometeorological field data shows that linear

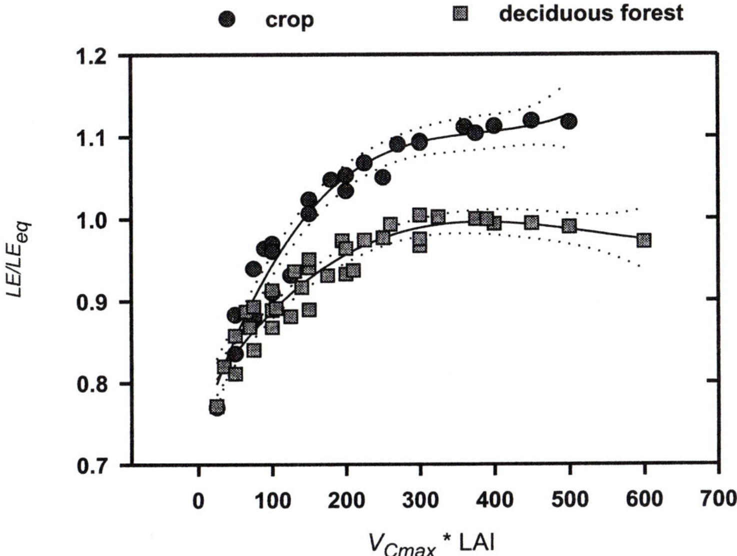

Figure 7.6 Relationship between the ratio LE/LE_{eq} and the product V_{Cmax} LAI (see text for explanation).

regressions on Q_p account for over 80% of the variance in Pc. However, these data seem to be concentrated among cases over closed, well-watered crop canopies. On the other hand, another body of data, derived from forests and sparse vegetation, reveals a markedly nonlinear response between Pc and absorbed radiation (Ruimy et al. 1995).

We propose an explanation for linear and nonlinear dependencies of canopy photosynthesis with respect to absorbed sunlight, on the basis of the interactive influences of LAI and photosynthetic capacity, as quantified by V_{Cmax}. Leaf area influences canopy photosynthesis by defining a canopy's ability to absorb more sunlight, by increasing the population of photosynthetic organs, and by shading the soil and reducing its heating and respiration. Photosynthetic capacity, on the other hand, governs the magnitude of photosynthesis that is attainable under given light conditions.

Calculations generated by a coupled canopy photosynthesis/micrometeorology model (Baldocchi and Meyers 1998) show that the canopy photosynthesis-light response curve of vegetation with high photosynthetic capacity (V_{Cmax} = 100 μmol m^{-2} s^{-1}) is saturated when a canopy is sparse (LAI = 1) and it is linear when the canopy is closed (LAI = 5) (Figure 7.7).

In recent years, a growing body of theoretical (Wang et al. 1991) and experimental (Hollinger et al. 1994; Baldocchi and Harley 1995; Leuning et al. 1995) data has shown that the slope of the relationship between canopy photosynthesis and available sunlight is affected by whether the sky is clear or cloudy. Rates of net ecosystem CO_2 exchange during clear conditions are diminished by more than half when compared to those observed under cloudy skies and similar photon flux densities. The effect is manifested through the effect of sky conditions on the distribution of light through the canopy and on the energy budget of leaves. Under cloudy skies, incoming sunlight is more isotropic (Ross 1981; Myneni et al. 1992), so light transmits deeper into the canopy (Ross 1981). The process allows more light to reach deeper into the canopy and illuminate typically sunlight-deficient leaves. On sunny days, the photosynthetic rates of upper sunlit leaves tend to be light saturated. They also experience a higher heat load, which enhances respiration, so it lowers net photosynthesis (Baldocchi and Harley 1995). These results have a major implication on the application of simple multiplicative relationships for estimating global NPP.

Time averaging also affects the relationship between canopy-scale CO_2 exchange and available solar energy. Theoretical calculations by Leuning et al. (1995) and a literature survey by Ruimy et al. (1995) suggest that canopy photosynthesis is a linear function of absorbed solar radiation when summed over the course of a day. In terms of light use efficiency, it has been shown (Ruimy et al. 1995) that the apparent quantum yield is 0.035 and 0.023 mol mol^{-1} for forests and crops canopies, respectively, which is lower than the value of leaves of C_3 plants of 0.05 mol mol^{-1}. In general, when all vegetation types are grouped into one class, the apparent quantum yield becomes 0.044 mol mol^{-1}, which is closer to leaf value. Thus, in general, the apparent quantum yield of canopies is smaller than that of single leaves.

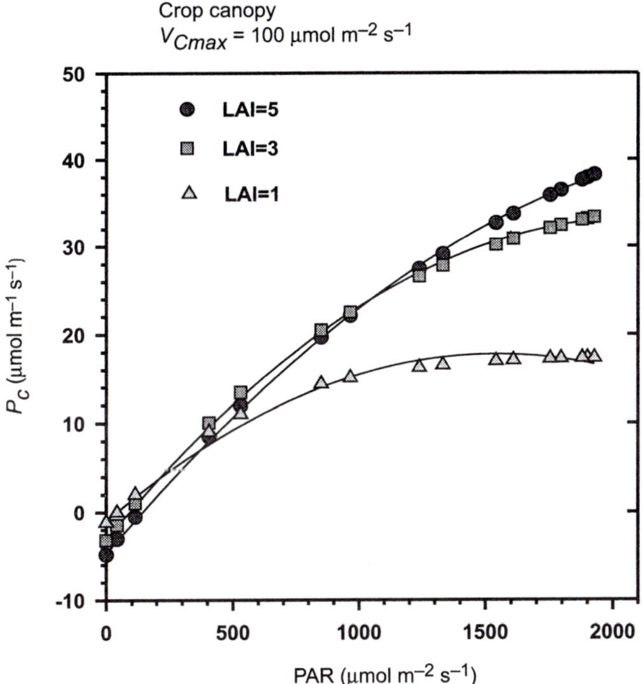

Figure 7.7 Simulated light response curves of canopy CO_2 exchanges (P_c) versus photosynthetic photon flux density (PAR) for a crop canopy with LAI = 5 at different levels of V_{Cmax}.

Nitrogen

A linkage between carbon and nitrogen budgets is manifested through the dependency of photosynthetic capacity on a leaf's nitrogen content (Wong et al. 1985; Field and Mooney 1986; Field 1991) and on the rate a leaf takes up nitrogen (Woodward and Smith 1994). Photosynthesis is linked to nitrogen because the enzyme Rubisco, a nitrogen-rich compound, catalyzes carboxylation. In combination, the link of nitrogen to water is mediated through stomata. Since stomatal conductance is linked to photosynthesis (as shown previously), it is also correlated with nitrogen content (Schulze et al. 1994; Kelliher et al. 1994). In general, at the leaf level, a linear relationship between nitrogen content and photosynthetic capacity is observed. The same property is observed along a light gradient within the canopy, where a decline of nitrogen content accompanied by a decline of photosynthetic capacity is shown. From these observations, it has been suggested (Hirose and Werger 1987; Hirose et al. 1988) that, given a fixed amount of nitrogen available to leaves, plants allocate nitrogen in order to optimize total canopy photosynthesis. The optimization theory (Field and Mooney 1986) suggests that any form of investment in resource acquisition (e.g., nitrogen) must be counterbalanced by a gain in a given function (e.g., photosynthesis). This explains

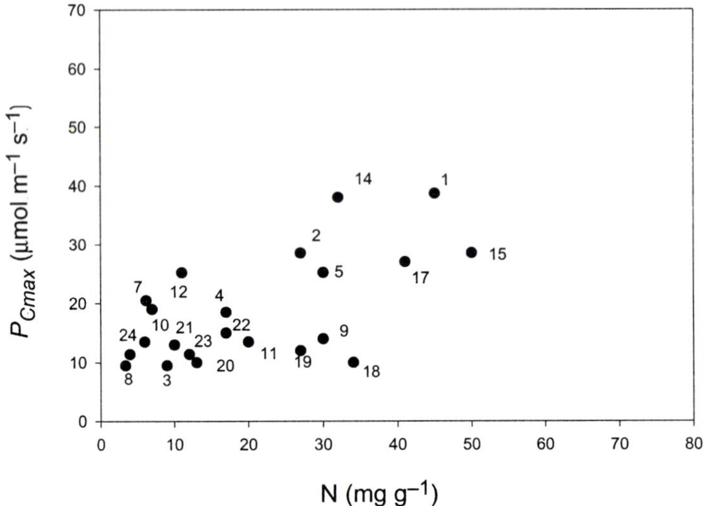

Figure 7.8 Relationship between nitrogen concentration (N, mg g^{-1}) and maximum canopy CO_2 fluxes (P_{Cmax}), µmol CO_2 m^{-2} s^{-1}) measured by the eddy covariance technique across different vegetation types (cf. legend in Figure 7.4).

why, for example, a low level of nitrogen corresponds to a limited photosynthetic capacity. There have been a number of studies showing the occurence of near optimization of resource use at the leaf level; however, few data are available at the canopy scale, despite the review compiled by Schulze et al. (1994), which mainly focuses on the effects of nitrogen concentration on stomatal conductance.

We have reviewed other ecosystem data, including those reported by Schulze et al. (1994), with the objective of determining to what extent the linear dependence between nitrogen concentration and photosynthesis, demonstrated valid at leaf level, occurs across a range of vegetation types. The results are presented in Figure 7.8, indicating that the limitation observed in nutrient resources has a pronounced impact on ecosystem carbon gain.

Phenology and Seasonality Effects on Fluxes

The view of functional properties described above implicitly applies to features of ecosystems that are vegetation type dependent in a static view of ecosystem processes. We can consider these as mean properties. However, ecosystems show a strong seasonality in terms of both phenology and the impact of climate factors. Leaf development is a strong determinant of carbon and water fluxes, while climate affects carbon and water fluxes by modifying some of the key functions we have discussed so far.

A major difference among vegetation types with regard to seasonality and phenology concerns leaf duration and aging, as seen in evergreen, annual, or deciduous habits.

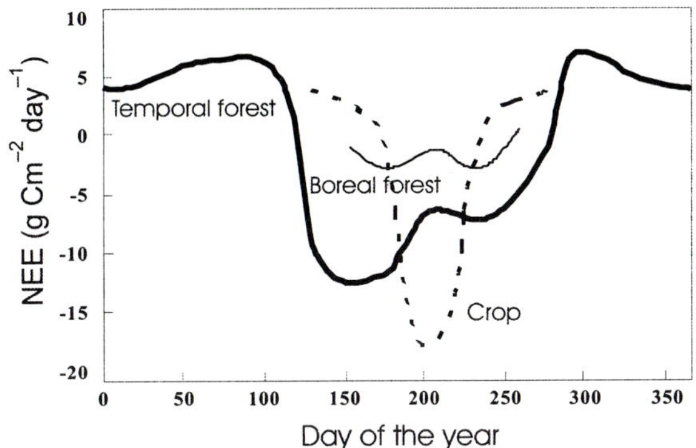

Figure 7.9 Yearly trend of net ecosystem exchange (NEE, gC m^{-2} day^{-1}) measured by the eddy covariance technique over a temperate oak forest (thick line), a cereal crop (broken line), and a boreal forest (thin line). The sign follows the micrometeorological convention, where a minus represents an exchange going from the atmosphere to the ecosystem.

An example of the annual trend of carbon fluxes for the three different vegetation types (Baldocchi and Amthor 1999) is presented in Figure 7.9. The reported example indicates changes in maxima, duration, and their timing, which are specific for the vegetation type. Phenology also has a strong influence on the annual carbon balance of ecosystems. For example, a 10 g C m^{-2} day^{-1} difference can be observed between leafless and full leaf states of a deciduous forest canopy. Since leaf out can vary 10 to 30 days at a site or by differences in latitude, this can cause a 100–300 g C m^{-2} difference in annual net ecosystem carbon exchange.

The effect of seasonality can shift overall ecosystem properties from one extreme to another. In Figure 7.10, an example is shown of how a low fertility Californian grassland can be atmospherically decoupled during its peak of vegetation development, when water resources are readily available, and can become extremely coupled when *Rs* is strongly reduced in response to leaf aging and climate (Valentini, Gamon et al. 1995).

TOWARD A FUNCTIONAL CLASSIFICATION OF TERRESTRIAL ECOSYSTEMS

Traditionally, vegetation has been described by taxonomic features, such as species composition, which has also been used to illustrate functional properties of the biomes. Here we make an attempt to aggregate functional properties, as directly determined by biosphere–atmosphere exchanges, in order to achieve a "functional" classification of the ecosystems. For this purpose, we present in Figure 7.11 a principal component analysis of a number of experiments where a set of variables have been considered to rank terrestrial ecosystems (see Table 7.2). The selected variables have been chosen as indicators of overall ecosystem properties: canopy resistances for both

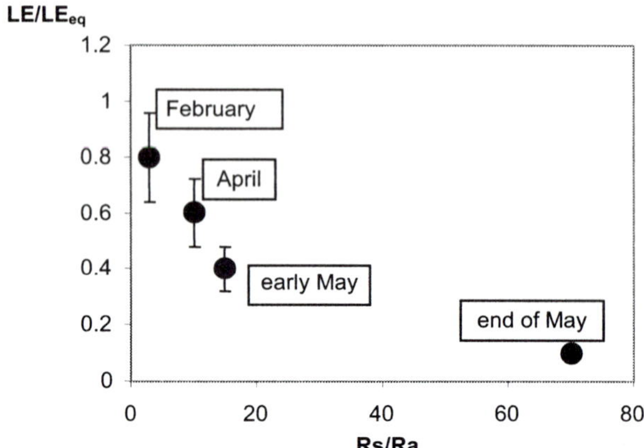

Figure 7.10 Changes in the ration of LE/LE_{eq} as a function of Rs/Ra during the season passing from nonlimiting (February) to limiting (end of May) water conditions on a low production Californian grassland.

water and carbon fluxes, the ratio of evapotranspiration to the equilibrium value and Bowen ratio for water processes, nitrogen content of leaves for the nitrogen cycle, and maximal assimilation for carbon gain. Resistances and canopy assimilation are minimal and maximal values, respectively, while the other factors are typical values for the vegetation community.

The graph shows how the various ecosystems group according to the variance explained by the first two principal components. The first component accounts for 86% of the total variance. The inclusion of a second component allows an explanation of 97% of the total variance. It is possible to distinguish three groups (α, β, γ). Group α

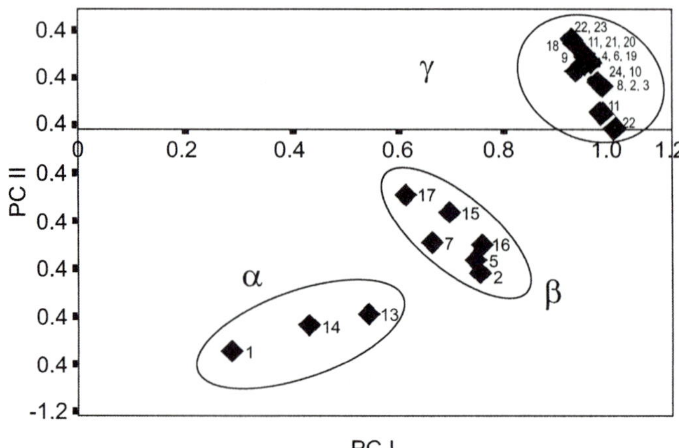

Figure 7.11 Principal component analysis of the data presented in Table 7.2. The three functional groups, α, β, and γ, explain 86% of the total variance.

Table 7.2 Functional properties of various terrestrial ecosystems.

Ecosystem	Rc (s m^{-1})	LE/LE_{eq}	Ra (s m^{-1})	N (mg g^{-1})	LAI	Beta	A_{max} (μmol m^{-2} s^{-1})	Ref.erence
Triticum aestivum (wheat)	20	1.5	80	38.65	2.6	0.1	45	Baldocchi (1994)
Zea mays (corn)	50	1.2	60	28.5	1.7	1	27	
Pinus taeda	200	0.6	5	9.5	2.5	6	9	Baldocchi and Vogel (1996)
Quercus alba	100	0.95	20	18.5	4.9	0.5	17	Verma et al. (1986)
Grassland /FIFE	67	1.25	83	25.2	2.8	0.25	30	Stewart and Verma (1992)
Grassland/FIFE dry	498	0.65	83	25.2	1.7	4	11	
Bog (Central Siberia)	71	1.049	100	20.5	1.7	0.6	6.2	Valentini, unpubl. (1996 expedition in Central Siberia)
Pinus sylvestris (Central Siberia)	555	0.225	35	9.5	0.5	2.9	3.4	
Fagus sylvatica	200	0.5	10	14	4.5	2	30	Valentini et al. (1996)
Quercus ilex dry	312	0.37	10	13.5	3	2.8	6	Valentini et al. (1991)
Quercus ilex	47	1.08	10	13.5	3	1.1	20	
Larix gmelinii	430	0.6	8	19	3	1.5	7	F. Kelliher, unpublished
Glycine max (soybean)	33	0.92	54	45	3.5	0.25	23	Anderson et al. (1984)
Alfalfa	20	1.4	34	38	2	0.18	32	McGinn and King (1990)
Zea mays (corn)	40	1.3	18	28.5	3.4	0.64	50	
Solanum tuberosum (potato)	60	1.235	65	38	2.2	0.11	10	Baldocchi and Shankar (1995)
Hevea Brasilianensis (rubber)	30	1.18	8	27	4.3	0.2	41	Monteney et al. (1985)
Picea sitkensis	80	0.47	6	10	7.8	2.2	34	Jarvis (1994)
Pinus sylvestris (U.K.)	120	0.9	5.5	12	4.3	3.5	27	Stewart (1988)
Pseudotsuga menziesii	63	0.75	1	10	3.5	1	13	Price and Black (1990)
Eucalyptus maculata	71	0.69	4	13	3.5	1	10	Leuning and Attiwell (1978)
Tropical Rainforest (Rondonia)	63	0.96	21	15	4	0.43	17	Grace et al. (1995)
Cerrado	100	0.82	21	11.4	1	0.67	12	Miranda et al. (1996)
Cerrado dry	312	0.55	21	11.4	0.4	1.5	4	

consists primarily of agricultural crops, which are the ones showing the highest photosynthesis rates. The β group is constituted by grasslands, bogs including also some crops (potato), and tropical plantation forests (rubber). The third group, γ, includes mostly forests.

In some respects the principal component analysis reflects the taxonomic differences of the broad vegetation types (crops, grassland, forests). However there are exceptions, and more remarkably, is the inclusion of stressed grassland into the γ group, indicating a shift of functional properties according to the seasonal stage. Similarly, the tropical plantation is placed in the β group, which otherwise includes short structure vegetation. Considering group γ, both the *Quercus ilex* forest (wet) and the tropical rainforest seem to form a sub-cluster, indicating a close similarity, while the *Quercus ilex* dry forest belongs to the forest clusters.

This exercise is limited to a small number of data sets and a few parameters. However, it shows remarkable differences in functional properties among different communities and that the taxonomical description is not always adequate to explain the role of vegetation cover in land-surface interactions.

FEEDBACK WITH THE ATMOSPHERE

Given the importance of functional properties for understanding biosphere–atmosphere interactions, how do the the properties discussed so far affect regional and global climate? A large body of evidence is currently available in the literature (Sellers 1991) clearly showing the increasing importance of vegetation exchange on the results of global circulation models. Less emphasis has so far been given to the effects of land use/cover changes at regional scale. Recently Pielke et al. (1991, 1998) pointed out the importance of vegetation feedbacks with the atmosphere at regional scale, particularly on the formation of clouds and in determining precipitation patterns. Other studies concerned the Amazonian basin, where a reduction of precipitation due to deforestation has been demonstrated (Shukla et al. 1990).

The knowledge of surface fluxes, particularly energy partitioning at the surface, allows estimations of the convective boundary-layer growth and hence the conditions relevant for cloud formation. Such measurements are now planned for Central Siberia, where the landscape is dominated by an alternation of forested areas (*Pinus sylvestris* L.) and bogs, extending for several thousands of kilometers from the Urals to the Yenisei River, to see how patterning in surface properties is linked to local climate. The forest is regenerated by frequent fires, which are dependent on thunderstorms due to convective clouds formation.

As an example, the Bowen ratio of two contrasting canopies of the Central Siberian region is presented in Figure 7.12. The bog ecosystem has a Bowen ratio close to 1, while the sparse canopy of an 18-year-old *Pinus sylvestris* forest exhibits a ratio of 4, indicating high dissipation of energy through sensible heat. Calculations of the PBL height associated with these two formations provide for the bog area 1600 m and for the forest a value close to 2550 m. The high sensible heat flux over forest is the driving

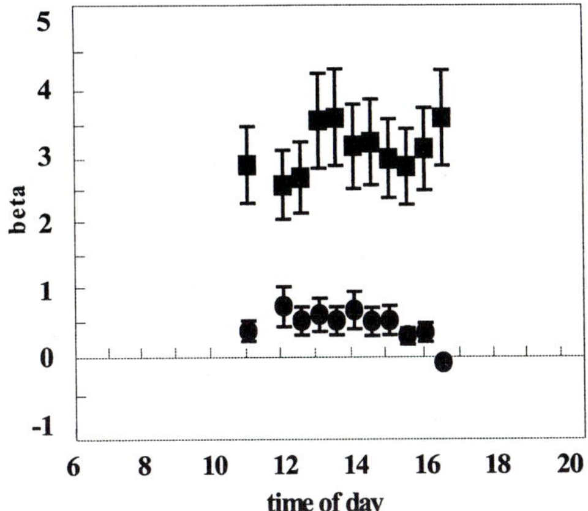

Figure 7.12 Daily trend of the Bowen ratio (beta) for a bog ecosystem (closed circles) and a pine forest (closed squares) in Central Siberia.

mechanism for the vertical transport of water vapor in the atmosphere, which is mainly evaporated by the bog areas. In this way thunderstorms may be generated, while fires maintain the ecological succession dynamics of the vegetation. A scheme for these processes is shown in Figure 7.13.

The energy partitioning over the forest is mainly determined by limiting soil fertility. Indeed the lack of adequate nitrogen supply and a small soil organic layer have a profound impact on the physiology of photosynthesis. However, since carbon, water, and nitrogen cycles are intimately related, biological limitations such as nitrogen availability can influence the atmospheric exchanges and the extent of energy partitioning at the surface, including the CBL growth rates.

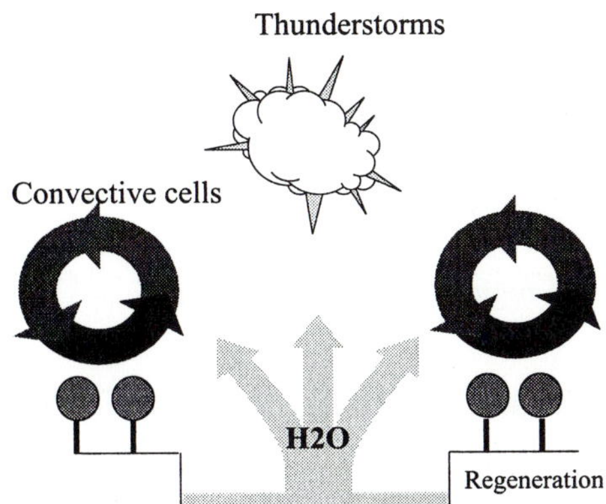

Figure 7.13 Possible mechanism for convective cloud formation and precipitation of the Siberian landscape leading to a closed water cycle. Thunderstorms generate fires and interact with the ecology of forest succession.

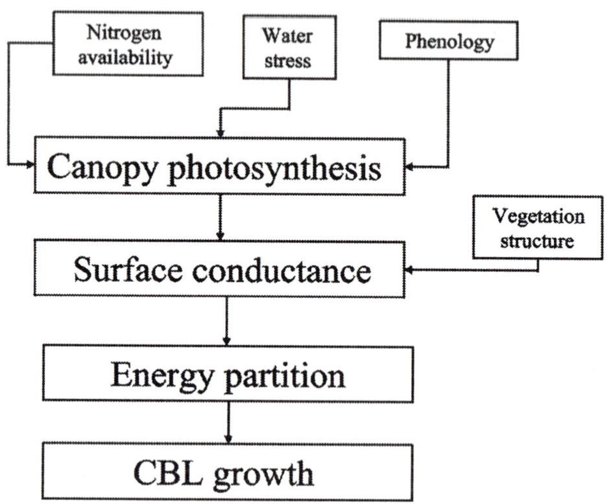

Figure 7.14 A scheme of biospheric controls of the energy partition at the surface.

Is Ecology Driving the Climate System?

The connection of water, carbon, and nitrogen cycles and their interactions provide a basic framework for understanding the potential feedbacks with the atmosphere. Most of the relationships we discussed intrinsically contain a strong biological control, which is dependent on functional type and environment. A possible consequence of the strong coupling among ecological processes is the effect of the biotic control on land–atmosphere interactions. Figure 7.14 is presented as a conceptual scheme of biological control of surface energy exchanges. Biological constraints, such as nitrogen availability, have impacts on maximal carbon assimilation, which in turn is related to the maximal canopy conductance, which regulate the extent of energy partitioning at the surface and, hence, the CBL growth.

Land use/cover changes are, therefore, a strong driving variable for understanding climate processes both at regional and global scales.

DO LAFUs EXIST? SYNTHESIS AND CONCLUSIONS

The relations we have discussed recognize properties that are invariant across scale changes from leaves to canopies. Examples are the assimilation–conductance relationship, which has been demonstrated valid for leaves and canopies. Similarly, the effect of nitrogen availability on the maximal assimilation of carbon also scaled from single leaves to the ecosystem. More importantly, we have seen that functional properties not only scale across spatial dimensions (leaf versus canopy) but are maintained across various biomes. Controls on evapotranspiration are scaled across a range of biomes from crops to forests, the maximal canopy assimilation and conductances scale also among different vegetation types, according to their productivity and the interactions with the environment.

We have also shown that such biodiversity of functions can be ranked, and it is possible to discriminate groups of ecosystems (α, β, γ LAFUs). Such groups represent broad vegetation categories (crops, grassland, and forests), but exceptions occur due to seasonality. Physical climate modifies time-integrated functional properties and leads to grouping other than that which would be assumed solely from morphological and taxonomical relationships.

The biodiversity of function in terrestrial ecosystems, as viewed from the standpoint of LAFUs, can have a significant impact on regional and global climate through a series of feedbacks, which in turn are important for biological processes at the Earth's surface.

REFERENCES

Amthor, J.S., M.L. Goulden, J.W. Munger, and S.C. Wofsy. 1994. Testing a mechanistic model of forest-canopy mass and energy exchange using eddy correlation: Carbon dioxide and ozone uptake by a mixed oak-maple stand. *Austral. J. Plant Physiol.* **21**:623–651.

Anderson, D.E., S.B. Verma, and N.J. Rosenberg. 1984. Eddy correlation measurements of CO_2, latent heat and sensible heat fluxes over a crop surface. *Boundary-Layer Meteorol.* **29**:263–272.

Arya, S.P. 1988. Introduction to Micrometeorology. San Diego: Academic.

Baldocchi, D.D. 1994. A comparative study of mass and energy exchange rates over a closed C^3 (wheat) and open C^4 (corn) crop. I. The partitioning of available energy into latent and sensible heat exchange. *Agr. For. Meteorol.* **67**:191–220.

Baldocchi, D.D., and J.S. Amthor. 1999. Canopy photosynthesis: History, measurements and models. In: Terrestrial Global Productivity: Past, Present and Future, ed. J. Roy, B. Saugier, and H. Mooney. San Diego: Academic, in press.

Baldocchi, D.B., and P.C. Harley. 1995. Scaling carbon dioxide and water vapor from leaf to canopy in a deciduous forest: Model testing and application. *Plant Cell Env.* **18**:1157–1173.

Baldocchi, D.D., B.B. Hicks, and T.P. Meyers. 1988. Measuring biosphere–atmosphere exchanges of biologically related gases with micrometeorological methods. *Ecology* **69**:1331–1340.

Baldocchi, D.D, and T.P. Meyers. 1998. On using eco-physiological, micrometeorological and biogeochemical theory to evaluate carbon dioxide, water vapor and gaseous deposition fluxes over vegetation. *Agr. For. Meteorol.* **90**:1–26.

Baldocchi, D.D., and R.K. Shankar. 1995. Intra-field variability of scalar flux densities across a transition between a desert and an irrigated potato field. *Boundary-Layer Meteorol.* **76**:109–136.

Baldocchi, D., R. Valentini, S. Running, W. Oechel, and R. Dahlman. 1996. Strategies for measuring and modelling carbon dioxide and water vapor fluxes over terrestrial ecosystems. *Glob. Change Biol.* **2**:159–168.

Baldocchi, D.D., and C.A. Vogel. 1996. Energy and CO_2 flux densities above and below a temperate broad-leaved forest and a boreal pine forest. *Tree Physiol.* **16**:5–16.

Ball, J.T., I.E. Woodrow, and J.A. Berry. 1988. A model predicting stomatal conductance and its contribution to the control of photosynthesis under different environmental conditions. In: Progress in Photosynthetic Research, ed. J. Biggins, pp. 221–224. Dordrecht: Nijhoff.

Björkman, O. 1981. Responses to different quantum flux densities. In: Encyclopedia of Plant Physiology, ed. O.L. Lange, P.S. Nobel, C.B. Osmond, and H. Ziegler, vol 12A, pp. 57–107. Plant Physiological Ecology I. Berlin: Springer.

Bonan, G.B. 1993. Importance of leaf area index and forest type when estimating photosynthesis in boreal forests. *Remote Sens. Env.* **43**:303–314.

Bonan, G.B. 1995. Land–atmosphere interactions for climate change system models: Coupling biophysical, biogeochemical and ecosystem dynamical processes. *Remote Sens. Env.* **43**:303–314.

Budyko, M.I. 1982. The Earth's Climate: Past and Future. International Geophysics Series 29. London: Academic.

Charlson, R.J. 1992. The Atmosphere. In: Global Biogeochemical Cycles, ed. S.S. Butcher, R.J. Charlson, G.H. Orians, and V.G. Wolfe, pp. 213–238. London: Academic.

Collatz, G.J., J.T. Ball, C. Grivet, and J.A. Berry. 1991. Physiological and environmental regulation of stomatal conductance, photosynthesis and transpiration: A model that includes a laminar boundary layer. *Agr. For. Meteorol.* **54**:107–136.

Cowan, I.R. 1982. Regulation of water use in relation to carbon gain in higer plants. In: Encyclopedia of Plant Physiology, ed. O.L. Lange, P.S. Nobel, C.B. Osmond, and H. Ziegler, pp. 589–613. Physiological Plant Ecology I. Berlin: Springer.

Cowan, I.R. 1986. Economics of carbon fixation in higher plants. In: On the Economy of Plant Form and Function, ed. T.J. Givnish, pp. 133–170. Cambridge: Cambridge Univ. Press.

Denning, A.S., J.G. Collatz, C. Zhang, D.A. Randall, J.A. Berry, P.J. Sellers, G.D. Colello, and D.A. Dazlich. 1996. Simulation of terrestrial carbon metabolism and atmospheric CO_2 in a general circulation model. 1. Surface carbon fluxes. *Tellus* **48B**:521–542.

Falge, E., R.J. Ryel, M. Alsheimer, and J.D. Tenhunen. 1997. Effects of stand structure and physiology on forest gas exchange: A simulation study for Norway spruce. *Trees* **11**:436–448.

Farquhar, G.D., S. von Caemmerer, and J.A. Berry. 1980. A biochemical model of photosynthetic CO_2 assimilation in leaves of C_3 species. *Planta* **149**:78–90.

Field, C.B. 1991. Ecological scaling of carbon gain to stress and resource availability. In: Integrated Responses of Plants to Stress, ed. H.A. Mooney, W.E. Winner, and E.J. Pell, pp. 35–65. San Diego: Academic.

Field, C., J.A. Berry, and H.A. Mooney. 1982. A portable system for measuring carbon dioxide and water vapor exchange of leaves. *Plant Cell Env.* **5**:179–186.

Field, C., and H.A. Mooney. 1986. The photosynthesis–nitrogen relationship in wild plants. In: On the Economy of Plant Form and Function, ed. T.J. Givnish, pp. 25–55. Cambridge: Cambridge Univ. Press.

Fowler, D., and J. Duyzer. 1989. Micrometeorological techniques for the measurement of trace gas exchange. In: Exchange of Trace Gases between Terrestrial Ecosystems and the Atmosphere, ed. M.O. Andreae, and D.S. Schimel, pp. 189–207. Dahlem Workshop Report LS 47. Chichester: Wiley.

Gholz, H. 1982. Environmental limits on aboveground net primary production, leaf area and biomass in vegetation zones of Pacific Northwest. *Ecology*. **63**:469–481.

Goulden, M.L., J.W. Munger, S.-M. Fan, B.C. Daube, and S.C. Wofsy. 1996. Measurements of carbon sequestration by long term eddy covariance: Methods and critical evaluation of accuracy. *Glob. Change Biol.* **2**:169–182.

Grace, J., et al. 1995. Fluxes of carbon dioxide and water vapour over an undisturbed tropical forest in south-west Amazonia. *Glob. Change Biol.* **1**:1–12.

Hirose, T., and M.J.A. Werger. 1987. Maximizing daily canopy photosynthesis with respect to the leaf nitrogen allocation pattern in the canopy. *Oecologia* **72**:520–526.

Hirose, T., M.J.A. Werger, T.L. Pons, and J.W.A. van Rheenen. 1988. Canopy structure and leaf nitrogen distribution in a stand of *Lysmachia vulgaris* L. as influenced by stand density. *Oecologia* **77**:145–150.

Hollinger, D.Y., et al. 1994. Carbon dioxide exchange between an undisturbed old-growth temperate forest and the atmosphere. *Ecology* **75**:134–150.

Jarvis, P.G. 1994. Capture of carbon dioxide by a coniferous forest. In: Resource Capture by Crops, ed. J.L. Monteith, R.K. Scott, and M.H. Unsworth, pp. 351–374. Nottingham: Nottingham Univ. Press.

Jarvis, P.G., and K.G. McNaughton. 1986. Stomatal control of transpiration: Scaling up from leaf to region. *Adv. Ecol. Res.* **15**:1–49.

Kelliher, F.M., R. Leuning, M. Raupach, and E.D. Schulze. 1994. Maximum conductances for evaporation from gobal vegetation types. *Agr. For. Meteorol.* **73**:1–16.

Kimmins, J.P. 1987. Forest Ecology. New York: Macmillan.

Lean, J., and D.A. Warrilow. 1989. Simulation of the regional climatic impact of Amazon deforestation. *Nature* **342**:411–413.

Leuning, R. 1990. Modelling stomatal behavior and photosynthesis of Eucalyptus grandis. *Austral. J. Plant Physiol.* **17**:159–175.

Leuning, R., and P.M. Attiwell. 1978. Mass, heat and momentum exchange between a mature Eucalyptus forest and the atmosphere. *Agr. Meteorol.* **19**:215–241.

Leuning, R., F.M. Kelliher, D.G.G. dePury, and E.D. Schulze. 1995. Leaf nitrogen, photosynthesis, conductance an transpiration: Scaling from leaves to canopies. *Plant Cell Env.* **18**:1183–1200.

McGinn, S.M., and K.M. King. 1990. Simultaneous measurements of heat, water vapour and CO_2 fluxes above alfalfa and maize. *Agr. For. Meteorol.* **49**:331–349.

Miranda, A.C., H.S. Miranda, J. Lloyd, J. Grace, J.A. McIntire, P. Meir, P. Riggan, R. Lockwood, and J. Brass. 1996. Carbon dioxide fluxes over a cerrado sensu stricto in central Brazil. In: Amazonian Deforestation and Climate, ed. J.H.C. Gash, C.A. Nobre, J.M. Roberts, and R.L. Victoria, pp. 353–364. New York: Wiley.

Moncrieff, J., R. Valentini, S. Greco, G. Seufert, and P. Ciccioli. 1997. Trace gas exchange over terrestrial ecosystems: Methods and perspectives in micrometeorology. *J. Exp. Bot.* **48**:1133–1142.

Monteith, J.L. 1975. Vegetation and the Atmosphere, vol. 2. New York: Academic.

Monteney, B.A., J.M. Barbier, and C.M. Bernos. 1985. Determination of the energy exchange of a forest-type culture: *Hevea brasilianensis*. In: The Forest–Atmosphere Interaction, ed. B.A. Hutchinson and B.B. Hicks, pp. 211–233. Dordrecht: Reidel.

Myneni, R.B., G. Asrar, D. Tanre, and B.J. Choudhury. 1992. Remote sensing of solar radiation absorbed and reflected by vegetated land surfaces. *IEEE Trans. Geosci. Remote Sens.* **30**:302–314.

Pielke, R.A., R. Avissar, M.R. Raupach, H. Dolman, X. Zeng, and S. Denning. 1998. Influence of short and long-term ecosystem dynamics on weather and climate. *Glob. Change Biol.* **4**: 461–475.

Pielke, R.A., G. Dalu, J.S. Snook, T.J. Lee, and T.G.F. Kittel. 1991. Nonlinear influence of mesoscale land use on weather and climate. *J. Clim.* **4**:1053–1069.

Price, D.T., and T.A. Black. 1990. Effects of short term variation in weather on diurnal canopy CO_2 flux and evapotranspiration of a juvenile douglas-fir stand. *Agr. For. Meteorol.* **50**:120–159.

Raupach, M.R., and J.J. Finnigan. 1988. Single-layer evapotranspiration models are incorrect but useful, whereas multilayer models are correct but useless: Discuss. *Austral. J. Plant Physiol.* 1**5**:715–726.

Ross, J. 1981. The Radiation Regime and Architecture of Plant Stands. The Hague: Dr. W. Junk.

Ruimy, A., P.G. Jarvis, D.D. Baldocchi, and B. Saugier. 1995. CO_2 fluxes over plant canopies and solar radiation: A literature review. *Adv. Ecol. Res.* **26**:1–68.

Salati, E., and P.B. Vose. 1984. Amazon Basin: A system in equilibrium. *Science* **225**:1–29.

Schimel, S.D. 1995. Terrestrial ecosystems and the carbon cycle. *Glob. Change Biol.* **1**:77–91.

Schulze, E.-D., F.M. Kelliher, C. Körner, J. Lloyd, and R. Leuning. 1994. Relationships between maximum stomatal conductance, ecosystems surface conductance, carbon assimilation rate and plant nitrogen nutrition: A global ecology scaling exercise. *Ann. Rev. Ecol. Syst.* **25**:629–660.

Sellers, P. 1991. Modelling and observing land surface–atmosphere interactions on large scales. In: Land Surface–Atmosphere Interactions for Climate Modeling, ed. E.F. Wood, p. 474. London: Kluwer Academic.

Sellers, P.J., Y. Mintz, Y.C. Sud, and A. Dalcher. 1986. A simple biosphere model (SiB) for use within general circulation models. *J. Atmos. Sci.* **43**:505–531.

Shukla, J., C. Nobre, and P. Sellers. 1990. Amazon deforestation and climate change. *Science* **247**:1322–1325.

Stewart, J.B. 1988. Modelling surface conductance of pine forest. *Agr. For. Meteorol.* **43**:19–35.

Stewart, J.B., and S.B. Verma. 1992. Comparison of surface fluxes and conductances at two contrasting sites within the FIFE area. *J. Geophys. Res.* **97**:8623–8628.

Suyker, A.E., and S.B. Verma. 1993. Eddy correlation measurements of CO_2 flux using a closed-path sensor: Theory and field tests against an open-path sensor. *Boundary-Layer Meteorol.* **64**:391–407.

Tanner, C.B., and T.R. Sinclair. 1983. Efficient water use efficiency: Research or re-search ? In: Limitations of Efficient Water Use in Crop Production, ed. H. Taylor, W. Jordan, and T. Sinclair, pp. 1–27. Madison, WI: Am. Soc. of Agronomy.

Tenhunen, J.D., J.F. Reynolds, S. Rambal, R. Dougherty, and J. Kummerow. 1989. A physiologically-based growth simulator for drought adapted woody plant species. In: Biomass Production by Fast-Growing Trees, ed. J.S. Pereira and J.J. Landsberg, pp. 135–168. NATO ASI Series, Applied Science 166. Dordrecht: Kluwer Academic.

Thornley, J.H.M. 1976. Mathematical Models in Plant Physiology. London: Academic.

Valentini, R., D.D. Baldocchi, and S. Running. 1997. The IGBP-BAHC global flux network initiative (FLUXNET): Current status and perspectives. *Glob. Change Newsl.* **28**:14–16.

Valentini, R., P. De Angelis, G. Matteucci, R. Monaco, S. Dore, and G.E. Scarascia Muguozza. 1996. Seasonal net carbon dioxide exchange of a beech forest with the atmosphere. *Glob. Change Biol.* **2**:199–208.

Valentini, R., J.A. Gamon, and C.B. Field. 1995. Ecosystem gas exchange in a California serpentine grassland: Seasonal patterns and implications for scaling. *Ecology* **76(6)**:1940–1952.

Valentini, R., G. Scarascia Mugnozza, P. De Angelis, and R. Bimbi. 1991. An experimental test of the eddy correlation technique over a Mediterranean macchia canopy. *Plant Cell Env.* **14**:987–994.

Valentini, R., G. Scarascia Mugnozza, P. De Angelis, and G. Matteucci. 1995. Coupling water sources and carbon metabolism of natural vegetation at integrated time and space scale. *Agr. For. Meterol.* **73**:297–306.

Verma, S.B., D.D. Baldocchi, D.R. Anderson, D.R. Matt, and R.E. Clement. 1986. Eddy fluxes of CO_2, water vapor and sensible heat over a deciduous forest. *Boundary-Layer Meteorol.* **36**:71–91.

Vitousek, P.M., H.A. Mooney, J. Lubchenko, and J.M. Melillo. 1997. Human domination of Earth's ecosystems. *Science* **277**:494–499.

Walter, H. 1983. Vegetation of the Earth. New York: Springer.

Wang, Y.P., P.G. Jarvis, and C.M.A. Taylor. 1991. PAR absorption and its relation to above-ground dry matter production of sitka spruce. *J. Appl. Ecol.* **28**:547–560.

Waring, R.H., and W.H. Schlesinger. 1985. Forest Ecosystems: Concepts and Management. Orlando: Academic.

Williams, M., D.N. Rastetter, M.L. Fernades, M.L. Goulden, S.C. Wofsy, G.R. Shaver, J.M. Melillo, J.W. Munger, S.M. Fan, and K.J. Nadelhoffer. 1996. Modelling the soil–plant–atmosphere continuum in a Quercu -Acer stand at Harvard Forest: The regulation of stomatal conductance by light, nitrogen and soil/plant hydraulic properties. *Plant Cell Env.* **19** :911–927.

Wofsy, S.C., M.L. Goulden, J.W. Munger, S.-M. Fan, P.S. Bakwin, B.C. Daube, S.L. Bassow, and F.A. Bazzaz. 1993. Net exchange of CO_2 in a mid-latitude forest. *Science* **260**:1314–1317.

Wong, S.C., I.R. Cowan, and G.D. Farquhar. 1979. Stomatal conductance correlates with photosynthetic capacity. *Nature* **282**:424–426.

Wong, S.C., I.R. Cowan, and G.D. Farquhar. 1985. Leaf conductance in relation to rate of CO_2 assimilation. II. Effects of short-term exposure to different photon flux densities. *Plant Physiol.* **78**:826–829.

Woodward, F.I. 1987. Climate and Plant Distribution. Cambridge: Cambridge Univ. Press.

Woodward, F.I., and T.M. Smith. 1994. Predictions and measurements of the maximum photosynthetic rate, Amax at the global scale. In: Ecophysiology of Photosynthesis, ed. E.D. Schulze and M.M. Caldwell, pp. 491–509. New York: Springer.

Wullschleger, S.D. 1993. Biochemical limitations to carbon assimilation in C_3 plants — A retrospective analysis of the A/Ci curves from 109 species. *J. Exp. Bot.* **44**:907–920.

8

Effects of Landscape Pattern and Topography on Emissions and Transport

F.X. MEIXNER[1] and W. EUGSTER[2]

[1] Biogeochemistry Dept., Max Planck Institute for Chemistry, P.O. Box 3060, D–55020 Mainz, Germany
[2] Institute of Geography, University of Bern, Hallerstrasse 12, CH–3012 Bern, Switzerland and Dept. of Plant Ecology, University of Bayreuth, D–95440 Bayreuth, Germany

ABSTRACT

Landscape pattern (land use and natural patchiness) and complex topography strongly influence regional variation in emissions of trace gases and pollutants as well as in atmospheric transport processes. This leads to small-scale variation in the amount of (biogenic) emission and atmospheric deposition and in local concentrations of atmospheric constituents. This chapter addresses the most important processes that control trace gas emission and that govern atmospheric transport in nonflat and nonhomogeneous landscapes. The focus is on the controlling factors of biogenic emissions and on those transport and deposition processes that differ from idealized conditions in flat and homogeneous terrain.

INTRODUCTION

The steady increase of carbon dioxide and several other trace gases in our atmosphere will have serious consequences for the habitability of our planet (Ramanathan et al. 1987). Among these consequences are:

1. global warming due to increased atmospheric concentrations of greenhouse gases (e.g., CO_2, CH_4, N_2O),
2. destruction of the stratospheric ozone layer due to increased halogenated compounds and N_2O,
3. the increase of tropospheric ozone due to increased emissions of nitrogen oxides NO_x ($NO_x = NO + NO_2$), CO, and volatile organic compounds (VOCs), and

Integrating Hydrology, Ecosystem Dynamics, and Biogeochemistry in Complex Landscapes
Edited by J.D. Tenhunen and P. Kabat

4. changes in tropospheric cloud cover and stratospheric aerosol due to increased emissions of dimethyl sulfide (DMS) and carbonyl sulfide (COS).

While the global cycles of some of these trace gases are heavily affected by anthropogenic emissions (e.g., CO_2, CO, and NO_x), most are dominated by biospheric processes (occurring within soil, vegetation, waters, and animals). In this chapter, we focus on biogenic emissions, especially those from soils. Knowledge on how biogenic emissions vary with landscape pattern and/or topography is scarce. Therefore, we emphasize those emission controlling factors that may be affected most by landscape pattern and topography.

Both anthropogenic and biogenic emissions are generally nonequally distributed in complex landscapes. Furthermore, atmospheric transport in complex landscapes leads to regional- and small-scale differences in concentrations and deposition. We examine the potential effects of specific transport processes thought to have an influence at regional and local scales that result in differences between idealized (homogeneous landscape, flat terrain) and actual encountered conditions. In the case of atmospheric transport processes, patchy land-use pattern and complex terrain exhibit interactive effects. Therefore, we will not separate our discussion with respect to these two components and are well aware that some of our considerations also apply to patchy land-use patterns on a flat topography.

EMISSIONS OF TRACE GASES FROM SOILS AND VEGETATION

Many trace gases are produced and consumed within soils. Therefore, soils can contribute to the global budgets of CO, CH_4, COS, N_2O, NO, and DMS as both sources and sinks (Table 8.1a). The budget values and percent contributions in Table 8.1a are highly uncertain for a number of reasons (Conrad 1996). With a varying degree of reliability, fluxes of many trace gases can presently be measured with a variety of techniques, ranging from small-scale leaf enclosure to tower-based and airborne micrometeorological methods (Matson and Harriss 1995). Nevertheless, it is by far not trivial to estimate atmospheric budgets from fluxes measured at individual field sites (Andreae and Schimel 1989). Most uncertainties and problems can be traced to the fact that fluxes, determined by a vast diversity in mostly microbial processes, are highly variable with respect to time and space. The problems are not necessarily solved by integration of fluxes over larger areas and longer time periods, since "each individual flux event is caused by deterministic processes that change in a nonlinear way when conditions even change slightly" (Conrad 1996). This is also valid for the variation in biogenic emissions from plants (Table 8.1b), especially for those of volatile organic compounds (nonmethane VOCs). There are several hundreds of biogenic VOCs known, but those of atmospheric relevance comprise the isoprenoids, alkanes and alkenes, organic acids (e.g., formic and acetic acids), carbonyl compounds (e.g., aldehydes), alcohols, as well as esters and ethers (e.g., methanol, ethanol, ethylacetate). The global source strength of the most prominent biogenic VOC compounds,

Table 8.1a Contribution of soils to the global cycles of atmospheric trace gases. After Conrad (1996); Meixner (1994).

Trace Gas	Ambient Concentration (ppb)	Lifetime (days)	Total Budget (Tg y^{-1})	Annual Increase (%)	Contribution of Soils as Source (%)	Contribution of Soils as Sink (%)
CO	70–170	100	2600	1.0	1	15
CH_4	1700	4000	540	~0.8	60	5
COS	0.5	1500	2.3	0	23? [a]	23? [a]
N_2O	310	60,000	15	0.2–0.3	70	?
NO	0.1–20	1	60	?	20	?
DMS	~0.1	~0.9	38	?	~0.1	0

[a] Inferred from Andreae and Crutzen (1997); however, any contribution of soils as a source of COS is presently under discussion.

Table 8.1b Contribution of plants to the global cycles of atmospheric trace gases.

Trace Gas	Ambient Concentration (ppb)	Lifetime (days)	Total Budget (Tg y^{-1})	Annual Increase (%)	Contribution of Plants as Source (%)	Contribution of Plants as Sink (%)
CO [e]	70–170	100	2600	1.0	6	0
VOCs [c]	0.1–100	0.004–2	900–1400	?	80	1?
COS [d]	0.5	1500	2.3	0	0	40[b]
NO [f]	0.1–20	1	60	?	4	?
DMS [a]	~0.1	~0.9	32–83	?	5–45	0

[a] Andreae and Jaeschke (1992)
[b] Andreae and Crutzen (1997)
[c] Kesselmeier and Staudt (1999)
[d] Kesselmeier et al. (1997)
[e] Seiler and Conrad (1987); Schade (1997)
[f] Wildt et al. (1996)

isoprene and monoterpenes, is estimated to range between 300–980 Tg y^{-1}, which is similar to and above the global source strength of CH_4 (540 Tg y^{-1}; Table 8.1a). The carbon loss due to biogenic VOC emissions is generally considered not to exceed some few percent of plant assimilated carbon (cf. Kesselmeier and Staudt 1999); occasionally it may reach 10%–20%. However, biogenic VOC emissions comprise more than two thirds of global VOC emissions, which play a crucial role in atmospheric chemistry, particularly for tropospheric oxidant formation, including ozone.

Mechanisms of Trace Gas Production and Consumption in Soils

Soil processes can be classified into chemical (abiotic) and microbial (biotic) processes. Additional soil enzymatic processes, which are very important for the sink strength of atmospheric hydrogen (H_2), but unimportant for CO, CH_4, COS, N_2O, NO, and DMS, are not discussed here. Chemical processes for COS production and consumption are likely but unknown; those for NO can either not be quantified at present, or their importance (e.g., NO oxidation by O_2) is limited by the requirement of unrealistic high NO mixing ratios in soil air (Conrad 1996). The production of CO in soils is dominated by chemical processes, namely the thermal decomposition of humic acids and other organic material. However, the contribution of soil emissions to the global CO budget is marginal (Table 8.1a). Microorganisms are responsible for most production and consumption processes in soils. In their oxidative and reductive metabolisms, trace gases could act as growth substrates and/or co-metabolites, or they are considered to be stoichiometric and other products. For a comprehensive review on microorganisms and relevant metabolic mechanisms, the reader is referred to Conrad (1996). From this review, we can learn that production and consumption, and hence exchange of trace gases between soils and atmosphere, are predominantly controlled on the microscopic level, i.e., the level of microorganisms' metabolism. However, in the context of how landscape pattern and topography may affect biogenic trace gas emissions, a higher scale of organization and controls must be considered.

Controls of Emission Pathways from Soils by Functional Groups and Processes

Considering production and consumption of trace gases from soils, Davidson and Schimel (1995) prefer so-called functional groups rather than species of microorganisms, since (a) common modes of metabolism exist among many groups of microbes and (b) knowing the predominant functional group facilitates identification of factors which control rates of production and consumption. Because of limited space, we confine the following discussion to functional groups, processes, and chemical and physical properties that are important for CH_4, N_2O, and NO emissions from soils.

Methane Production

Methane-producing bacteria (methanogens) are a group of the archaeobacteria, which predominantly produce CH_4 via cleaving acetate to CO_2 and CH_4, and reducing CO_2 with H_2. Acetate (as substrate for methanogens) and H_2 may originate from the fermentation of simple substrates (sugars, amino acids, etc.), which are released directly by plants or from soil organic matter breakdown. Most methanogens operate within a temperature range of about 20°–40°C and are very responsive to temperature changes within this range. Methanogens are extremely sensitive to oxygen and reactive oxygen species: methane production in soils occurs under anaerobic and highly reducing conditions in the absence of NO_3^-, SO_4^{2-}, or ferric ion. Therefore, rice paddies (rich in organic matter), wetlands (tropical, temperate, and arctic or boreal), marine sediments, the digestive tract of many animals and insects, as well as anaerobic sewage digestors are prominent methane production sites. Significant CH_4 production was also observed in (generally aerobic) upland soils as a consequence of rain events saturating the soil for a more or less extended period. However, CH_4 production in soils should not be confused with CH_4 emission from wetland soils: more than 80% of the diffusive CH_4 flux from the anaerobic part of the soil is oxidized in the oxic layer of freshwater wetlands. However, ebullition flux of CH_4 (and other trace gases contained in the bubbles) from inundated and nonvegetated wetlands does not seem to be affected by the CH_4-oxidizing bacteria (Conrad 1996). If wetland soils are covered by appropriate vegetation, the transport through the aerenchyme of those plants is the dominating emission pathway to the atmosphere (e.g., Davidson and Schimel 1995).

Methane Oxidation

Methane-oxidizing bacteria (methanotrophs) are a diverse group of aerobic bacteria, but they are ubiquitous and can be found in all types of soils. They oxidize CH_4 primarily by (a) the enzyme CH_4 monooxygenase (methanotrophic bacteria) and (b) the enzyme NH_3 monooxygenase (nitrifying bacteria). O_2 is required and is an obligate reagent in the first reaction of the oxidation pathway. Thus, any soil that is porous enough to allow the diffusion of atmospheric O_2 into it has the potential to act as a net consumer of CH_4 (as the majority of aerated upland soils). In those ecosystems where seasonal inundation occurs (rice paddies, tropical floodplains, arctic or boreal wetlands), the position of the water table may be the primary factor regulating the magnitude and even the sign of the CH_4 exchange with the atmosphere.

Denitrification

In denitrification (an anaerobic process), bacteria utilize nitrate (NO_3^-) and nitrite (NO_2^-) for their growth and reduce them (via N_2O and NO) to N_2. The absence of oxygen results from either high soil water content or large respiration and oxygen consumption rates in the soil. Readily oxidizable organic carbon is a requirement for most denitrifying bacteria (heterotrophs). Denitrification occurs over a wide temperature range (maximum rate at ~65°–70 °C), roughly doubling with every 10 K increase. However, the ability of specific denitrifying bacteria to produce and consume N_2O and NO was shown (Conrad 1996).

Nitrification

Nitrification involves the biological oxidation of nitrogen, typically the oxidation of soil ammonium (NH_4^+) to NO_3^- (with NO_2^- as intermediate), but there are also bacteria which oxidize NH_4^+ to NO_2^- and NO_2^- to NO_3^-. These bacteria are generally chemoautotrophic and require only CO_2, H_2O, O_2, and either NH_4^+ or NO_2^- for growth. If NH_4^+ or urea ($(NH_2)_2CO$) is present, nitrification will occur in well-aerated soils. As a by-product of NH_4^+ oxidation, nitrifying bacteria produce N_2O and NO. While it is unknown which biochemical pathway is the most important for NO, the N_2O production is most likely via reduction of NO_2^- by NH_4^+ oxidizers. It has been shown that NO production may be dominated by nitrification in a particular soil and by denitrification in another one (Conrad 1996).

Bidirectional Fluxes and the Compensation-Concentration Concept

As discussed above, trace gas production and consumption processes occur simultaneously in a given (upland) soil; consequently, bidirectional fluxes have been observed under laboratory and field conditions. A conceptual explanation is given in Figure 8.1: the relative distribution of trace gas-producing vs. trace gas-consuming soil crumbs in upland soil determines the magnitude of corresponding (gross) fluxes from and to the surface (caveat: only net fluxes can be measured with techniques commonly applied in the field). In aggregated soils, soil crumbs may be considered as units of trace gas metabolism (in nonaggregated soils, sand grains covered with a microbial biofilm may play that role). Soil crumbs are usually heterogeneous regarding their aerobic–anaerobic metabolism, since even (generally well) aerated upland soils may also contain anoxic microniches (Conrad 1996).

The compensation-concentration concept is based on simultaneous trace gas production and consumption in a given bulk soil sample and on the observation that consumption rates are a function of trace gas concentration, whereas production rates are not. The compensation concentration is then "the concentration at which the consumption rate reaches the same value as the production rate so that the result of both processes is zero flux" (Conrad 1994). Given the (gross) production rate P and the pseudo first-order uptake constant k of a specific trace gas, then the net release rate R is given as $R = P - km$ (where m is the atmospheric trace gas concentration), and the compensation concentration is given by $m_c = P/k$. The net release rate is linked to the flux F between the soil and the atmosphere by the diffusion within the soil. This was first described by the conceptual model of Galbally and Johansson (1989) and can be formulated as $F = R/k\sqrt{(k\rho_s D)} = (m_c - m)\sqrt{(k\rho_s D)}$, where ρ_s is the bulk density of the dry

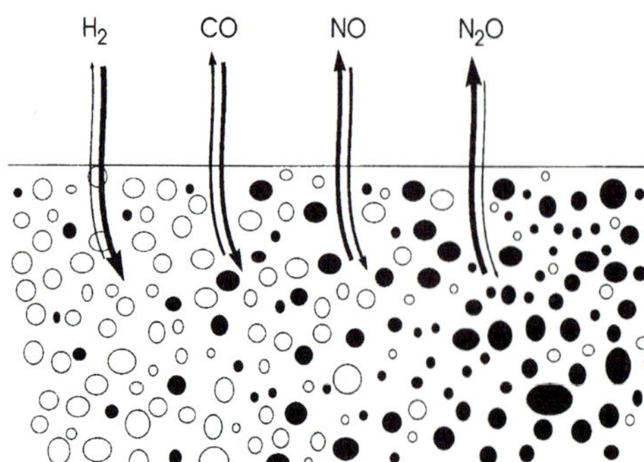

Figure 8.1 Conceptual scheme of H_2, CO, NO, and N_2O bidirectional fluxes between upland soil and the atmosphere as a result of the relative amount of soil entities (e.g., soil crumbs) that either release trace gas into the soil air (solid symbols) or take it up (open symbols). Figure taken from Conrad (1996).

soil and D the diffusivity of the trace gas in the soil. It is evident that (a) the magnitude of the flux F increases with the difference between the compensation and ambient concentrations, and (b) F will result in a net emission from the soil if $m_c > m$ and in a net uptake if $m_c < m$. Sensitivity analysis has shown that F strongly depends on k (Remde et al. 1993).

For the application of the compensation concept, it is essential that production and consumption processes are more or less homogeneously distributed in the soil layer (sample) under investigation. In the case of CH_4, there is obviously a substantial vertical separation of both processes in most soils; no CH_4 compensation points were reported so far (Conrad 1994, 1996). Generally, the compensation concentration may be considered as a critical variable of the regulation of trace gas surface exchange, if (a) there is a dynamic change of compensation concentration in the soil and/or high fluctuations of ambient concentrations, and (b) at least a partial overlap of ranges of compensation and ambient concentration (see Table 8.2). Under these premises, CO and NO are likely candidates for the compensation-concentration concept, whereas COS and N_2O are usually not. For techniques and methods how to determine compensation point concentrations, the reader is referred to Conrad (1994).

Nitrogen Availability

Considering the microbial processes which determine the production and consumption of CO, CH_4, N_2O, and NO in soils, it becomes obvious that any biological process controlling nitrogen and carbon availability (and their interrelation) is of fundamental importance for emission fluxes. Several attempts have been made to describe local, regional, and global CH_4, N_2O, and NO emission patterns on the basis of carbon availability, gross N mineralization, denitrification, nitrification, decomposition, and soil carbon–nitrogen fluxes (see Bartlett and Harris 1993, Potter et al. 1996, and references therein). Numerous significant correlations between N_2O and NO fluxes and the NO_3^- or NH_4^+ soil concentration exist in literature, but they were shown to be very site specific, and there is no consistent trend across the different studies (Davidson and

Table 8.2 Overlapping and nonoverlapping ranges of ambient and compensation concentrations. After Conrad (1994); completed by data from Meixner (1994).

Trace Gas	Ambient Concentration (ppb)	Compensation-Concentration (ppb)
CO	70–170	~5–1200
COS	0.5	10–200
N_2O	310	ca. 500
NO	< 0.1–20	0.3–75 (aerobic) 1600–2200 (anaerobic)

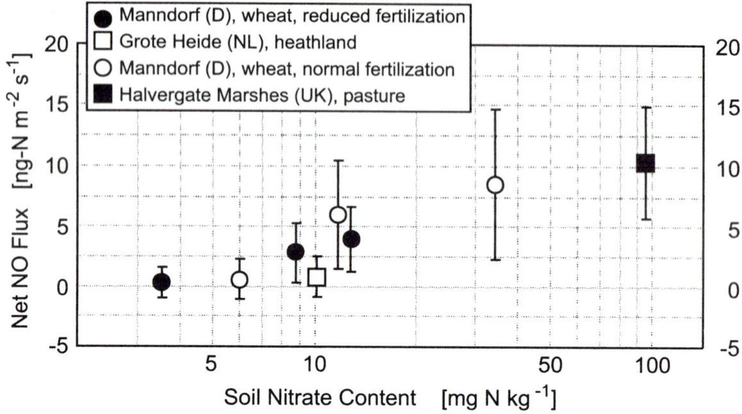

Figure 8.2 The net NO flux from different European ecosystems as a function of soil nitrate content: Halvergate (UK) September 1989; Manndorf (D, Lower Bavaria) May, June, July, August 1990; Grote Heide (NL) May 1991. Data are from Ludwig (1994).

Schimel 1995). Conversely, intersite predictions of NO emission from actual soil NO_3^- are less significant, but large-scale differences between ecosystem types are usually reflected by accumulation of soil NO_3^-, for which an example is given in Figure 8.2.

Soil Water Content and Gaseous Diffusivity

Soil water content is intimately related to CH_4 production and oxidation, as well as to nitrification and denitrification for two important reasons (Davidson and Schimel 1995): (a) the substrate supply for soil microorganisms (e.g., NH_4^+ for nitrifying and acetate for methanogenic bacteria) is accomplished by diffusion of the substrates in soil water films, and (b) water in soil pores is the dominant controller of gaseous diffusion in soils. Since O_2 and CH_4 have to enter soil water films prior to their consumption by microorganisms, this phase change may also be rate limiting. The water-filled pore space (WFPS), the ratio of volumetric soil water content to total porosity of the soil, was shown to be the most suitable among various expressions of soil water, since WFPS is largely comparable among soils of different texture.

Sixty percent WFPS seems to be the optimal soil water content for aerobic processes, and > 80% WFPS for anaerobic processes. Considering N_2O and NO sources and source partition, it was suggested that (a) if WFPS < 60%, nitrification is more important than denitrification and the N_2O:NO emission ratio is < 1, while (b) if WFPS >60%, denitrification overrides nitrification and N_2O:NO > 1 (Davidson and Schimel 1995). Under completely anaerobic conditions, N_2 may be the dominant end product of denitrification. Potter et al. (1996) calculated soil biogenic N_2O and NO release from global patterns of gross mineralization and scaled them by relative emission rates, which are dependent on the transient water-filled pore space (Figure 8.3). For

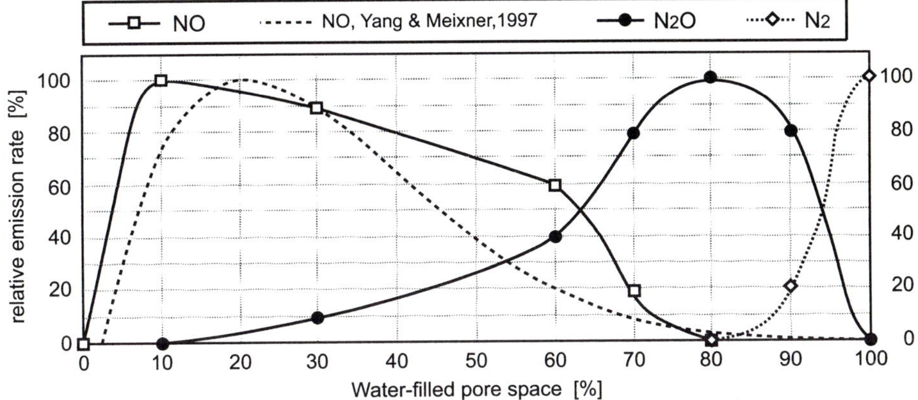

Figure 8.3 The influence of water-filled pore space on the relative emissions of NO, N_2O, and N_2 from (upland) soils. After Potter et al. (1996); completed by recent data of Yang and Meixner (1997).

many soils, field capacity is around 60% WFPS, which according to Figure 8.3, separates ranges of dominance of nitrification versus denitrification and NO vs. N_2O production. The shape of relative N_2O and NO emission curves, a result of two opposing processes, the substrate diffusion limit (towards low WFPS) and the O_2 diffusion limit (towards high WFPS) is a more general issue of soil water content versus soil microbial activity (Skopp et al. 1990).

Bartlett and Harriss (1993), having analyzed global data sets of CH_4 uptake of upland soils from the subarctic to the tropics, found a striking similarity of uptake rates regardless of ecosystem type and/or latitude. Consequently, they suggested that CH_4 uptake on a global scale is controlled primarily by rates of diffusion rather than by more variable biological factors, which meanwhile received confirmation (see Davidson and Schimel 1995, Conrad 1996, and references therein). Diffusion of CH_4 into the soil therefore largely controls CH_4 oxidation rates by controlling the supply of substrate to microbial populations. Whereas geomorphological and edaphic factors (soil texture) are determining the diffusion characteristics of a given soil, WFPS (which can be related to soil diffusivity) is also an important parameter to scale CH_4 uptake in upland soils. Since substantial rates of CH_4 production are only observed in fully saturated soils, WFPS will certainly not be an appropriate parameter for its description. However, water table height is the more useful measure of soil water in studying CH_4 production. For temperate wetlands, water table height was found to explain 62% of the CH_4 flux variability (see Bartlett and Harriss 1993). As soon as the water table drops below the soil surface, CH_4 emission rate will decrease rapidly: this is caused by the combined effect of reduced CH_4 production (the anaerobic soil volume becomes smaller) and increasing CH_4 consumption in the increasing layer of aerobic surface soil.

Soil Temperature

Production and consumption of CH_4, N_2O, and NO have frequently been reported to be temperature dependent. For instance, an approximate doubling of the NO emission rate for each 10 K rise in (soil surface) temperature was found (Andreae and Schimel 1989). This is due to the fact that rates of enzymatic processes usually increase exponentially with temperature, as long as other factors (substrate or moisture availability) are not limiting. Indeed, complete temperature control of short-term (diel) NO emission was observed for subtropical savanna soils immediately after the first rains which terminated the dry season (Meixner et al. 1997). Within temperate wetlands there is a general correspondence between CH_4 emission and temperature in virtually all seasonal flux studies (Bartlett and Harriss 1993). However, the attempt to correlate CH_4 emissions from temperate wetlands with soil temperature for spatial extrapolation failed in most cases. As far as northern wetlands are concerned, the situation tends to improve. That is, statistically significant correlations between flux and environmental variables are generally found between CH_4 flux and soil temperature (Bartlett and Harriss 1993). An example is given in Figure 8.4. While there is still considerable variation within each data set, slopes of the individual relationships are of surprising similarity.

Mechanisms of VOC and NH_3 Emissions from Plants

From a biochemical perspective, isoprene and monoterpenes belong to the isoprenoids which all originate from isopentenylpyrophosphate (IPP). Biosynthesis of IPP ("mevalonate pathway") requires activated substrates (e.g., acetyl-CoA), chemical energy

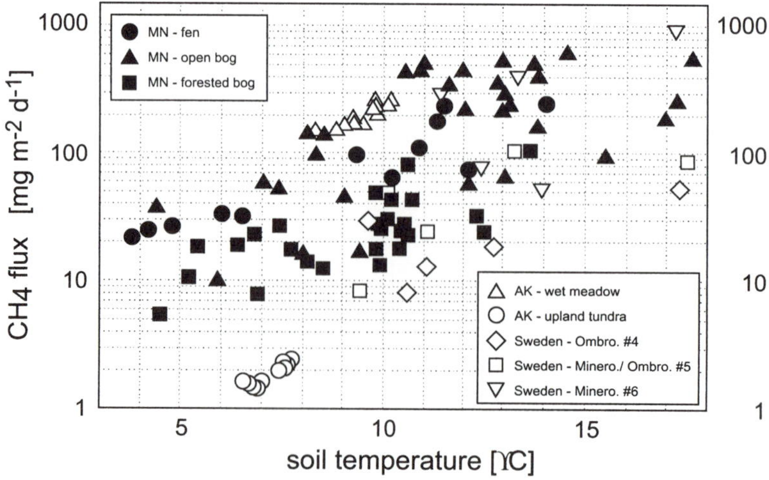

Figure 8.4 CH_4 emission fluxes and soil temperatures from northern wetlands (after Bartlett and Harriss 1993). Measurement depth of soil temperatures: –0.1 m (Minnesota, MN), –0.15 m (Alaska, AK; an average of those from –0.1 and –0.2 m), and –0.02 m (Sweden).

(ATP), and reduction equivalents (NADPH, NADH). Isomerization of IPP leads finally to the substrate of the key enzyme isoprene synthase. Very recently, an alternative pathway ("Rohmer pathway") of isoprenoid biosynthesis, which avoids the acetyl-CoA-pool, was reported (cf. Kesselmeier and Staudt 1999). Flowers and fruits release VOCs, but the majority of VOCs is emitted from green leaves. The actual production of terpenes occurs within leaf plastids or cytosol. Ammonia exchange with vegetation is related to the glutamine synthetase/glutamate synthase (GS/GOGAT) cycle, the dominant pathway of NH_4^+ assimilation in plants. In leaf tissues, NH_4^+ is continuously produced in substantial amounts during photorespiration, NO_3^- reduction, protein turnover (e.g., senescence-induced protein degradation), and lignin biosynthesis (cf. Schjørring et al. 1998). Dissolved NH_4^+ is found in the aqueous interface between the leaf and the atmosphere, which is the apoplast (consisting of the water contained within the cell wall of the mesophyll cells). GS and GOGAT activities are critical to avoid (toxic) NH_4^+ accumulation in leaf tissues.

Controls of Biogenic Emission from Plants

Isoprene and Monoterpenes

Basically, one has to distinguish between terpenes which, after biosynthesis, will be emitted from leaves more or less immediately and will be stored in corresponding pools prior to emission. If the inner leaf is the location of actual production or storage pools, then the release of terpenes is most likely via leaf stomata. However, a general stomatal control of the emission of isoprene and monoterpenes could not be observed until now; diffusion through the leaf's cuticle is likely. If glands and hairs on the leaf surface are the location of the actual production or the storage pools, then stomatal control is nonexistent and emissions will solely be controlled by leaf surface temperature and permeability. It was shown that emissions of both isoprene and monoterpenes strongly depend on temperature. The emission of isoprene is generally also controlled by photosynthetic active radiation (PAR) (Figure 8.5). While responses of isoprene emission and CO_2 assimilation are similar with regard to PAR, they differ for temperature. Isoprene response to temperature exhibits an optimum between 35°–40°C. For explanation of the observed radiation and temperature dependencies, it is assumed that isoprene as well as monoterpene synthesis, if located in the leaf's chloroplast, is (a) coupled to the availability of products of photosynthesis and (b) limited by the temperature-dependent activity of the enzyme isoprene synthase (cf. Kesselmeier and Staudt 1999). According to Guenther (1997), there are four major factors controlling natural VOC emissions: landscape average (species-specific) emission potential ε (e.g., in $\mu g\ g^{-1}\ s^{-1}$), foliar density D (e.g., in $g\ m^{-2}\ g_{dry\ mass}^{-1}$), an emission activity factor γ (nondimensional), and a second emission activity factor δ (nondimensional) to account for longer-term (> 1 h) controls of emission variations. Emission fluxes (e.g., in $\mu g\ m^{-2}\ s^{-1}$) can then be modeled by $F = \varepsilon\ D\ \gamma\ \delta$. The generalized behavior of the short-term emission activity factor γ is given in Figure 8.5. For appropriate modeling,

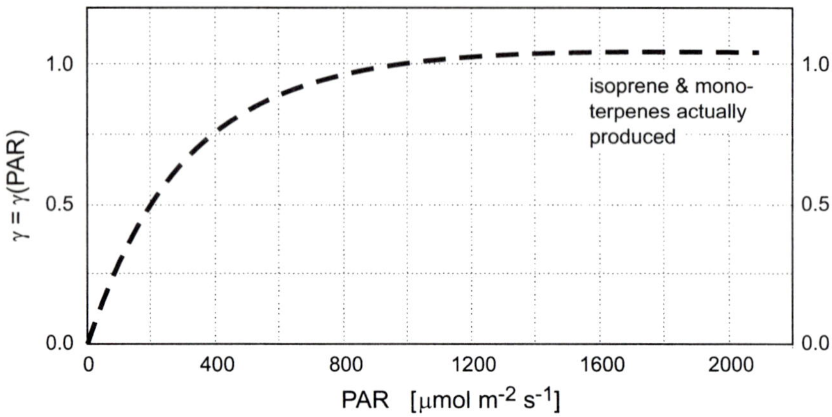

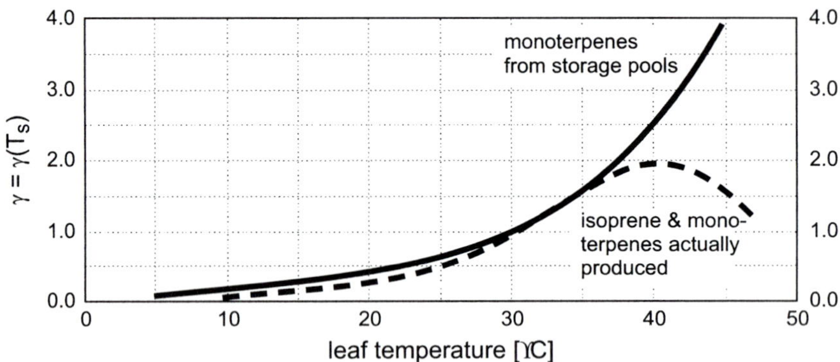

Figure 8.5 Generalized behavior of nondimensional activity factors $\gamma = \gamma\,(T_S)$ and $\gamma = \gamma\,(PAR)$ for biogenic emissions of isoprenoids. Emission activity factors are used to scale plant species specific emission potentials to short-term variations of leaf surface temperature (T_S) and photsynthetic active radiation (PAR). Figures are taken from Kesselmeier and Staudt (1999).

in-canopy profiles of temperature and radiation must be considered. Other factors (δ), which may influence emission activity, may comprise seasonal effects (growth environment, leaf age), phenological events, leaf VOC concentration and nitrogen content, water status, insect herbivory, disease, physical injury, and others (Guenther 1997). Typical landscape-scale variations of foliar density D and problems of their deduction (from remote sensing data) are addressed elsewhere in this volume. The emission potential ε accounts for genetic controls of isoprene and monoterpene production. It represents the emission rate per unit foliar mass for a specific plant species normalized to leaf surface temperature $T_S = 30°C$ and PAR = 1000 μmol m^{-2} s^{-1}. Available emission potentials of isoprene and monoterpenes, determined by cuvette and/or micrometeo-rological techniques, were repeatedly reviewed (cf. Kesselmeier and Staudt 1999; Guenther 1997) and vary over three orders of magnitude for different species of trees and bushes. This simply reflects the diversity of underlying production mechanisms, which even may vary among taxonomically and ecologically similar species.

Ammonia

If gaseous NH_3 concentration in substomatal cavities is higher than ambient NH_3 concentration, plant leaves will emit NH_3, otherwise plants will take up NH_3 from the atmosphere. Therefore, the NH_3 exchange with plants is basically bidirectional. The NH_3 concentration, which is in equilibrium with apoplast NH_4^+, is considered as the leaf compensation concentration of NH_3 exchange (Schjørring et al. 1998). Its mixing ratio ranges between 0.5 and 20 ppb (also the range of ambient NH_3 concentrations), and it is mainly controlled by apoplastic NH_4^+ concentration, pH, and leaf temperature. The latter also affects physiological processes generating or assimilating NH_4^+ in the leaf tissues (see above). The importance of leaf temperature is demonstrated in Figure 8.6: under controlled environmental and nutritional conditions, increasing leaf temperature switches the NH_3 exchange status from deposition to emission. Diffusion into and out of the plant leaves was found to be entirely controlled by stomatal opening. However, it was observed that the leaf NH_3 conductance varies with the nitrogen status of the plant (Schjørring et al. 1998). Furthermore, the plant's nitrogen uptake (from the root medium in the form of NO_3^- or NH_4^+) affects the strength of NH_3

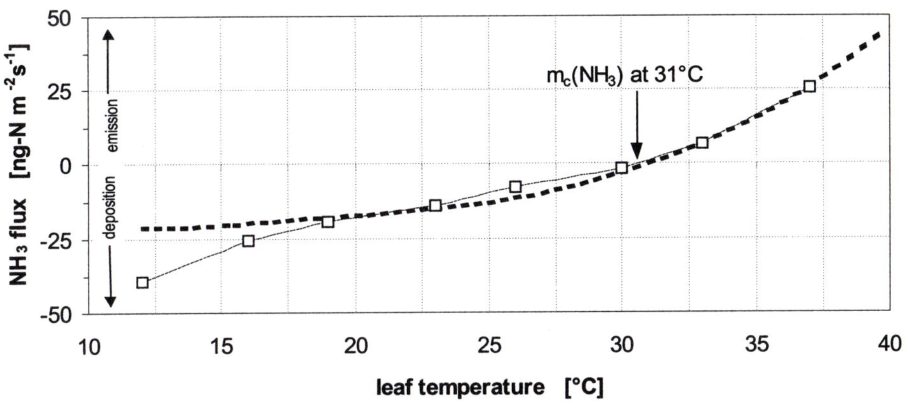

Figure 8.6 The effect of increasing leaf temperature in *Brassica napus* plants exposed to 15 ppb NH_3 at $T_{air} = 25$ °C, PAR = 500 μmol m^{-2} s^{-1}, and r. h. = 20%. Plants were growing at a high nitrogen supply and (duplicate) measurements were performed during the vegetative growth state. The dashed line is deduced from relevant thermodynamical equilibria at ambient NH_3 concentration 15 ppbv (a temperature-independent leaf NH_3 conductance of 0.12 mol m^{-2} s^{-1} was assumed). Figure taken from Schjørring et al. (1998).

emission: NH_4^+-absorbing plants emit considerably more NH_3 than those which absorb NO_3^-. Assimilation, (internal) transport and turnover of nitrogen change dramatically over the various stages of plant development: a bimodal pattern of the NH_3 compensation point of cereals is generally found (with maxima around tillering and grain filling, separately).

Nitric Oxide

Very recently, biogenic NO emission from a variety of plant species was observed by Wildt et al. (1996) during laboratory fumigation experiments at low ambient NO concentrations. For the plants studied so far (sunflower, sugar cane, soybean, corn, rape, spruce, spinach, and tobacco) they suggest an NO compensation point around 1 ppb. The nature of plant physiological processes that emit NO is almost unknown. However, a first-order relationship between rates of daytime NO emission and uptake rates of CO_2 was observed (3×10^{-6} mol of NO emitted per mole CO_2 taken up). Based on this relation, the global strength of NO emission by plants may be on the order of 2.4 Tg N y^{-1}.

Landscape Pattern Control over Biogenic Emissions: Some Examples

Soil water, due to its outstanding role in plant physiology, substrate (trace gas precursor) production, and diffusivity of trace gases could certainly be considered as the most important controlling factor among those discussed above. However, temporal and spatial distribution of (surface) soil water results from microclimatic and hydrological conditions in the landscape, which are usually modified by land forms (which comprise the landscape). In turn, the type and intensity of key factors for plants and soils (and hence, patterns of soils and plants at a site) are largely controlled by these land form-influenced environmental conditions. Therefore, biogenic emissions of trace gases, particularly those from soils, should be prone to characteristic variations of land forms best studied along toposequences (e.g., Figure 8.7).

The inserted table in Figure 8.7 summarizes CH_4 oxidation processes and CH_4 transport processes described above under the sections *Methane Production* and *Methane Oxidation*. These are mainly controlled by (seasonal) inundation, changes in water table height, and soil moisture content; as these hydrological quantities vary with landscape–land form pattern, CH_4 exchange will also vary. While the (surface) soil moisture content (and soil temperature) will determine the intensity of CH_4 deposition within ecosystem type 6, special emphasis should be put on type 5 (Figure 8.7). In this system, the water table height may switch parts of or the entire (periodically drying) wetland ecosystem from a significant CH_4 source to a significant CH_4 sink (Bartlett and Harriss 1993). For northern boreal peatlands, spatial scales of the variability of CH_4 exchange were recently reported by Waddington and Roulet (1996): seasonally averaged CH_4 emissions ranged over up to three orders of magnitude between hummocks and hollows (3 m scale), ridges, lawns and pools (100 m scale), and on the land-form scale (500 m).

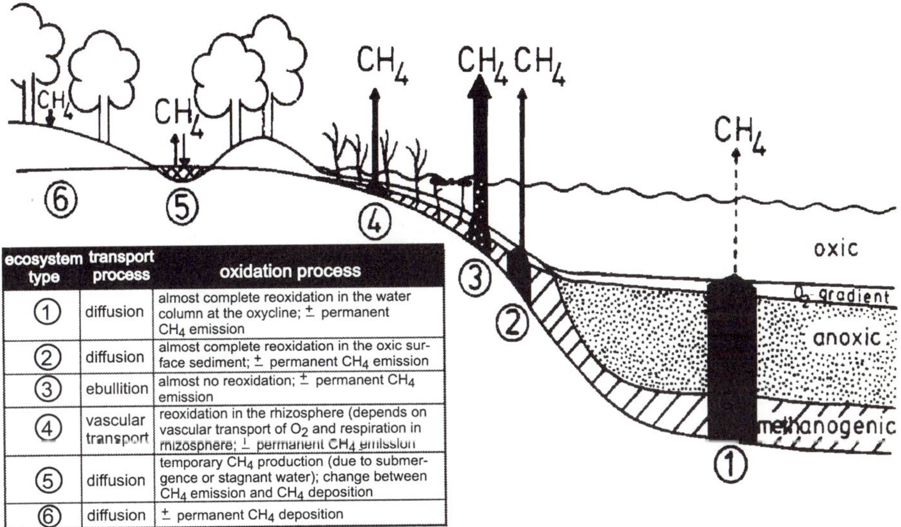

ecosystem type	transport process	oxidation process
1	diffusion	almost complete reoxidation in the water column at the oxycline; ± permanent CH_4 emission
2	diffusion	almost complete reoxidation in the oxic surface sediment; ± permanent CH_4 emission
3	ebullition	almost no reoxidation; ± permanent CH_4 emission
4	vascular transport	reoxidation in the rhizosphere (depends on vascular transport of O_2 and respiration in rhizosphere; ± permanent CH_4 emission
5	diffusion	temporary CH_4 production (due to submergence or stagnant water); change between CH_4 emission and CH_4 deposition
6	diffusion	± permanent CH_4 deposition

Figure 8.7 Soil transport and oxidation processes involved in CH_4 cycling between terrestrial ecosystems and the atmosphere (Conrad 1989).

For N_2O emissions from cultivated soils, high spatial variability of soil microbial activity is usually assumed to avoid the development of predictive relationships between factors which impact a given microbial process. However, Corre et al. (1996) showed that such relationships can be derived if an appropriate spatial sampling scheme is used and those statistical measures for data evaluation, which are resistant to the effects of few extreme values, are applied. Some of their results are summarized in Figure 8.8. With the exception of the pasture site, a significant pattern on the landscape scale (50–250 m) was observed, showing higher N_2O emissions on the footslope than on the shoulder complexes. As suggested by the inserted schematic in Figure 8.8, the lower N_2O emission activity associated with the shoulder compared with the footslope complex reflects the influence of topographically induced water redistribution and better internal drainage conditions of the soils on the upper than on the lower landscape conditions. Corre et al. (1996) conclude that, during their study in the Canadian Black soil zone, only a discrete number of factors controlled the N_2O flux variability, namely, soil-parent sediment, vegetation–land-use pattern, climate, and topography. They presented a concept of links between large-scale controllers and proximal factors of N_2O emissions, which is shown in Figure 8.9.

As already mentioned in the section *Soil Water Content and Gaseous Diffusivity* (see also Figure 8.3), the partition of gaseous nitrogen emissions between NO, N_2O, and N_2 depends strongly on the actual soil moisture content. If we accept general validity of the observations by Corre et al. (1996), we would expect higher NO emissions

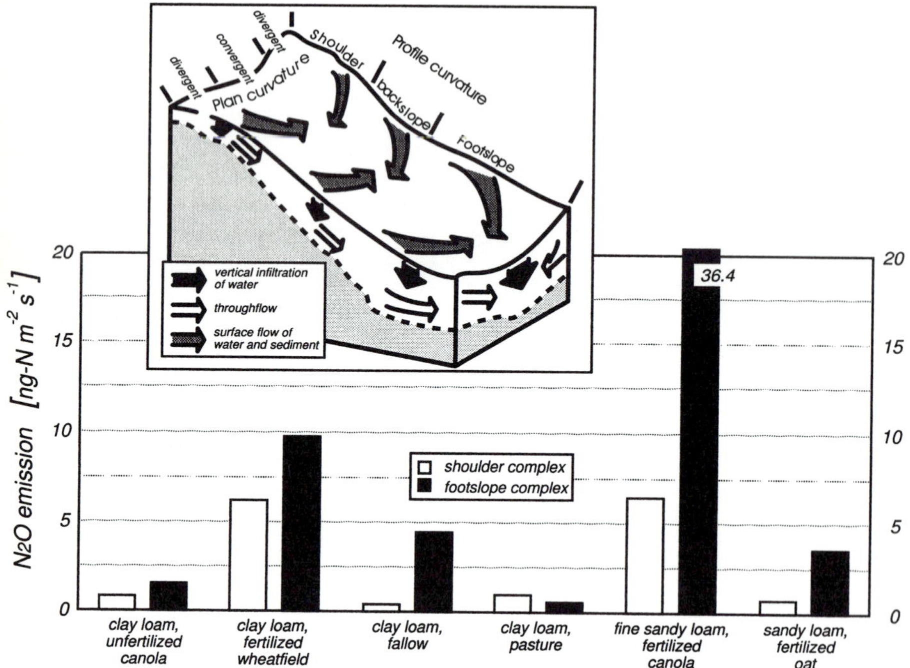

Figure 8.8 Landscape patterns of N_2O emissions (May, 1993, to May, 1995) from the black soil zone (Udic Boroll soils) near St. Louis, Saskatchewan, Canada (data are taken from Corre et al. 1996). Basic land form classification (□ = shoulder complex, ■ = footslope complex) are from Pennock et al. (1994). The insert shows a summary diagram of probable water movement and concentration associated with different landform elements in a hillslope system (from Pennock et al. 1987).

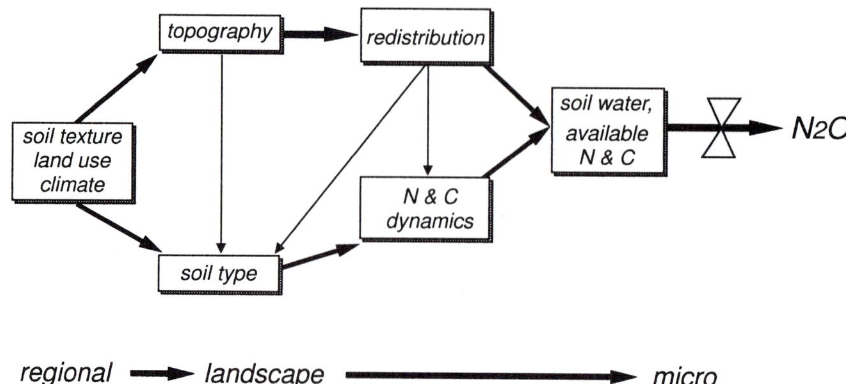

Figure 8.9 Conceptual links between large-scale controllers and proximal factors of N_2O emissions (after Corre et al. 1996).

from shoulder than from footslope complexes of areas of similar geomorphological characteristics. For NO emissions, it also has to be emphasized that the controlling processes are most likely confined to the uppermost soil layers (0–3 cm; cf. Remde et al. 1993, Meixner 1994). Therefore, diurnal effects of surface soil temperature and wetness may result in more short-term variations of NO emissions as compared to N_2O emissions. Also, unlike N_2O, NO is highly reactive in the atmosphere and converts (by reaction with ozone) rapidly to NO_2, which in turn is photolyzed back to NO on a similar time scale under daylight conditions (cf. Remde et al. 1993). However, within vegetation canopies, biogenic NO from soil will partly be converted to NO_2, since shadowing by vegetation elements will more or less suppress NO_2 photolysis. Uptake of NO_2 by vegetation elements is small but still an order of magnitude larger than that for NO (cf. Meixner 1994). As a result, not all of the NO emitted from soils may escape to the atmospheric boundary layer (ABL). First estimates of the so-called "canopy reduction" of soil NO emission range within some ten percent; quantification on local and/or regional scales is lacking.

TRANSPORT AND DEPOSITION PROCESSES IN COMPLEX LANDSCAPES

Transport processes in complex landscapes with variable topography and small- to large-scale differences in land use are strongly affected by local wind systems, which develop owing to differential surface heating and differences in surface moisture (see Pielke et al., this volume). In general, the transport of a trace gas in the atmosphere ends with the deposition to the Earth's surface or with chemical depletion (which is not considered here). In this chapter we do not address all transport and deposition processes in detail but restrict ourselves to the most relevant ones for complex landscapes.

Sea- and Land-breeze Systems and the Effect of Large-Scale Differences in Land Use

We start our discussion of the transport processes at the largest scale considered in the present context, which is less than 100 km in the horizontal direction. On this scale, the most spectacular transport processes are related to the photochemical smog problems, e.g., those of Los Angeles or Athens. The daytime sea-breeze wind system transports pollutants from the seashore into the urban area, where they are trapped by local topography (mountain barrier), while at night, when the land surface becomes colder than the sea surface, the trapped pollutants are blown back to the sea without the possibility of rapid turbulent dispersion into the free troposphere, as would be the general case in flat and level terrain.

Although topography is the main factor in this example, Zhong and Doran (1995) showed that regional variation in the land-use pattern might even have a stronger effect on atmospheric transport than topography alone, even in complex terrain. In a modeling exercise on the regional distribution of (potential) temperature and wind fields over a rural area in the northwestern U.S.A., they demonstrated that model runs considering actual topography and different land use (steppe vs. irrigated farmland or open water surface) represented observed fields more closely when topography was neglected than when land use was neglected (Figure 8.10).

However, the question of how large areas of different land-use pattern must be to have a significant influence on the regional scale still remains open. For instance, aircraft transect flights addressing the transition from dry to irrigated farmland (Pielke and Avissar 1990; Mahrt et al. 1994) showed that the transition of (potential) temperature was gradual (rather than stepwise) and extended over a horizontal distance of more than 10 km (at flight levels of 140 m and more above ground; Pielke and Avissar 1990) to a few km (at flight levels of 30 m; Mahrt et al. 1994). Thus, for most areas of the world where land-use patterns change on the smaller scale of a few tens to hundreds of meters, the importance of such land-use differences seems to lie primarily in the alteration of turbulent exchange with the surface within the surface layer.

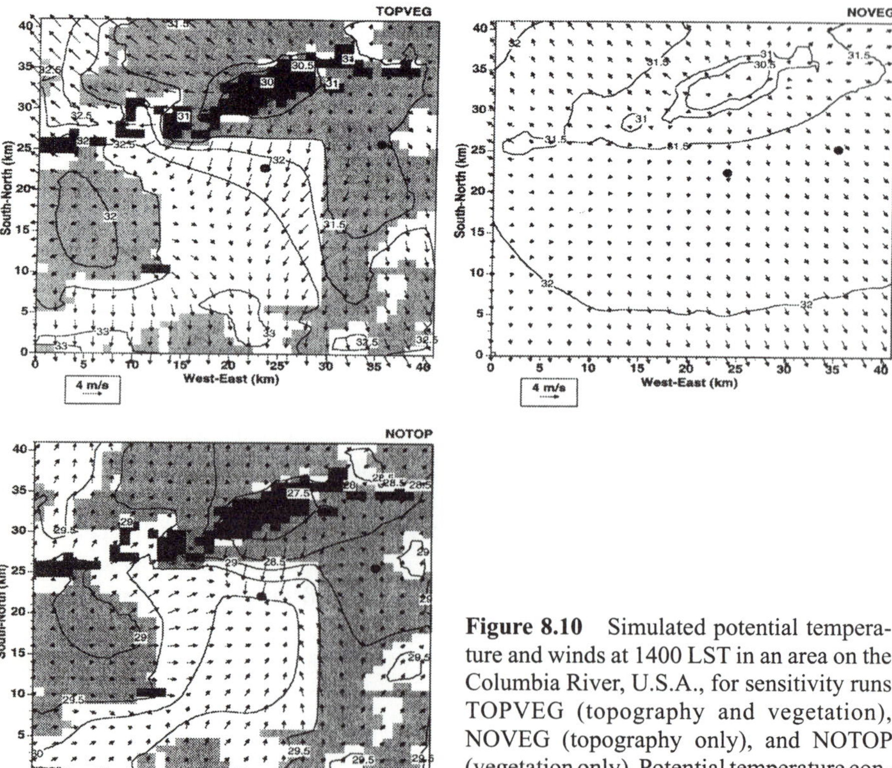

Figure 8.10 Simulated potential temperature and winds at 1400 LST in an area on the Columbia River, U.S.A., for sensitivity runs TOPVEG (topography and vegetation), NOVEG (topography only), and NOTOP (vegetation only). Potential temperature contours are in degrees Celsius. Figure taken from Zhong and Doran (1995).

Valley and Slope Wind Systems and the Recirculation of Trace Substances

Valley and slope wind systems were initially described for deep and long mountain valleys. There, a persistent diel wind direction pattern with frequently high wind speeds (typically on the order of 10–20 m s^{-1} in valleys like the Kaligandaki valley of Nepal) can be observed not only by the educated meteorologists. The importance of valley and slope wind systems for the transport and recirculation of atmospheric constituents was recently shown by Allwine and Whiteman (1994) for the Grand Canyon region. Although with less severe restrictions on dispersion and transport due to the local topography, similar influences of landscape patterns and topography can be expected in moderately complex terrain. This is due to the interaction between the lateral topographic restrictions of the air volume in which trace gases are dispersed and the height of the ABL. Usually, a (temperature) inversion is topping the ABL within which most surface-emitted trace gases are trapped owing to the very weak turbulent exchange across the inversion layer. For moderately complex topography with differences in elevation less than 400 m, the top of the noon and early afternoon ABL is typically higher than the peaks of local topography. Therefore, the restrictions on lateral dispersion due to topography are limited to late afternoon, nighttime and early morning periods, when the ABL height may be less than the peak heights of local topography. In deep mountain valleys even the maximum daytime ABL height may stay below the mountain crests, which significantly reduces the atmospheric volume into which surface emitted trace gases can be dispersed. The simple concept of the topographic amplification factor (TAF) (Whiteman 1990) illustrates this for air temperature, but it is also applicable to the transport of trace substances. In complex terrain, there is a smaller volume of air inside a box with the bottom defined by the terrain elevation, compared to a box of the same size over flat terrain. Additionally, the real surface of the topography inside the complex terrain box is larger than the corresponding surface of the flat terrain. Given a constant incoming solar radiation flux density, this causes the larger surface to heat a smaller volume, and hence the air temperatures are higher over complex terrain than over flat topography. By analogy, the concentration of a trace gas emitted from complex terrain should be, in principle, higher than over flat terrain, as the larger surface area would emit the trace gas into a smaller volume of air. However, we showed above under *Soil Temperature* that surface (soil) temperature may be a major influencing factor for the biogenic emission of many trace gases and the response to it is generally exponential. On the other hand, there is the dependence of surface temperatures on slope angle and orientation (e.g., Swift 1976). Therefore, it cannot be said a priori that a complex topography will directly result in higher overall emissions only owing to its larger surface: some of the area in complex terrain would certainly have significantly lower surface temperature owing to its exposure away from the sun, which may reduce overall emissions as compared to a flat plain with the same projected surface area. Soil moisture is probably the *most* important factor for emissions, especially for methane (water table controls whether CH_4 is emitted or taken up by soils). Surface soil moisture on slopes is not only controlled by radiative input and thus exposure of the slope, but also by subsurface flows of water. If subsurface flow is absent, then the slopes exposed towards the sun are drier and warmer compared with those exposed away from the sun, potentially leading to a moisture limitation of soil emission processes or to a temperature limitation in the second case.

Air-mass Blocking and Cavities

An important factor for atmospheric transport in complex terrain is air-mass blocking in front of cavities or behind obstacles, owing to deceleration of atmospheric flow. This can be seen on all scales from whole mountain systems down to exposed single hills, forest edges, and buildings, when the atmosphere is stably stratified. Depending on the strength of atmospheric stability and the size of the obstacle, atmospheric flow will be either around (small Froude numbers, *Fr*, very stable stratification with weak winds) or over obstacles (high *Fr*, slightly stable stratification with moderate to strong winds), or a combination of both. On the upwind side (upwind cold-air blocking, low *Fr*) or on the downwind side (lee-side rotor cavity, intermediate *Fr*), trace gases may accumulate and lead to both higher concentrations and higher surface deposition than the regional average. Additionally, wind direction and wind speed can differ considerably in blocked air mass owing to the decoupling from the atmosphere aloft. This is often the case in valleys when synoptic winds are perpendicular to the valley (Whiteman and Doran 1993). In such cases, the atmosphere inside the valley may be decoupled from the atmosphere aloft, which can have significant impact on the direction of atmospheric transport and vertical dispersion at the interface between the blocked air near the surface and the rest of the ABL aloft. The general pattern based on the Froude number is the following: during weak wind conditions with (very) stable stratification, air pollutants accumulate upwind of an obstacle; while at higher wind speed and decreased stability, air pollutants tend to trap behind the obstacle in a cavity.

Orographic Effects: The Seeder–Feeder Mechanism

The picture gets even more complicated in the presence of clouds. Fowler et al. (1995) found that the concentration of major pollutant ions (Cl^-, SO_4^{2-}, NO_3^-, NH_4^+) in the cloud water is related to the relative proximity of a site to marine or anthropogenic sources of aerosol. However, they also found that concentrations in orographic clouds (e.g., "cap clouds" around summits, generated by the forced ascent of moist air on a mountain barrier) exceed those in frontal (upwind) clouds by a factor of five to ten, leading to 20%–50% higher wet deposition at summit sites compared to low ground locations. This important influence of topography on transport and deposition of dissolved pollutants is called the seeder–feeder mechanism: rain droplets from frontal clouds (seeder) in higher levels are incorporated in orographical cap clouds (feeder), where pollutants accumulate before they are wet deposited to the surface. In addition to the dissolved pollutants received from the seeder clouds, orographical clouds may scavenge trace gases and aerosol particles out of the upwind air mass, thus further influencing the transport and wet deposition of such constituents in complex landscapes. Such orographic effects may shift the ratio between dry and wet deposition towards increased wet deposition.

Forest Edges, Changes in Displacement Height, and the Smooth-to-rough Transition

Along forest edges and wherever surface roughness suddenly increases or a considerable step in the height of roughness elements occurs (i.e., where the so-called aerodynamic displacement height increases), there is an aerodynamic effect that may lead to enhanced dry deposition of trace gases and particles, as well as of enhanced interception of fog droplets at some (short) distance from the edge line. This is due to a rotating turbulence element which is created at the edge. It will lead to persistent downdrafts at a downwind distance of the turbulence element's average size from the edge line. This is also the range with highest dynamic pressure, where most windfelled trees can be found and agricultural crops are most likely bent down at some distance from the field edge after the passage of a thunderstorm. Draaijers et al. (1994) performed a field experiment at eight forest edges and showed that current deposition models underestimate deposition of acidifying compounds to Dutch forests by 10% just owing to edge effects. Although their focus was on forest edges, this phenomenon is not restricted to forest edges alone but also to any similar condition in a landscape. As the total length per unit area of edge lines (between surfaces of different roughness-displacement height) increases with the complexity of landscape pattern and vegetation mosaic, we expect biased estimates of both transport and exchange of trace substances in complex terrain, if the basic assumptions in the corresponding modeling approach is based on homogeneous and flat surface. We think that this may generally lead to a systematic underestimation of local deposition and thus to an overestimation of horizontal transport distances in complex landscapes.

Rough-to-smooth Transition and Other Roughness Length Effects

Together with atmospheric stability (i.e., the thermal structure of the ABL), the aerodynamic roughness of a surface determines the turbulent structure (the vertical turbulent exchange) and the vertical wind profile (the vertical shape of horizontal transport) in the ABL. There is an important difference of total surface roughness between homogeneous terrain and patchy landscapes, where very frequent transitions from smooth to rough surfaces and vice versa occur. Over horizontally homogeneous terrain, the aerodynamic roughness of the surface can be described by only the geometry and density of the vegetation elements growing on the surface or by the size of small-scale structures of nonvegetated surfaces (e.g., wave height over water surfaces). However, in complex terrain, the total aerodynamic roughness of the landscape is a composite of local surface roughness plus the roughness of the (bare) topography, which is generally a few orders of magnitude larger than the local surface roughness. Thus, aerodynamic roughness is not scale-invariant. Models will use different values, depending on spatial resolution and thus on the scale of processes which are explicitly modeled. Due to the fact that turbulence is generated almost immediately at a smooth to rough transition, the decay of this turbulence at the rough to smooth transition is governed by the internal properties of the flow, the dissipation rate for kinetic energy, which has a much longer time constant than the generation of turbulence. Thus, the atmosphere above a landscape with frequent changes in surface roughness has a higher grade of turbulence than above a homogeneous landscape, and the effect of mechanic turbulence created over complex topography adds to this landscape-influenced turbulence. The consequences for transport of constituents in the atmosphere are identical to those already stated for the forest edge effect: over complex landscapes we expect increased turbulence, which leads to increased deposition and thus to decreased horizontal transport distances compared with over horizontally homogeneous landscapes. Because the concentration of trace gases in the atmosphere is extremely low, there is no back-pressure effect induced by increased deposition and decreased transport distances over rough surfaces that would lead to an increase or reduction of trace gas emissions.

Laminar-turbulent Transfer Resistance

So far, our description of transport processes considered turbulent and mesoscale scales, which are roughly three orders of magnitude more efficient for dispersion of constituents in the atmosphere than molecular diffusion. However, very close to the surface, around any vegetation element, object, obstacle, etc. there is a quasi-laminar layer of air, which is only a few tens of micro- or millimeters depth. Transfer across this quasi-laminar layer may be an essential control for the exchange (emission and deposition) of trace substances. Given the rough guess of the efficiency of turbulent versus molecular diffusion, one can estimate that it should take an atmospheric constituent as long to cross a 1 mm quasi-laminar layer as it would take it to travel 1 m in the turbulent ABL. The depth of this quasi-laminar layer depends mainly on wind speed and other micrometeorological parameters, but also on the shape and size of the object.

Internal Boundary Layers, Fetch and Footprint, and the Urban Heat Island

Changes in landscape pattern are almost always associated with changes in surface roughness, albedo, surface moisture, plant activity, and thus surface energy balance. Such changes induce new internal boundary layers owing to changes in mechanic turbulence production (especially at the smooth-to-rough transition) and thermal convection (due to changes in sensible heat flux). The urban heat island is an example of such an internal boundary layer resulting from increased sensible heat flux from urban areas (where surface water availability is low and thus evapotranspiration is small). Raupach (1993) has shown that there are three kinds of heterogeneity to be considered:

1. microscale heterogeneity between adjacent surface patches for which advection occurs in the surface layer;

2. mesoscale heterogeneity for which the ABL is well-mixed, but advection occurs on the scale of the ABL; and
3. macroscale heterogeneity for which advection is negligible and surface patches are energetically independent.

Currently there are two experimental concepts used by micrometeorologists to deal with inhomogeneities in landscape:

1. Flux measurements are made in the surface layer as close to the canopy as possible to get measurements that are representative of a distinctive surface or vegetation type due to its small footprint (area that influences the flux measurement) (Schmid and Oke 1990).
2. Flux measurements are made on a tall tower 10–100 m or with an aircraft to get flux measurements that represent an average of a large area and thus integrate over all the patches represented in the footprint of the measurement.

The remaining key question, however, is: What is the footprint of a measurement and how representative are any of the flux measurements taken with either measuring concept? The concepts of instrument fetch (the upwind distance, which must be homogeneous to allow equilibration of the atmospheric flow with that surface) and the footprint (the contributing surface area for a measurement) are both based on homogeneous and idealized conditions. For example, a change in roughness is most often also in direct combination with a change in displacement height. However, footprint models neglect differences in displacement height, not to mention complex topography. This may lead to severe errors in estimation of footprint areas in complex landscapes. However, there is some general agreement that, although imperfect, such footprint models are a valid tool to assess the problem of spatial representativeness of flux measurements obtained over nonhomogeneous and nonflat terrain.

The independent variables in the Flux Source Area Model by Schmid and Oke (1990) are the ratio between measuring height and surface roughness (z/z_0), the atmospheric stability (z/L), and the turbulence parameter for lateral turbulence (σ_v/u_*) which determines the width of the footprint. While large z/z_0 lead to large footprints, any increase of z_0 due to landscape pattern and complex terrain automatically results in the reduction of the footprint size, while lateral dispersion typically increases over such landscapes, and thus, the flux footprint area gets wider. In general, it seems possible to extend atmospheric flux measurements, which are best described for horizontally homogeneous and flat landscape, to complex terrain and patchy landscape, if they are accompanied by careful investigation of the flux footprint in order to know what flux measurements obtained in nonideal terrain actually represent. In most cases, a good starting point is to investigate the dependence of a flux measurement on wind direction (e.g., Jarvis et al. 1997). Such an analysis reveals many features that can be directly explained by local heterogeneities within the footprint area. Although the size of the footprint area may not be accurately estimated by a footprint model that was designed for homogeneous topography, the order of magnitude is expected to be correct even in complex landscape and is an indication of where to search for surface heterogeneities that may explain the measured dependence of flux on wind direction. A further refinement would include a classification of the measuring conditions according to atmospheric stability, which strongly influences the horizontal extent of the footprint area.

Effects of Atmospheric Stability: Katabatic Drainage Flow and Cold Air Lakes

We cannot make a priori statements about the effect of landscape pattern and topography on emission and transport which would generally be valid for any and all types of complex landscapes. Atmospheric stability influences all of the above listed transport processes, and atmospheric stability depends on (a) net radiative input (which varies with time of day and season); (b) how this energy is partitioned into sensible, latent, and ground heat flux at the surface (the fraction converted to sensible heat flux directly affects atmospheric stability); and (c) mechanical turbulence driven by the wind shear in the surface layer. All of these properties are influenced by complex terrain and a vegetation mosaic in a nonlinear way such that their influence on atmospheric stability and transport must be studied separately for each unique study area of the world. Even well-known and widely accepted models for processes such as katabatic drainage flow (bringing cold air and its constituents from higher elevations into basins and valleys during nighttime) do not necessarily lead to identical results in similar landscapes. In one case, drainage flow may lead to deep cold air lakes with surface fog, while in another case there is no fog and cold air is not pooling within the study domain. This may be due to differences in orientation of the complex landscape: in south-going valleys of the northern hemisphere, the daytime up-valley wind is generally stronger than the nocturnal down-valley wind, while a valley with identical shape and topographic relief that is opening to the north may show a weaker daytime up-valley wind and a stronger nocturnal down-valley wind, a difference that only results from exposure differences, not differences in landscape patterning, topographic relief, or the vegetation mosaic.

OPEN QUESTIONS, POINTS OF CONTROVERSY, RESEARCH NEEDS

Our description of effects of landscape and topography on emissions and transport addresses a very wide range of scales for the processes involved, with the smallest scale for the microbial processes described under the section **EMISSIONS OF TRACE GASES FROM SOILS AND VEGETATION**, and a larger scale for the atmospheric processes in **TRANSPORT AND DEPOSITION PROCESSES IN COMPLEX LANDSCAPES**, where we emphasize the small-scale processes with a more detailed description. This simply reflects the fact that, unlike for transport (as well as for water and CO_2 exchange), research on the biogenic emission of most trace gases is comparably young and still needs basic mechanistic studies for understanding. In contrast, research on turbulent transports has reached a more advanced state of scientific understanding, thanks to all the well-funded research that was carried out in the past few decades in the fields of aircraft engineering and atmospheric dispersion of radioactive substances that would occur following a nuclear strike. To link small-scale heterogeneity with the large-scale atmospheric processes is also the key issue. We believe, that despite some obvious success of previous large-field experiments, there are still plenty of open questions to be addressed here.

Under the consensus that the microscopic level of microorganisms' metabolism exerts any *initial* control over biogenic emissions (Conrad 1996), the open questions that still remain are:

1. What are the consequences of the vast microbial diversity in plants and soils for higher levels of organization and control?
2. Can we — as for energy, CO_2 and H_2O (see Valentini et al., this volume) — define any spatial unit(s) on which biogenic emission of a given trace gas shows a coordinated and specific response to environmental factors?
3. What is the appropriate level of control (microbial community, soil crumbs, vegetation elements, (whole) species, patch, biome) where the development of functional relationships (for extrapolation to larger spatial and temporal scales) can start?

There is some discussion (not even controversy) to which extent the diversity of microorganisms matters in terms of trace gas emissions. While Conrad (1996) favors microbial species diversity as an important factor in ecosystem function, Schimel (1995) asks whether the microbial diversity and community structure in soil have any consequences at the ecosystem level. However, there is growing interest with respect to functional groups of microbes (rather than on species of microorganisms) for identification of factors which control rates of production and consumption (Davidson and Schimel 1995). Simple concepts on how to deal with the effects of nonhomogeneous and nonflat landscapes do not exist, neither for biogenic emissions nor for atmospheric transport. Major points of controversy in this respect exist on where and how to do experimental measurements in order to get representative data sets for an extended area (e.g., a region). The most widespread concept to date is to select locations that are only disturbed by as few heterogeneities and as little terrain relief as is unavoidable in a given region. A second concept, which is finding more acceptance within the scientific community, is to view landscape heterogeneity and complex terrain issues as a challenge, based on the understanding that this is the real world. Another concept, although it utilizes a better theoretical base so far, does not address the real situation of large areas of the Earth's land surface. The drawback, however, is the need for additional data that are not required in studies carried out in flat and homogeneous landscapes. However, we believe that the scientific community has already begun to develop the right tools for understanding complex landscapes, such that the expenses needed for measuring additional data can be scientifically justified. Two key questions addressing these scientific tools were supplied by one of the reviewing workshop participants:

1. How can the relationships between large-scale landscape pattern, small-scale spatial heterogeneity, and emission of trace gases be quantitatively related?
2. How can spatial variability be handled by appropriate experimental designs?

We think that the most promising approach to answer the first question is spatial explicit modeling with a dynamic atmospheric model that incorporates a mosaic of small-scale land use (and thus trace gas emission sources) for each model grid cell with a resolution that suits the subgrid scale heterogeneity of complex landscapes. We suggest that the required grid resolution, depending on the application, should be 1 km^2 (or finer) and the mosaic within each grid cell must be able to resolve surface heterogeneities on the scale of 10×10 m^2 plots. An approach to develop the methodology addressing the second question was presented by Eugster et al. (1997). The general idea is similar to what has been practiced in climatology for decades: long-term (anchor) stations should provide flux measurements against which short-term measurements from various localities in a complex landscape can be cross-referenced to address spatial variability. Furthermore, as in the field study of Corre et al. (1996), spatial sampling schemes for biogenic emissions, which are tailored to the particular local situation, should be developed; and those statistical measures for data evaluation, which are resistant to the effects of few extreme values, should be applied. Difficult and unresolved problems, however, lie in the statistical integration of such spatiotemporal data sets. Wikle and Cressie (1998) presented a statistical approach for measurements obtained from research vessels, and we expect that similar approaches will be extended to land surfaces and, especially, complex landscapes in the near future.

Research needs may be summarized as:

- continuing field experiments on biogenic emission of trace gases from carefully selected key ecosystems of the globe *and* laboratory experiments on the corresponding soils and plants;
- finding more generally applicable theories on where and how to measure nonideal landscapes in order to be able to receive accurate regional estimates;
- separating local problems from general problems within different types of complex landscapes;
- defining the time lags and scales of processes that are affected by landscape pattern and topography (e.g., inundations, pollutant-rich cold-air lakes); and
- improving knowledge on how land-use changes, anthropogenic activities (in the widest sense), climate variability, and climate change interact with each other on the landscape scale (human dimensions are not addressed herein).

Clearly, for biogenic emissions, solutions to methodological problems will promote field studies focusing on increasingly complex landscapes, and the complementary laboratory experiments, under well-defined conditions, will increase knowledge of underlying mechanisms and quantify controlling factors, provided the applicability of laboratory results to field conditions is guaranteed.

ACKOWLEDGMENTS

We acknowledge generous support received from the Universities of Bern and Bayreuth (WE), and the Max Planck Institute for Chemistry, Mainz (FXM) to work on this paper. Many thanks to Ms. Carol Strametz for improving the language and style of this paper considerably.

REFERENCES

Allwine, K.J., and C.D. Whiteman. 1994. Single-station integral measures of atmospheric stagnation, recirculation, and ventilation. *Atmos. Env.* **28**:713–721.

Andreae, M.O., and P.J. Crutzen. 1997. Atmospheric aerosols: Biogeochemical sources and role in atmospheric chemistry. *Science* **276**:1052–1058.

Andreae, M.O., and W.A. Jaeschke. 1992. Exchange of sulfur between biosphere and atmosphere over temperate and tropical regions. In: Sulfur Cycling on the Continents Systems and Wetlands, ed. R.W. Howarth, J.W.B. Stewart, and M.V. Ivanov, pp. 27–61. Chichester: Wiley.

Andreae, M.O., and D.S. Schimel, eds. 1989. Exchange of Trace Gases between Terrestrial Ecosystems and the Atmosphere. DahlemWorkshop Report LS 47. Chichester: Wiley.

Bartlett, K.B., and R.C. Harriss. 1993. Review and assessment of methane emissions from wetlands. *Chemosphere* **26**:261–320.

Conrad, R. 1989. Control of methane production in terrestrial ecosystems. In: Exchange of Trace Gases between Terrestrial Ecosystems and the Atmosphere, ed. M.O. Andreae and D.S. Schimel, pp. 39–58. DahlemWorkshop Report LS 47. Chichester: Wiley.

Conrad, R. 1994. Compensation concentration as critical variable for regulating the flux of trace gases between soil and atmosphere. *Biogeochemistry* **27**:155–170.

Conrad, R. 1996. Soil microorganisms as controllers of atmospheric trace gases (H_2, CO, CH_4, OCS, N_2O, and NO). *Microbiol. Rev.* **60**:609–640.

Corre, M.D., C. van Kessel, and D.J. Pennock. 1996. Landscape and seasonal patterns of nitrous oxide emissions in a semiarid region. *Soil Sci. Soc. Am. J.* **60**:1806–1815.

Davidson, E.A., and J.P. Schimel. 1995. Microbial processes of production and consumption of nitric oxide, nitrous oxide and methane. In: Biogenic Trace Gases: Measuring Emissions from Soil and Water, ed. P.A. Matson and R.C. Harriss, pp. 327–357. Oxford: Blackwell.

Draaijers, G.P.J., R. van Ek, and W. Bleuten. 1994. Atmospheric deposition in complex forest landscapes. *Boundary-Layer Meteorol.* **69**:343–366.

Eugster, W., J.P. McFadden, and F.S. Chapin III. 1997. A comparative approach to regional variation in surface fluxes using mobile eddy correlation towers. *Boundary-Layer Meteorol.* **85**:293–307.

Fowler, D., I.D. Leith, J. Binnie, A. Crossley, D.W.F. Inglis, T.W. Choularton, M. Gay, J.W.S. Longhurst, and D.E. Conland. 1995. Orographic enhancement of wet deposition in the United Kingdom: Continuous monitoring. *Water Air Soil Poll.* **85**:2107–2112.

Galbally, I.E., and C. Johansson. 1989. A model relating laboratory measurements of rates of nitric oxide production and field measurements of nitric oxides from soils. *J. Geophys. Res.* **94**:6473–6480.

Guenther, A. 1997. Seasonal and spatial variations in natural volatile organic compound emissions. *Ecol. Appl.* **7**:34–45.

Jarvis, P.G., J.M. Massheder, S.E. Hale, J.B. Moncrieff, M. Rayment, and S.L. Scott. 1997. Seasonal variation of carbon dioxide, water vapor, and energy exchanges of a boreal black spruce forest. *J. Geophys. Res.* **102**:28,953–28,966.

Kesselmeier, J., P. Schröder, and J.W. Erisman. 1997. Exchange of sulfur gases between biosphere and the atmosphere. In: Transport and Chemical Transformation of Pollutants in the Troposphere, ed. P. Borrel, P.M. Borrel, T. Cvitas, K. Kelly, and W. Seiler, vol. 4, Biosphere–Atmosphere Exchange of Pollutants and Trace Substances, ed. J. Slanina, pp. 176–198. Heidelberg: Springer.

Kesselmeier, J., and M. Staudt. 1999. Emissions of biogenic volatile hydrocarbons — A review. *J. Atmos. Chem.*, in press.

Ludwig, J. 1994. Untersuchungen zum Austausch von Stickoxiden zwischen Biosphäre und Atmosphäre. Ph.D. diss., Univ. of Bayreuth, Germany.

Mahrt, L., J.I. MacPherson, and R. Desjardins. 1994. Observations of fluxes over heterogeneous surfaces. *Boundary–Layer Meteorol.* **67**:345–367.

Matson, P.A., and R.C. Harriss. 1995. Biogenic Trace Gases: Measuring Emissions from Soil and Water. Oxford: Blackwell.

Meixner, F.X. 1994. Surface exchange of odd nitrogen oxides. *Nova Acta Leopoldina NF* **70(288)**:299–348.

Meixner, F.X., Th. Fickinger, L. Marufu, D. Serca, F.J. Nathaus, E. Makina, L. Mukurumbira, and M.O. Andreae. 1997. Preliminary results on nitric oxide emission from a southern African savanna ecosystem. *Nutr. Cyc. Agroecosys.* **48**:123–138.

Pennock, D.J., D.W. Anderson, and E. de Jong. 1994. Landscape-scale changes in indicators of soil quality due to cultivation in Saskatchewan, Canada. *Geoderma* **64**:1–19.

Pennock, D.J., B.J. Zebarth, and E. de Jong. 1987. Landform classification and soil distribution in hummocky terrain, in Saskatchewan, Canada. *Geoderma* **40**:297–315.

Pielke, R.A., and R. Avissar. 1990. Influence of landscape structure on local and regional climate. *Landsc. Ecol.* **4**:133–155.

Potter, C.S., P.A. Matson, P.M. Vitousek, and E.A. Davidson. 1996. Process modeling of controls on nitrogen trace gas emission from soils worldwide. *J. Geophys. Res.* **101**:1361–1377.

Ramanathan, V., L. Callis, R. Cess, J. Hansen, I. Isaksen, W. Kuhn, A. Lacis, F. Luther, J. Mahlman, R. Reck, and M. Schlesinger. 1987. Climate–chemical interactions and effects of changing atmospheric trace gases. *Rev. Geophys.* **25**:1441–1482.

Raupach, M.R. 1993. The averaging of surface flux densities in heterogeneous landscapes. In: Exchange Processes at the Land Surface for a Range of Space and Time Scales, ed. H.-J. Bolle, R.A. Feddes, and J.D. Kalma, pp. 343–355. IAHS Publ. 212. Oxfordshire: Intl. Assn. Hydrological Sciences (IAHS) Press.

Remde, A., J. Ludwig, F.X. Meixner, and R. Conrad. 1993. A study to explain the emission of nitric oxide from a marsh soil. *J. Atmos. Chem.* **17**:249–275.

Schade, G. 1997. CO Emissions from Degrading Plant Litter. Ph.D. diss., J. Gutenberg Univ., Mainz, Germany.

Schimel, J.P. 1995. Ecosystem consequences of microbial diversity and community structures. *Ecol. Stud. (Berlin)* **113**:239–254.

Schjørring, J.K., S. Husted, and M. Mattson. 1998. Physiological parameters controlling plant-atmosphere ammonia exchange. *Atmos. Env.* **32(3)**:507–512.

Schmid, H.P., and T.R. Oke. 1990. A model to estimate the source area contributing to turbulent exchange in the surface layer over patchy terrain. *Qtly. J. Roy. Meteorol. Soc.* **116**:965–988.

Seiler, W., and R. Conrad. 1987. Contribution of tropical ecosystems to the global budget of trace gases, especially CH_4, H_2, CO, and N_2O. In: The Geophysiology of Amazonia, ed. R.E. Dickinson, pp. 133–162. Chichester: Wiley.

Skopp, J., M.D. Jawson, and J.W. Doran. 1990. Steady-state aerobic microbial activity as a function of soil water content. *Soil Sci. Soc. Am. J.* **54**:1619–1625.

Swift, L.W. 1976. Algorithm for solar radiation on mountain slopes. *Water Res.* **12**:108–112.

Waddington, J.M., and N.T. Roulet. 1996. Atmosphere-wetland carbon exchange: Scale dependency of CO_2 and CH_4 exchange on the developmental topography of a peatland. *Glob. Biogeochem. Cyc.* **10(2)**:233–245.

Whiteman, C.D. 1990. Observations of thermally developed wind systems in mountainous terrain. In: Atmospheric Processes over Complex Terrain, ed. W. Blumen, pp. 5–42. Meteorological Monographs 40. Boston: Am. Meteorol. Soc.

Whiteman, C.D., and J.C. Doran. 1993. The relationship between overlying synoptic-scale flows and winds within a valley. *J. Appl. Meteorol.* **32**:1669–1682.

Wikle, C.K., and N. Cressie. 1998. A dimension-reduction approach to space-time Kalman filtering. Statistical Laboratory, Preprint No. 97–24. Ames, IA: Iowa State Univ.

Wildt, J., D. Kley, A. Rockel, P. Rockel, and H.J. Segschneider. 1996. Emission of NO from several higher plant species. *J. Geophys. Res.* **102**:5919–5927.

Yang, W.X., and F.X. Meixner. 1997. Laboratory studies on the release of nitric oxide from sub-tropical grassland soils: The effect of soil temperature and moisture. In: Gaseous Nitrogen Emissions from Grasslands, ed. S.C. Jarvis and B.F. Pain, pp. 67–71. Wallingford, U.K.: CAB International.

Zhong, S., and J.C. Doran. 1995. A modeling study of the effect of inhomogeneous surface fluxes on boundary-layer properties. *J. Atmos. Sci.* **52**:3129–3142.

Standing, left to right:
Dennis Baldocchi, Hans-Jürgen Bolle, Werner Eugster, Riccardo Valentini, Roger Pielke, Sr., Franz Meixner
Seated, left to right:

9

Group Report: How Is the Atmospheric Coupling of Land Surfaces Affected by Topography, Complexity in Landscape Patterning, and the Vegetation Mosaic?

M.R. RAUPACH, Rapporteur

D.D. BALDOCCHI, H.-J. BOLLE, L. DÜMENIL, W. EUGSTER, F.X. MEIXNER, J.A. OLEJNIK, R.A. PIELKE, SR., J.D. TENHUNEN, and R. VALENTINI

INTRODUCTION

The terrestrial and atmospheric components of the earth system are coupled by land–atmosphere exchanges of energy, matter (water, carbon dioxide, trace gases, aerosols, and other entities), and momentum. If one regards the coupled dynamics of ecosystems, water, carbon, and nutrients as the "metabolism" of the global terrestrial biosphere, then land–atmosphere exchanges can be thought of as the "breathing" of the biosphere. The discussions leading to this report were concerned with the present state and future directions of research on land–atmosphere exchanges, in the context of studies of the whole earth system.

While the earth system is governed at a basic level by natural (physical, chemical, biological, and ecological) principles, its behavior is also profoundly affected by another set of influences, namely those arising from human actions. Over the last decade, much attention has been devoted to predicting the consequences of one kind of human-induced perturbation on the earth system, the effects of changes in atmospheric

Integrating Hydrology, Ecosystem Dynamics, and Biogeochemistry in Complex Landscapes
Edited by J.D. Tenhunen and P. Kabat

composition through the enhanced-greenhouse mechanism. However, another kind of perturbation, anthropogenic changes in land cover, is already having major consequences on the behavior of the earth system at regional and smaller scales, and may well have global impacts comparable with those arising from atmospheric composition changes. Changes in land use, such as forest clearing or the use of dry land for irrigation, have profound effects not only on local and regional climates and ecosystems, but also on human well-being through consequences on water quantity and quality. In understanding and predicting the consequences of these changes, studies of land–atmosphere exchange processes play a significant part along with many other disciplines.

Bearing in mind this overall context, in this report we focus on the ways in which land–atmosphere exchanges are influenced by the complexity of real landscapes. This complexity arises not only from spatial variations (in soils and topography) and temporal fluctuations (associated with processes in the atmosphere, hydrosphere, or biosphere), but also from human-induced variations of land cover. Our broad goals are: (a) to clarify the current state of knowledge (but not by undertaking a review); (b) to identify areas where knowledge is weak; (c) to identify current research trends; and (d) to suggest future research directions and projects to address identified weaknesses.

In approaching this task, we adopted a few ground rules. First, we take a *connective view* in which land–atmosphere exchanges are regarded not in isolation but as parts of a much larger system in which their primary role is connective. The biosphere, hydrosphere, and atmosphere all influence each other dynamically through land–atmosphere fluxes, which appear as important (often dominant) terms in the budgets of water, energy, carbon, nitrogen, and other entities in both land and atmosphere. Second, we have tried to maintain *freedom from disciplinary allegiances.* Because of their connective role, it makes no sense to study land–atmosphere exchanges from any single disciplinary viewpoint. Third, we adopt *a global perspective with a regional focus.* Because of the issues raised by the influences of anthropogenic land cover changes on the earth system, we give primary attention to regional spatial scales of a few kilometers to a few hundred kilometers. This does not preclude attention to local-scale or global-scale questions.

Our report is organized around five questions:

1. For land–atmosphere exchanges, what is landscape complexity and when does it matter?
2. How should we best describe, conceptualize, and model land–atmosphere exchanges in complex landscapes, in the context of broader models of the earth system?
3. What are the key feedbacks and interactions in this coupled system?
4. What are the limits on predictability imposed by nonlinearities and other factors?
5. What are the most appropriate observational strategies for land–atmosphere exchange studies?

LANDSCAPE COMPLEXITY AND LAND–ATMOSPHERE EXCHANGES

Even on a land surface as simple as an extensive, uniform, closed vegetation canopy on level ground, land–atmosphere exchanges of energy and matter involve complex interactions among radiative, aerodynamic, physiological, hydrological, and biogeochemical processes. Knowledge of these processes and interactions has accumulated slowly but steadily over several decades. Relative to the state of knowledge represented by the landmark "Vegetation and the Atmosphere" volumes published over twenty years ago (Monteith 1975, 1976), there has been major progress in many areas: see, for instance, recent reviews in Steffen and Denmead (1988), Kalma and Sivapalan (1995), Michaud and Shuttleworth (1997), Pielke et al. (1998), and Raupach (1998). This body of knowledge is by now encapsulated in a large number of SVAT (Soil–Vegetation–Atmosphere Transfer) schemes that model the biophysical processes governing land–atmosphere exchanges of energy and matter on a discrete "patch" of landscape (Chen et al. 1997). Practically all SVAT formulations rest on the key assumption that the patch under consideration is uniform and large enough to neglect the influence of neighboring patches, so that the transfers of energy and matter in both soil and canopy layers can be described in terms of fluxes and gradients only in the vertical (more generally, surface-normal) direction.

In the context of land–atmosphere exchanges, the phrase "landscape complexity" can denote *spatial complexity, temporal complexity,* and *process complexity.* Spatial complexity denotes the ways in which landscapes depart from the idealization of simple vertical transfers on a uniform patch: major examples are provided by sparse vegetation cover, patchiness (horizontally nonuniform surface cover on level terrain), and topographic variation (hills, valleys, mountains, escarpments). Temporal complexity denotes situations in which unsteadiness in time is an important feature. Process complexity denotes the suite of processes occurring on practically any landscape that includes not only the biophysical processes governing energy and water exchanges but also hydrological, biogeochemical, physiological, and ecological processes, together with the couplings or feedbacks causing all these processes to interact. Clearly, spatial, temporal, and process complexity can all occur together, though one or other may predominate in any instance.

Spatial Complexity

Sparse canopies, patchy landscapes, and hills can all affect wind flows and the land–atmosphere exchanges of entities such as heat, water vapor, or trace gases. The essence of this form of complexity is that land–atmosphere exchanges can no longer be described adequately using a one-dimensional (vertical) formulation. Spatial complexity can have several implications:

1. wind velocities may be horizontally nonuniform and perhaps even recirculating;

2. scalar (temperature, water vapor, or trace gas) spatial distributions may be strongly influenced by horizontal transfers;
3. parameterizations of terms in governing equations may need to adopt more general forms than those applicable in horizontally uniform situations;
4. the radiation field may be affected significantly; and
5. there may be complex interactions between spatial and chemical processes (an example of the combined effects of spatial and process complexity).

Budgets for scalar entities, such as heat, water vapor, or trace gases, can provide a useful test for whether spatial complexity is significant or not in particular circumstances. Consider the mass of scalar $\int Cdz$ in a column between two arbitrary heights z_1 and z_2, where C is the scalar concentration. The conservation equation for Cdz (a quantity with the dimension of mass per unit horizontal area) is:

$$\frac{\partial\left[\int Cdz\right]}{\partial t} = F(z_1) - F(z_2) - \frac{\partial\left[\int UCdz\right]}{\partial x} - \frac{\partial\left[\int VCdz\right]}{\partial y} + \int Pdz - \int kCdz\,, \qquad (9.1)$$

where U and V are the mean velocity components in the horizontal (x and y) directions, $F(z)$ is the scalar flux in the vertical (z) direction, P is the production rate due to biogeochemical or other sources, and kC is the destruction rate (assumed to follow first-order kinetics with rate constant k). Simplifications in Equation 9.1 include the use of a coordinate system in which the mean vertical velocity W is zero, and neglect of horizontal turbulent fluxes. Because of integration over the column from z_1 to z_2, vertical flux divergence appears as the difference $F(z_1) - F(z_2)$.

Spatial complexity appears in Equation 9.1 in several ways: first, through the horizontal divergence terms $\partial[\int UCdz]/\partial x$ and $\partial[\int UCdz]/\partial y$. These terms can be significant (relative to other terms) for two reasons: (a) when the velocity field itself is nonuniform or perhaps even recirculating as in flows around hills, in sparse canopies, and in convectively driven mesoscale circulations; or (b) when the surface flux distribution is not horizontally uniform (even if the velocity field is horizontally homogeneous) because of landscape patchiness induced by variations in vegetation properties such as height, leaf area, or stomatal behavior, or soil characteristics such as depth, texture, or nutrient status. The significance of these forms of spatial complexity can be quantified by investigating the relative order of magnitude of $\partial[\int UCdz]/\partial x$ and $\partial[\int UCdz]/\partial y$ in Equation 9.1, using scale analyses or other means.

A second way in which spatial complexity can appear in Equation 9.1 is through the parameterizations of flux, production, and destruction terms which are always required in practice to solve a set of equations like 9.1 in a model. Parameterizations used in uniform flows may be quantitatively inaccurate in nonuniform flows and may need to be made more sophisticated. For example, atmospheric models treat both horizontal and vertical fluxes as the sum of resolvable and subgrid-scale components. The resolvable horizontal and vertical fluxes are represented explicitly, using conservation equations like Equation 9.1, and other physical principles. However, without exception to our knowledge, vertical subgrid-scale fluxes are parameterized using theory developed for horizontally homogeneous landscapes, in which horizontal nonuniformities are absent, while subgrid-scale horizontal fluxes are neglected altogether. It is clear that these assumptions are questionable in spatially complex situations.

Third, spatial complexity occurs when chemical processes — represented in Equation 9.1 by source-sink terms — interact with spatial processes to cause the behavior of the system to change substantially from the situation in spatially simple systems. This can happen, for example, through the height-integrated destruction term $\int kCdz$ which will often not be equal to $k\int Cdz$, as in the interaction of NO, NO_2, and O_3 in a deep canopy (Meixner and Eugster, this volume). Usually, such a simplification is appropriate only when chemical time constants ($1/k$) are much longer than time constants for diffusive transport.

Fourth, spatial complexity can appear in, and through, the radiation field. Examples are provided by the radiation field in sparse canopies and the radiative effects of topography. Radiation interacts with Equation 9.1 through its effects on surface scalar fluxes which appear as boundary conditions, and also through its direct contributions to source-sink terms, for instance through radiative warming or cooling in the heat budget, or through photochemical processes in the budgets of photochemically active species.

These consequences of spatial complexity can be treated explicitly using three-dimensional models. Alternatively, it may be appropriate to recover a one-dimensional formulation by taking appropriate horizontal averages over spatial heterogeneities, but such an operation amounts to parameterizing the effects of spatial complexity in a larger-scale description. The averaging may preserve the form of the governing equations, as in the case of linear averages of vertical exchange fluxes, or it may require different model forms at different scales, as in the case of infiltration and runoff (see Wood, this volume).

Temporal Complexity

Complexity is also added when processes are nonsteady state. Quantitatively, this is important when the storage term $\partial[\int UCdz]/\partial t$ in Equation 9.1 is significant in comparison with other terms, after averaging over an appropriate time. Let us consider examples at short, medium, and long time scales.

At *short* time scales (less than diurnal), temporal complexity can have significant implications especially for flux measurements. An example arising directly from the column storage term in Equation 9.1 (the term on the left-hand side) is that of nocturnal CO_2 fluxes, especially in tall vegetation canopies. At night, CO_2 produced by autotrophic and heterotrophic respiration accumulates inside the canopy volume from which it is exported to the overlying atmosphere in intermittent "flushes" when the wind speed occasionally increases, and also when convection commences in the morning. During periods of rapidly changing external conditions (around sunrise and sunset, under changing cloud shadows or during wind shifts), measurement complications arise not only from storage changes in the atmospheric column but also because of the nonapplicability of conventional (steady and one-dimensional) parameterizations such as flux-gradient relationships. At *medium* time scales (from subseasonal to a few years), temporal complexity influences land–atmosphere exchanges through phenological, biogeochemical, hydrological, and climatic processes. Examples include the seasonal patterns imposed on land–atmosphere exchanges by leaf dynamics in deciduous forests, the effects of soil dry-down after intermittent rainfall, the dynamics of the vegetative and soil stores of carbon and nitrogen, and effects of climate forcing through mechanisms such as El Niño events. Land–atmosphere exchanges are also modulated by processes at *long* time scales (from decades to paleoclimatic scales), an example being the interaction of deep freezing and thawing cycles with trace gas emissions in boreal and arctic ecosystems.

Process Complexity

Land–atmosphere exchanges involve a broad range of processes and phenomena, extending far beyond the biophysical processes governing energy and water exchanges (the domain of the traditional SVAT) to include hydrological, biogeochemical, physiological, and ecological processes. Interactions among these processes occur on a very wide range of temporal and spatial scales. Figure 9.1 indicates schematically the range of processes involved, and some of the major interactions between them. Processes taking place at (or just above and just below) the land surface include:

1. radiative processes, important both for their role in the energetic forcing of land–atmosphere exchanges and also their implications for remote sensing;
2. turbulent transfers of heat, water vapor, trace constituents, and momentum;
3. short-term physiological processes, such as vegetative carbon assimilation and microbial and root respiration in the soil;
4. flow and transport of heat, water, and nutrients in the soil;
5. seasonal-scale vegetation developmental processes, influencing properties such as vegetation height, leaf area index, albedo, and root depth and distribution;
6. biogeochemical processes involved in the land–atmosphere exchange of trace gases including nitrogenous gases, methane, and volatile organic compounds (VOCs);
7. biogeochemical processes determining the stores of nutrients such as N, P, K, and S in soil, detrital, and plant pools.

In addition, land–atmosphere exchanges are intimately connected with many processes occurring in other parts of the earth system, such as the atmosphere high above the surface or the water stored in aquifers, rivers, and lakes. Some major examples are:

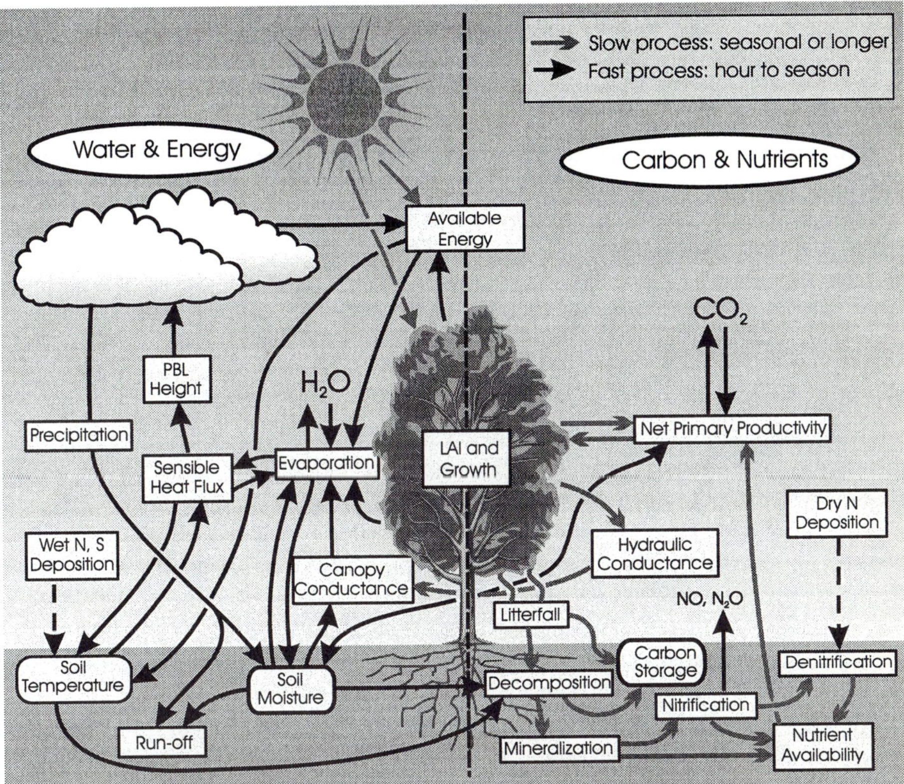

Figure 9.1 Schematic representation of processes and interactions at patch scale on a vegetated land surface.

8. development of the atmospheric planetary boundary layer and boundary-layer clouds, with implications for radiation, temperature, and humidity at the land surface;
9. development of deep cumulus convection and convective rainfall over land;
10. development of mesoscale circulations in heterogeneous landscapes, including convective "chimneys" connecting the free atmosphere with the land surface;
11. surface and subsurface lateral water movement and river runoff;
12. processes linked with direct human intervention on the landscape, including cropping, grazing, forestry, and urban land use.

The group found two concepts to be of great help in coming to grips with the task of conceptualizing and modeling this vast array of phenomena: first, the use of hierarchical approaches for describing and modeling complex systems, and second, the formal identification of feedbacks and interactions. These are the subjects of the next two sections.

CONCEPTUALIZING AND MODELING LAND–ATMOSPHERE EXCHANGES

In many respects, land–atmosphere exchanges form a major crossroads through which the atmosphere, hydrosphere, and biosphere interact. This set of interactions can be described at a number of levels of detail ranging from simple and highly aggregated to complex and highly detailed, forming a hierarchy of levels of description. In modeling terms, this hierarchy corresponds to a range of possible levels of complexity of process parameterization.

Hierarchical Description of Complex Systems

The idea of a hierarchy of levels of description has links with formal hierarchy theory, which is much more familiar in the environmental life sciences than in the environmental physical sciences. Hierarchy theory was developed within the framework of general systems theory in the 1960s, as a theory of complex systems that have a large number of interactive components (Simon 1962; Allen and Starr 1982; O'Neill 1988; Wu and Loucks 1995). The viewpoint is that complex systems generally have a high degree of redundancy in organization and are hierarchically structural. Representing such complex systems as hierarchical systems with a limited number of discrete levels (or scales) can lead to great simplification and better understanding. This characteristic of complex systems, that is, being able to be simplified into an ordered system containing only a small number of components with nonsignificant loss of information, is called near-decomposability (Simon 1962). The concept of a hierarchy of organizational levels has been identified as one of the key characteristics of a complex *adaptive* system (Waldrop 1992, p. 145), and is associated with the ability of a complex adaptive system to form coherent, self-reinforcing structures at several levels, with structures at one level acting as building blocks for structures at the next higher level (as in cells, organs, individuals, communities, ecosystems, biomes).

Hierarchical systems have some common properties regardless of their detailed content. Higher (more aggregated) levels exert constraints on lower (more detailed) levels, and lower levels provide the mechanisms for higher levels. Thus, in general, to understand a certain phenomenon of a complex system, one usually needs to look at three adjacent hierarchical levels: the focal level characteristic of the phenomenon, the next level above (the constraints), and the next level down (the mechanisms).

Implications for Land–Atmosphere Exchange Processes

These ideas have implications for conceptualizing and modeling land–atmosphere exchanges and their role in the earth system in general. First, the choice of an appropriate level of description is important because it is a means of focusing on what matters in any given situation, and dismissing irrelevant aspects of the system. Second, the role of hierarchical levels adjacent to the level of focus is clarified: more aggregated, larger-scale, slower levels provide constraints, and more detailed, smaller-scale, faster levels provide mechanistic understanding and process information. There is thus a general flow of information about constraints from more aggregated to more detailed levels in the hierarchy, and a flow of information about processes from more detailed to more aggregated levels. Models of a particular process at detailed levels tend to have many parameters, while models of the same process at more aggregated levels tend to have few parameters that embody information from more detailed parts of the hierarchy of description.

The application of these ideas can be clarified by means of several examples from different areas. The first is the modeling of energy and water exchanges in a SVAT. For each of the key processes in a traditional SVAT, a range of models at different levels of detail and parameterization is now available (Chen et al. 1997; Raupach 1998). For instance, soil water stores may be treated with a single layer (bucket) scheme, a two-layer scheme, or a multi-layer scheme involving the full numerical solution of Richards' equation for water movement in unsaturated soils. The physiological processes associated with the stomatal or surface conductance can be represented by single value of the conductance, a maximum conductance modified by environmental "stress functions," or a full assimilation model in which photosynthetic carbon uptake, stomatal conductance and intercellular CO_2 concentration are treated as linked dependent variables to be determined by three coupled equations (an approach that fundamentally links energy, water, and carbon exchanges). Similar ranges of models exist for aerodynamic and radiative processes. The point is not that more sophisticated models are more "correct" than simpler ones; indeed, excessive complication in parameterization can lead to problems of nonrobust calibration when parameters must be determined empirically (Franks et al. 1997). Rather, the points are that (a) the application of the SVAT determines the appropriate level of sophistication for its various process descriptions, consistent with the requirement that parameters need to be determined unambiguously; and (b) simple process descriptions, where their use is appropriate, can often utilize information from more detailed descriptions in the form of guidance on process behavior and constraints on parameters.

A second, closely related example is the modeling of the interactions between physiological and biogeochemical processes which enter the coupled water, carbon, and nitrogen cycles. These are now represented explicitly in many SVAT schemes, at varying levels of process resolution (partly determined by the number of distinct carbon and nitrogen pools recognized by the model) and at a wide range of temporal resolutions from subdiurnal to monthly or more. Well-known examples include CENTURY, initially developed for grasslands (Parton et al. 1993); the sequence of PnET models, describing forests at several different levels of detail (Aber et al. 1996, 1997); and models for global net primary production (NPP) which utilize remotely sensed data, including BIOME-BGC (Running and Hunt 1993), CASA (Potter et al. 1993, 1997), and 3PG (Landsberg and Waring 1997). Broadly speaking, the order of these instances reflects progressively more aggregation of the pools of carbon (and nitrogen in most cases) and progressively less process resolution.

A third, more specific example concerns the practical scaling of detailed physiological information to coarser levels in a study of fire dynamics in chaparral (Reynolds et al. 1993). Let L be the level of interest, with $L = 2$ being the leaf level (time scales of seconds), $L = 1$ the whole-plant level (time scales of hours to days), and $L = 0$ the community level (time scales of months). At $L = 2$, leaf photosynthesis was modeled as a function of enzyme concentrations, electron transport, light absorption, and related variables, and nitrogen uptake as a function of uptake kinetics, root length, and similar variables. At $L = 1$, these predictions were used as input to drive the carbon and nitrogen balance of the whole plant. From this, a growth rate (G) was obtained for a wide range of scenarios of external environmental drivers:

$$G = G(\mathrm{CO}_2, \mathrm{nitrogen}, \mathrm{light})\,. \tag{9.2}$$

A functional form for this relationship was then obtained and used directly in a community growth model ($L = 0$) for the biomass W_i of species i:

$$\frac{dW_i}{dt} = GW_i\left(1 - \frac{W_i}{W_{\max}}\right). \tag{9.3}$$

Equation 9.3 incorporates new information that is appropriate only at the community level, such as mortality, recruitment, and competition. In this way, physiological information is transferred up through a hierarchy of processes. This example also indicates how ecosystem responses and land–air exchanges need to be viewed interactively. Through land–atmosphere fluxes, changes in biomass, leaf area, and vegetation cover of the kind modeled by Equation 9.3 have implications for local climate and precipitation which feed back in turn on the ecosystem responses. These feedbacks are discussed in more detail in the next section.

A final example is the parameterization of physical processes in global climate models (GCMs), for instance the encapsulation of line spectral models for atmospheric transmission in broad-band models such as the Goody model (Houghton 1973). Broad-band radiation models are now widely used in GCMs to provide descriptions of atmospheric radiative transfer processes, greatly reducing computational load and encapsulating the basic physics with a few parameters. These radiative processes provide the primary forcing for the enhanced greenhouse effect, by linking perturbations in the thermal radiative properties of the atmosphere with changes in concentrations of radiatively active gases (CO_2, chlorofluorocarbons, CH_4, and others). However, the enhanced greenhouse effect does not depend on radiative processes alone but also on interactions with many other processes in the climate system, including cloud and water vapor feedbacks, interactions with oceanic and terrestrial biospheric processes, and more. To describe the overall complex system, GCMs include numerous encapsulated descriptions of physical processes similar to broad-band radiation models.

Implications for Spatial Aggregation into Land–Atmosphere Functional Units

The question now arises of the optimal level of aggregation at which to study land–atmosphere exchanges in complex landscapes. This will correspond with an optimal level in the hierarchy of processes governing land–atmosphere exchanges. Since we are concerned with the functional properties of landscapes in terms of energy and mass exchanges with the atmosphere, it is natural to imagine a complex landscape as a mosaic of *land–atmosphere functional units* (Valentini et al., this volume). Around any point, the land–atmosphere functional unit may be defined as "the largest possible patch of landscape which, through mass and energy exchanges with the atmosphere, shows a coordinated and specific response to environmental factors." The term "coordinated" means a response which reflects in a unique functional form the integrated effects of the biospheric processes determining land–atmosphere exchanges. This functional form can be either *invariant* or *emergent*. Processes of biophysics, biochemistry, and physiology which scale almost seamlessly from individual leaves to whole canopies are called *invariant* properties across these scales: for example, leaf and ecosystem photosynthesis have the same responses in terms of nitrogen. Processes which show a different behavior when scaled from single organs to the community lead to *emergent* properties at ecosystem level which are not clearly apparent at individual level. An example is provided by the light response curve of photosynthesis, which always saturates at leaf level but often does not saturate at ecosystem level. The term "specific" means that the energy and mass exchange properties of a land–atmosphere functional unit can be described in terms of aggregated (whole-system) parameters which represent specific responses at species or biome level. This aggregation is in terms of functional properties, not taxonomic classifications.

Landscape functional units may be defined from viewpoints other than land–atmosphere exchanges: for instance, remote sensing, hydrological function, or ecosystem function. Consideration of alternative definitions has been taken up by several groups at this Workshop. The relationship between alternative definitions of landscape functional units opens a research issue, since these definitions rest on views of landscape functional organization. The real issue is not one of mere definition but of the ways that different kinds of process (natural processes in the atmosphere, hydrosphere, and biosphere, or human-induced interventions such as land-cover changes) interact to produce evolving landscape organization and structure.

FEEDBACKS AND INTERACTIONS

We have already alluded to the numerous interactions and feedbacks which link land–atmosphere exchanges with other processes in the earth system. The behavior of the whole system is emergent, being determined at least as much by the nature of these interactions as by the behavior of individual or isolated processes. Examples of important interactions include:

- short-term biophysical feedbacks between stomatal conductances, surface temperature, and atmospheric thermal stability;
- the feedback between plant water use and soil water status;
- the feedback between surface radiation and the surface sensible and latent heat fluxes;
- feedbacks between the surface energy balance, the state and development of atmospheric boundary layer, and boundary-layer clouds;
- the feedback between soil moisture and locally generated convective rainfall, which may be enhanced by landscape patchiness or topography;
- interactions between soil temperature, soil moisture, and trace gas fluxes, including nitrogenous gases, methane, and VOCs;
- interactions between plant growth, soil moisture, and biogeochemical nutrient cycling;
- interactions between plant growth, ecosystem development, and fire behavior (frequency and intensity).

A much longer list can easily be constructed. Qualitative diagrams indicating many of the possible interactions have appeared in numerous reports produced by IGBP Core Projects and other agencies, for example IGBP (1993). However, following a recognition of the existence of these many interactions, the next step is an evaluation of their implications. Any particular interaction (such as one from the list above) can be characterized by five properties:

1. Is it a one-way or two-way interaction?
2. Is it linear or nonlinear?
3. Does it represent a positive or a negative feedback?
4. Is it strong or weak, relative to other influences on the system?
5. Is it short-term or long-term?

Again we take some examples, intended to illustrate the linked effect of *groups* of interactions. The first is the set of linkages between water and biogeochemical processes. Water is a key requirement for most biogeochemical activity, including assimilation of carbon by plants, soil and plant respiration, VOC emissions, and emission of nitrogenous gases, for example. Land–cover changes (induced by climatic variations, anthropogenic land-use changes, and other disturbances such as fire) therefore affect all these biogeochemical processes through their effect on soil water status via plant transpiration. Furthermore, the effect of atmospheric nitrogenous gases is two-way because these gases account for a significant fraction of the nitrogen input into many natural ecosystems, so that disturbance in the atmospheric cycle of these gases has implications for nitrogen input and plant growth. Currently this effect is generally to increase N input and promote growth of certain species, leading potentially to shifts in ecosystem composition. There are also implications for N movement into aquifers and aquatic systems, with potential consequences for water quality.

The second example is the relationship between cumulus convection, land–atmosphere water fluxes, ecology, and fire behavior in ecosystems as diverse as boreal forest and savanna. Convective storms simultaneously provide a source of water stimulating vegetation growth and an ignition source from lightning, stimulating disturbance of the ecosystem by fire. In the case of the tree-grass mixture comprising savanna, this disturbance favors grass over trees. The linkage between convective storms and ecosystem composition is two-way, as the drastic changes of land-surface albedo and water availability following burning have a significant influence on convective development. These interactions are also clearly strongly nonlinear.

A third example is provided by the effect of terrestrial vegetation on hypothesized temperature rises associated with the enhanced greenhouse effect and atmospheric CO_2 concentration increases. Rising temperatures have a significant effect on stomatal aperture and thence on transpiration and surface energy balance. At warm temperatures this is a two-way interaction providing positive feedback to a temperature rise, offset partly by a negative feedback due to the role of thermal instability in modifying aerodynamic coupling between the surface and the atmosphere (Raupach 1998). The feedback is observed with a wide range of process models for stomatal physiology, though its exact magnitude is model-dependent. The study of Sellers et al. (1996) used a coarse-resolution GCM containing a relatively sophisticated representation of vegetation physiology to conclude that this feedback would increase terrestrial near-surface temperatures by substantially more than the global average near-surface temperature increase under doubled CO_2. In focusing on the direct effects of atmospheric composition on short-term vegetative physiology, this study did not include related longer-term feedbacks associated with responses in the vegetation itself (for instance in leaf area and nutrient status). However, the study effectively highlights the kinds of change in system behavior that can result from the incorporation of biospheric interactions into climate models.

A fourth example is the interaction of long-term temperature changes, thawing of arctic permafrost, and increased methane emissions in boreal and arctic ecosystems, providing a positive feedback on temperature through the role of methane as a greenhouse gas.

Our fifth and final example is the interaction between terrestrial systems, aerosols, and climate. Although most attention has focused on the role of the oceans as sources of aerosol-forming compounds like dimethyl sulfide (DMS), the terrestrial biosphere also has a significant role both as a source of aerosol-forming compounds and as an aerosol sink.

These examples together illustrate the major implications of incorporating land–atmosphere exchange processes into global climate models. Through sheer computational power, GCMs span many hierarchical levels simultaneously and carry relatively detailed representations of many processes up to global scale. Moreover, there is a trend for GCMs to incorporate progressively more processes relevant to the functioning of the earth system, including more detailed representations of terrestrial physiological, hydrological, biogeochemical, and ecological processes, with links to atmospheric behavior through increasingly realistic descriptions of land–atmosphere exchanges. In this way, processes that at one time were specified by *parameters* are increasingly being described by *state variables*. This is an important trend in the development of understanding of the coupled biosphere–climate system. It shifts the terrestrial biospheric components of the system, which used to be treated as noninteractive external influences, into their true place as fully interactive parts of the system, both influencing and being influenced by the system as a whole. Similar trends are evident in treatments of other parts of the climate system, including the oceans and the cryosphere.

LIMITS TO PREDICTABILITY

We have argued in the last section that the dynamics of vegetation and land-surface processes interact strongly with climate, and that many aspects of these interactions are strongly nonlinear. There are limits to the predictability of this coupled system, arising in two basic ways. A first, fundamental limit is imposed by the degree of nonlinearity of the interactions. If nonlinearities are strong enough, they induce unpredictability in the solutions on time scales associated with the time scales of the nonlinear terms, by amplifying infinitesimal uncertainties in initial conditions into large uncertainties in the predicted system state at some later time. Nonlinearities with long time scales (oceans, terrestrial biogeochemistry, and ecology) may induce long-term unpredictability. It should be emphasized that it is an open question whether unpredictability is in fact induced by realistic climate–biosphere interactions, but the potential is there and a research question exists.

The second limit to predictability arises from limits on our understanding and ability to represent quantitatively the behaviors of system components, particularly (for present considerations) the biophysical, hydrological, biogeochemical, and ecophysiological properties of the terrestrial biosphere. Although the equations governing climate physics (including the equations of fluid motion and radiative transfer) are well known, the equations governing the biological and ecological evolution of the terrestrial biosphere are not so unequivocally determined.

One implication of a limit to global climate predictability is that accurate forecasts of future regional climates (ensembles of weather regimes) are unattainable beyond a forecast time linked with the time scales of the major nonlinear interactions. Beyond this time, the concept of a deterministic climate prediction or forecast must be abandoned. Therefore, to address the long-term evolution of climate in response to human-induced perturbations of land cover and atmospheric composition, it is better to use a probabilistic approach based on the concepts of vulnerability, susceptibility, and risk assessment.

The principle is to treat a climate system model as a mapping between some given present state and an *ensemble* of future states (because of the unpredictability of the system). The distribution of this ensemble is determined both by the natural stochastic evolution of the climate, as captured by the model, and also by imposed forcings or perturbations such as changes in land cover and atmospheric composition. If a variety of scenarios is used to specify these forcings, a variety of output ensembles of future states will be obtained. The set of ensembles can then be compared with a postulated range of acceptable future states (such as a range of temperature or precipitation). This identifies the part of the ensemble of future states which lies wholly or partly outside of the range of acceptable future states, and enables a risk to be assigned to each scenario of imposed perturbations. Such a procedure avoids the need to assume that we can deterministically forecast the future regional climate and its interaction with other environmental factors. However, the inclusion of all relevant interactions is critical. Up to the present, this mapping has been carried out without including many of the two-way, nonlinear interactions discussed in this paper.

APPROPRIATE STRATEGIES FOR OBSERVATIONS AND EXPERIMENTS

As with other branches of environmental science, the investigation of land–atmosphere interactions involves making a series of hypotheses which are subject to test through observation or experiment. Before considering land–atmosphere interactions specifically, we make some general comments about the role of experiments and observations in testing hypotheses.

First, we must distinguish between testing a hypothesis (an activity), falsification of that hypothesis (a possible outcome of that activity), and "validation" of the hypothesis (arguably an unachievable goal because of its implication of permanent certainty).

Second, we may think of a hypothesis as a mapping between spaces of "causes" and "effects." The testing exercise involves comparing the mapping predicted by the hypothesis with an independently determined (usually observed) data set which measures or constrains some aspect of this mapping. It depends on circumstance whether the forward or inverse mapping is tested, and for the present considerations the choice of option is immaterial. The result of the test can be "mapping consistent with independent data," implying that the hypothesis is not falsified; "mapping inconsistent with data," implying that the hypothesis is falsified; or "don't know," in which case the test is inconclusive. We must be open to this third possibility. An excellent current example is provided by attempts to test predictions of temperature response to changing atmospheric CO_2 concentrations, where the available data is both inconsistent (for instance, surface and satellite temperature series) and subject to an unknown amount of noise due to natural climate variability.

Third, it is important to distinguish between simple hypotheses about individual processes, such as photosynthesis, turbulent transfer or ecosystem development, and compound hypotheses such as climate model predictions of temperature response to changing atmospheric CO_2 concentrations. Tests can falsify simple hypotheses, leading eventually to improved hypotheses. With the compound hypotheses formed by complex models, falsification is possible in principle but a common use of additional information from observations is to modify the model, for instance by improving parameter choices. This means that observations and process understanding (as represented within the model) are used together to provide *constraints on predictions*. The interaction between the two is much closer than in the classical approach of testing a simple hypothesis with independent observations.

Turning more specifically to land–atmosphere exchanges, the testing of hypotheses or models takes place in several ways. First, replicated laboratory or controlled field experiments provide tests of simple hypotheses (in the above sense) about individual processes, for instance, in plant physiology or fluid mechanics. Controlled field experiments can include observations under specially selected "near-ideal" conditions, or manipulative studies creating deliberate landscape variation. When well designed, these experiments provide rigorous tests of such hypotheses. However, laboratory or controlled field experiments cannot explore a wide range of the interactions which combine to determine land–atmosphere exchanges in nature.

Second, beginning in the early eighties, the land–atmosphere scientific community began the task of overcoming the limitations of the tightly controlled field experiment. The extension to regional spatial scales has been undertaken through a series of "large-scale" experiments at regional scales of around 100 km, generally conducted mainly in intensive campaigns of a few weeks or months duration. Among these are HAPEX-MOBILHY (France, 1984–1986); HAPEX–SAHEL (Niger, 1992–1993); FIFE (U.S.A., 1987, 1989); EFEDA (Spain, 1991, 1994); BOREAS (Canada, 1993–1995); NOPEX (Scandinavia, 1994–1997); HEIFE (China, 1990, 1994); OASIS (Australia, 1994–1995); several campaigns in the Amazon; and the planned LBA experiment in the Amazon basin. The findings and philosophy of these experiments are treated by Hutjes et al. (1999). Further experiments of this kind are still required, especially in topographically complex landscapes.

Third, more recently, attention has turned to the need for long temporal records, to gain a better perspective on the long-term behavior of landscapes and their mass and energy exchange properties, and their responses to long-term influences such as natural climatic variability. In the past, most of our knowledge of the carbon and water exchanges of vegetation has been based on measurements taken at leaf or branch level with enclosure chambers or portable cuvettes. With the development of micrometeorological methods, a bigger scale was achieved, passing from the leaf level to studies of overall ecosystem fluxes in a direct, nondestructive way. Historically this technique has been used primarily in intensive and short-term land-surface experiments, but new opportunities are arising for application at longer temporal scales. A new generation of long-term flux measurements from towers is being initiated, using developments in micrometeorological technology and theory which have made the study of the interactions between vegetation and atmosphere routine. Of particular importance has been the progressive improvement of eddy covariance methods through better instrumentation, which now makes possible the continuous measurement of carbon dioxide, water, energy, and momentum fluxes on a seasonal basis with hourly resolution. These long-term flux measurements (now being globally coordinated by the Fluxnet initiative) need to be supplemented with campaigns to provide regional-scale spatial information, and linked with studies of hydrological and biogeochemical processes.

Fourth, a critical role is played by remotely sensed observations from space. The record of visible, near-infrared and thermal-band imagery now extends back over two decades, a resource of inestimable value. These observations not only provide spatially resolved inputs to land–atmosphere exchange models at a global scale (e.g., through methods based on visible and near infrared vegetation indices) but are also capable of providing independent observations for model testing (e.g., through the thermal band). Long-term evaluations of satellite data are necessary to minimize instrumental and atmospherically induced errors and to separate short-term variations in observed land–surface properties from long-term trends. To evaluate inferences about land–atmosphere exchanges from remotely sensed data, long-term flux measurements (including both radiative and eddy fluxes) are essential at a set of representative sites. This is a key role for the long-term flux measurements mentioned next.

Fifth, another vital group of observations is the long-term surface weather record. This provides one means of detecting both regional and global climate change and variability. To include the influences on near-surface climate of changes in land use (urban–rural differences, forestation, agricultural and grazing changes), it is important to link such work with atmospheric modeling studies of the role of spatial complexity in land–atmosphere exchanges and regional climates. The use of surface weather records in heterogeneous regions also requires an assessment of the representative footprint of what the weather record is measuring, accounting for both microclimate and mesoclimate effects.

SUMMARY AND RECOMMENDATIONS

This report can be summarized as follows. Our starting point was a recognition that the earth system is being profoundly influenced by human actions, two major drivers being anthropogenic changes in land cover and atmospheric composition, and that land–atmosphere exchange studies have a vital role (along with many other disciplines) in understanding and predicting the consequences of these influences. We identified three kinds of complexity in land–atmosphere interactions, which must be embraced to meet these challenges: spatial, temporal, and process complexity. Next we discussed the concept of hierarchical levels of description of complex systems as a means of providing structure to the initially daunting complications raised by spatial, temporal, and process complexity. We identified the *land–atmosphere functional unit* as an optimal level of aggregation for studying land–atmosphere exchanges in complex environments. We considered the feedbacks and interactions that are the hallmarks of process complexity, highlighting the trend for processes that at one time were specified by *parameters*, passed between decoupled models, to be increasingly described by *state variables* in fully coupled models such as climate–biosphere models. Next we explored the implications of feedbacks and interactions, along with observational uncertainties, for limits to predictability. We emphasized the benefits of probabilistic approaches based on the concepts of vulnerability, susceptibility, and risk, rather than deterministic predictions. Finally, we discussed the role of observations and experiments, distinguishing between simple hypotheses which can be directly tested by observation and compound hypotheses such as complex, coupled models, where the relationship between model and observation is much closer and the two are often used jointly to provide constraints on predictions. We considered the different roles of five kinds of observation: specific controlled experimentation, large-scale experiments, long-term site observations, remote sensing, and the weather record.

Based on our discussions of research trends and knowledge gaps, six recommendations can be highlighted:

1. Because much of the behavior of the coupled climate–biosphere system is emergent, depending on interactions between system components rather than the behaviors of components in isolation, it is important to recognize the full range of interactions and feedbacks in coupled system models.
2. It is helpful to view components of the earth system, including land–atmosphere exchanges, as an organizational hierarchy in a single complex system. There is a need to identify the relationships between land–atmosphere functional units (the building blocks of the system from a land–atmosphere exchange viewpoint) and other types of landscape functional unit, such as remote sensing or ecosystem. A long-term goal is to understand the ways that different kinds of process interact to produce evolving landscape organization and structure.
3. Because of inherent limits to predictability, restrictions on data availability and imperfect knowledge of system behavior, predictions about the evolution of the climate–biosphere system need to use probabilistic approaches based on the concepts of vulnerability, susceptibility, and risk assessment, rather than on deterministic approaches.
4. The response of the system to perturbation (for instance in land use) needs to be tested over a wide range of possible forcings, recognizing the inherent limits to deterministic prediction. This kind of testing, within a probabilistic framework, provides sensitivity studies necessary for an integrated approach to the management of landscapes for long-term sustainability.
5. Processes that at one time were specified by *parameters*, passed between decoupled models, are increasingly being described by *state variables* in fully coupled models (for instance, climate–biosphere models). This trend is a necessary step in the exploration of the consequences of the interactions responsible for the complex behavior of the coupled climate–biosphere system.

6. To realize the spatial and temporal integrative potential of remote sensing from space, continuing interaction is necessary between the remote sensing community and communities studying land–atmosphere exchanges with models and ground-based observations. Long-term evaluations of satellite data are necessary to minimize errors and to separate short-term variations in observed land–surface properties from long-term trends. Long-term flux measurements (including both radiative and eddy fluxes) are essential at a set of representative sites, to provide ground-based observations for testing inferences about land–atmosphere exchanges and land-surface behavior from remotely sensed data.

REFERENCES

Aber, J.D., S.V. Ollinger, and C.T. Driscoll. 1997. Modelling nitrogen saturation in forest ecosystems in response to land use and atmospheric deposition. *Ecol. Mod.* **101**:61–78.

Aber, J.D., P.B. Reich, and M.L. Goulden. 1996. Extrapolating leaf CO_2 exchange to the canopy: A generalized model of forest photosynthesis compared with measurements by eddy correlation. *Oecologica* **106**:257–265.

Allen, T.F.H. and T.B. Starr. 1982. Hierarchy: Perspectives for Ecological Complexity. Chicago: Univ. of Chicago Press.

Chen, T.H., A. Henderson-Sellers, P.C.D. Milly, et al. 1997. Cabauw experimental results from the Project for Intercomparison of Land-surface Parameterization Schemes. *J. Clim.* **10**:1194–1215.

Franks, S.W., K.J. Beven, P.F. Quinn, and I.R. Wright. 1997. On the sensitivity of soil–vegetation–atmosphere–transfer (SVAT) schemes: Equifinality and the problem of robust calibration. *Agr. For. Meteorol.* **86**:63–75.

Houghton, J.T. 1973. The Physics of Atmospheres. Cambridge: Cambridge Univ. Press.

Hutjes, R.W.A., P. Kabat, S.W. Running, W.J. Shuttleworth, C.B. Field, B. Bass, M. Assunçao da Silva Dias, R. Avissar, A. Becker, M. Claussen, A.J. Dolman, R.A. Feddes, M. Fosberg, Y. Fukushima, J.H.C. Gash, L. Guenni, H. Hoff, P.G. Jarvis, I. Kayane, A.N. Krenke, C. Liu, M. Meybeck, C.A. Nobre, L. Oyebande, A. Pittman, R.A. Pielke, M.R. Raupach, S. Saugier, E.D. Schulze, P.J. Sellers, J.D. Tenhunen, R. Valentini, R.L. Victoria, and C.J. Vorosmarty. 1999. Biospheric aspects of the hydrologic cycle. *J. Hydrol.*, in press.

IGBP (International Geosphere–Biosphere Programme). 1993. Biospheric Aspects of the Hydrological Cycle (BAHC): The operational plan. Intl. Geosphere–Biosphere Program. Report No. 27. Stockholm: IGBP.

Kalma, J.D., and M. Sivapalan, eds. 1995. Scale Issues in Hydrological Modelling. Chichester: Wiley.

Landsberg, J.J., and R.H. Waring. 1997. A generalised model of forest productivity using simplified concepts of radiation-use efficiency, carbon balance and partitioning. *For. Ecol. Manag.* **95**:209–228.

Michaud, J.D., and W.J. Shuttleworth, guest eds. 1997. Aggregate Description of Land-Atmosphere Interactions. Spec. Iss., *J. Hydrol.* **190**:173–414.

Monteith, J.L., ed. 1975. Vegetation and the Atmosphere. 1. Principles. London: Academic.

Monteith, J.L., ed. 1976. Vegetation and the Atmosphere. 2. Case Studies. London: Academic.

O'Neill, R.V. 1988. Hierarchy theory and global change. In: Scales and Global Change, ed. R. Rosswall, G. Woodmansee, and P. Risser, pp. 29–45. New York: Wiley.

Parton, W.J., J.M.O. Scurlock, D.S. Ojima, et al. 1993. Observations and modelling of biomass and soils organic matter dynamics for the grassland biome worldwide. *Glob. Biogeochem. Cyc.* **7**:785–809.

Pielke, R.A., R. Avissar, M.R. Raupach, H. Dolman, X. Zeng, and S. Denning. 1998. Interactions between the atmosphere and terrestrial ecosystems: Influence on weather and climate. *Glob. Change Biol.* **4**:461–475.

Potter, C.S., J.T. Randerson, C.B. Field, et al. 1993. Terrestrial ecosystem production: A process model based on global satellite and surface data. *Glob. Biogeochem. Cyc.* **7**:811–841.

Potter, C.S., R.H. Riley, and S.A. Klooster. 1997. Simulation modelling of nitrogen trace gas emissions along an age gradient of tropical forest soils. *Ecol. Mod.* **97**:179–196.

Raupach, M.R. 1998. Radiative, physiological, aerodynamic and boundary-layer feedbacks on the terrestrial surface energy balance. *Glob. Change Biol.* **4**: 477–494.

Reynolds, J.F., D.W. Hilbert, and P.R. Kemp. 1993. Scaling ecophysiology from the plant to the ecosystem: A conceptual framework. In: Scaling Physiological Processes: Leaf to Globe, ed. J.R. Ehleringer and C.B. Field, pp. 127–140. San Diego: Academic.

Running, S.W., and E.R. Hunt. 1993. Generalization of a forest ecosystem process model for other biomes, BIOME–BGC, and an application for global-scale models. In: Scaling Physiological Processes: Leaf to Globe, ed. J.R. Ehleringer and C.B. Field, pp. 141–158. San Diego: Academic.

Sellers, P.J., L. Bounoua, G.J. Collatz, D.A. Randall, D.A. Dazlich, S.O. Los, J.A. Berry, I. Fung, C.J. Tucker, C.B. Field, and T.G. Jensen. 1996. Comparison of radiative and physiological effects of doubled atmospheric CO_2 on climate. *Science* **271**:1402–1406.

Simon, H.A. 1962. The architecture of complexity. *Proc. Am. Phil. Soc.* **106**:467–482.

Steffen, W.L., and O.T. Denmead, eds. 1988. Flow and Transport in the Natural Environment: Advances and Applications. Berlin: Springer.

Waldrop, M.M. 1992. Complexity: The Emerging Science at the Edge of Order and Chaos. London: Penguin.

Wu, J., and O.L Loucks. 1995. From balance of nature to hierarchical patch dynamics: A paradigm shift in ecology. *Qtly. Rev. Biol.* **70**:439–466.

10

The Role of Lateral Flow: Over- or Underrated

E.F. WOOD
Program in Environmental Engineering and Water Resources, Princeton University, Princeton, NJ 08544, U.S.A.

ABSTRACT

Hydrological research over the last forty years has demonstrated that downslope redistribution of soil water is an important hydrological mechanism, and contributes to spatial heterogenity of soil and vegetation observed across landscapes. The mechanisms that contribute to lateral flow, and which are reviewed in this chapter, include: (a) surface runoff production due to heavy precipitation and locally saturated areas, (b) macropores within the soil matrix, and (c) spatial variability of soil moisture. Recent results from model intercomparison studies show that, in general, hydrological models do a poor job in accurately predicting soil water drainage, lateral flows, and the prediction of baseflow versus surface runoff in total streamflow. Analyses provided in the chapter show the importance of accurately determining lateral flows and soil moisture to the estimation of evapotranspiration over heterogeneous landscapes, and therefore the long-term water balance.

INTRODUCTION

The convergence of lateral flow of water within a landscape results in the formation of streams and rivers. The erosional activity of these streams results in the evolution of the landscape and gives it its form (Horton 1945; Stahler 1957; Willgoose et al. 1992; Rodriguez-Iturbe and Rinaldo 1997). Within the fluvial geomorphology and hydrogeomorphology literature there is a well-recognized link between landscape form and climate. Even though it was outside the scope of this workshop, these links must serve as a backdrop to a discussion on lateral flows in complex landscapes and the reader must be aware of the larger perspective within which catchment runoff and lateral flow are embedded.

What is the importance of lateral flows in complex landscapes? Through a number of field studies during the 1960s, many of which are reviewed in Kirkby (1978) and in Anderson and Burt (1990), there was an increased recognition that subsurface flows

Integrating Hydrology, Ecosystem Dynamics, and Biogeochemistry in Complex Landscapes
Edited by J.D. Tenhunen and P. Kabat

play a significant role in the spatial variation of soil moisture within a catchment. During inter-storm periods, the subsurface redistribution of soil moisture with the soil properties sets up spatially variable initial conditions for infiltration (during rain events) and evaporation (between events). In addition, this field work found that for areas of relatively shallow soils, the dominant control on the subsurface flows is topography. Thus, the potential exists that lateral flows and the subsequent spatial variability in soil moisture can play a critical role in the water and energy fluxes at landscape scales.

In this chapter I will try to present the underlying hydrologic processes of water redistribution within complex terrain, the linkage to the spatial distribution of soil moisture, and subsequently the spatial water and energy balance dynamics within catchments; approaches to modeling lateral flows and subgrid soil moisture variability within catchments; and an assessment of the importance of these processes to soil–vegetation–atmospheric transfer (SVAT) modeling of surface water and energy fluxes. Finally, I will try to assess the complexity required by models to properly account for lateral flow and its effects.

SURFACE RUNOFF AND SUBSURFACE REDISTRIBUTION OF WATER

This section discusses the pathways for water at the surface and the redistribution of water below the surface soil zone. The principles of infiltration and vertical movement of moisture in the unsaturated zone are assumed to be familiar to the reader. If not, the reader is referred to Philip (1969) or Hillel (1980). In general, the hydrologic aspects of lateral flow and its role in catchment response to precipitation events has often been referred to as "runoff generation" in the hydrology literature, and there are a number of excellent reviews (e.g., Freeze 1974) and books which focus on the problem (see Kirkby 1978; Anderson and Burt 1990). In this section, the various pathways for subsurface water flow and runoff generation are briefly reviewed.

The problem appears deceptively simple: for any particular rainfall event, determine the portion of incident rainfall which reaches the stream to become discharge during, or immediately after, a rainfall event. This surface runoff is measured as the storm hydrograph. The residual rainwater is (often) assumed to infiltrate into the soil, redistribute either vertically or horizontally (down the hillslope), and either return to the atmosphere through evapotranspiration or contribute to streamflow over an extended period of time (base flow discharge.) It is important to be able to estimate the "effective rainfall," which is the rainfall volume equal to the volume of the stream hydrograph, and to know the time transformation between the effective rainfall and the stream hydrograph which is controlled by catchment characteristics that govern its retardation and diffusive effects. For the modeling of the land surface hydrology within land-atmospheric coupled systems, it has been shown from intercomparisons carried out within the Project for the Intercomparison of Land Parameterization Schemes

(PILPS) that modeling runoff volumes correctly is critical to accurate modeling of latent and sensible heat fluxes (e.g., Chen et al. 1997; Wood et al. 1997).

Runoff Mechanisms

There is a broad spectrum of processes that are involved in the generation of runoff and streamflow and in the redistribution of infiltrated water. Figure 10.1, from Beven (1986), presents the major mechanisms for runoff production. As discussed by Beven (1986), Sklash (1990), Wood et al. (1990) among others, the response of any particular catchment may be dominated by a single mechanism or by a combination, depending on the magnitude of the rainfall event, the antecedent soil moisture conditions of the catchment, and/or the heterogeneity in soil hydraulic properties. Thus, during any particular storm, different mechanisms may generate runoff from different parts of a catchment. Surface runoff from these (partial) contributing areas may be generated by either the *infiltration excess* mechanism (Figure 10.1a, b) where the rainfall rate exceeds the infiltration capacity of the soil, or from rainfall on areas of soil saturated by a rising water table even in high permeability soils. Such *saturation excess overland flow*, is represented as mechanisms in Figure 10.1c, d.

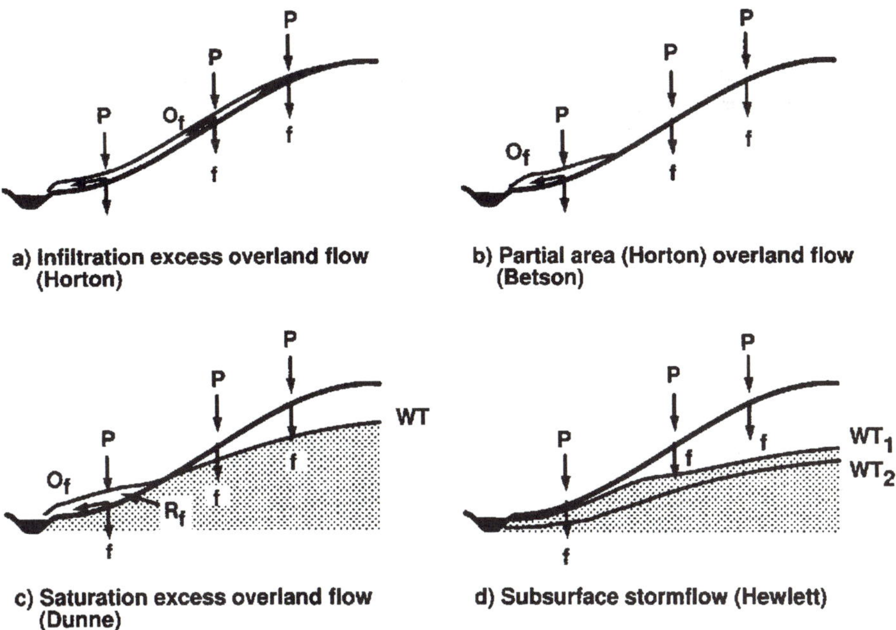

Figure 10.1 Mechanisms of runoff production (from Beven 1986).

The resulting hydrograph is often divided into a number of components depending on the pathway and time delay from the time the rain drop hits the ground surface to the time it enters the stream. These components are usually divided into the direct surface runoff, subsurface (or interflow) discharge, and groundwater baseflow discharge. Direct surface runoff results from rain falling on saturated surfaces (either onto partially saturated areas due to rain rates greater than the local infiltration rate or onto near-stream saturated areas resulting from emerging water tables). These saturated contributing areas expand and contract during and between rainfall events. Subsurface discharge from the unsaturated soil zone (often referred to as interflow) often arises from either flow through macropores in the upper soil (see section below on **Macropores and Redistribution of Soil Water**) or through downslope flow through the unsaturated soil matrix, often enhanced due to vertical anisotropic soil properties — especially the occurrence of an impeding soil layer to vertical infiltrating water (McCord et al. 1991).

It is also known, as a result of using natural tracers to determine the source of streamflow, that in many catchments a significant portion of the storm hydrograph is derived from subsurface water that is displaced from soil and groundwater by the incoming infiltrated rain water (e.g., Sklash et al. 1976; Stewart and McDonnell 1991; Yoshida et al. 1995; for a review of such tracer studies, see Sklash 1990). It is clear that the modeling of runoff needs to deal with the spatial patterns of antecedent moisture conditions, soil hydraulic properties, and rainfall intensities, with the expectations that under many conditions these patterns will be highly variable in space and time.

Macropores and Redistribution of Soil Water

Beven and Germann (1982) and Germann (1990) provide extensive reviews of macropores and their important role in subsurface water flow. Figures 10.2 and 10.3, taken from Beven and Germann (1982), provide a framework for understanding the role of macropores in infiltration and redistribution of water. Field studies have grouped macropores into four categories: pores formed by (a) soil fauna (earthworms, gophers, moles, etc.), (b) plant roots, (c) soil cracks and fissures, and (d) natural soil pipes. Land use has a significant role in the formation and changes in soil macropores, and in many catchments, it is widely accepted that the magnitude and shape of the storm hydrograph is controlled by subsurface flows due to macropore flow (Beven and Germann 1982). This suggests that macropores enhance both the vertical infiltration of water and the lateral downslope redistribution.

The results of the above field work can be summarized as follows: *Substantial hydrologic field work over the last 40 years has confirmed that downslope lateral flow is an important hydrologic process at hillslope-to-catchment scales.*

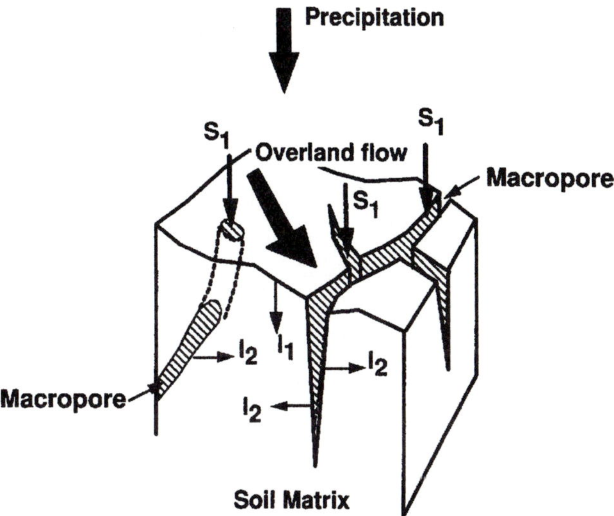

Figure 10.2 Schematic of water flows into soil with macropores: P, precipitation; I_1, infiltration into the matrix from the surface; I_2, infiltration into the matrix from the walls of the macropores; S_1, seepage into the macropores at the soil surface; S_2, flow within the macropores; O, overland flow (from Beven and Germann 1982).

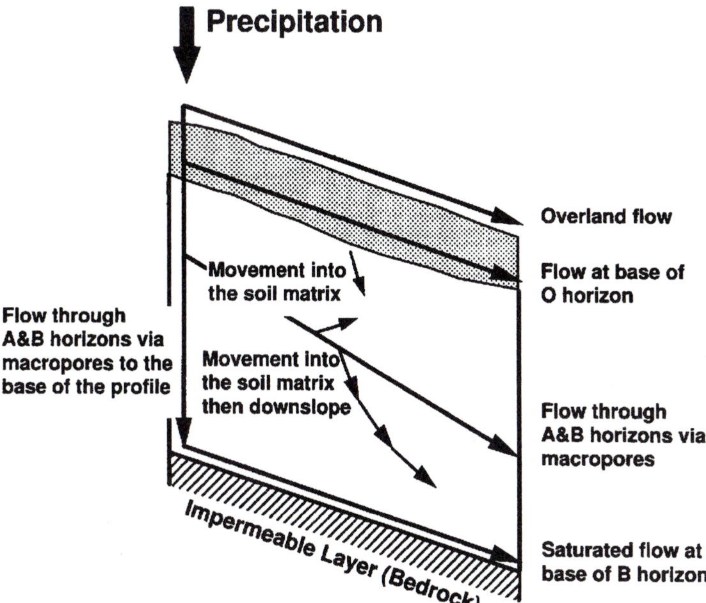

Figure 10.3 Flow paths for the movement of water through shallow soil with macropores (from Beven and German 1982).

EFFECTS OF LATERAL FLOW PROCESSES WITHIN WATERSHEDS

Observed Spatial Variability

During the last fifteen years, there has been an increasing awareness of the problems posed by the natural spatial heterogeneity found in catchment characteristics and meteorological inputs on runoff generation, soil moisture, and evapotranspiration. Some results for specific models are discussed in Wood et al. (1988), Beven et al. (1988), Goodrich (1990), and Grayson et al. (1992); reviews include Beven (1983b) and Wood (1997). Variability caused by precipitation variability (including partial precipitation over a catchment), topography, different soil, and vegetation types is readily apparent in spatial and temporal patterns in soil moisture, runoff, and evaporation. Even variations within seemingly homogeneous units (soil or vegetation) can also cause significant variations in soil moisture and its redistribution as well as the resulting fluxes of water (runoff and evaporation.)

To appreciate the natural variability in runoff that can occur, consider the field results observed by Hjelmfelt and Burwell (1984). Measurements of the runoff from forty 0.01 ha plots (adjacent to each other) were obtained over one season with storm totals ranging from 6–96 mm. The coefficient of variation for storm runoff volumes across the forty plots ranges from 0.071 to 1.09, showing dramatic variations in runoff over a very small area. In addition, there was considerable temporal scatter in the runoff ratios for individual plots. Thus, even in the smallest experimental plots, significant variability exists.

Variation in Soil Moisture

As discussed above, surface soil moisture plays a significant role in the infiltration of rainwater and the subsequent subsurface redistribution. There has been a significant number of investigations into how infiltration rates, hydraulic conductivity, porosity, soil moisture contents, and other soil characteristics vary in space (e.g., Nielsen et al. 1973; Bell et al. 1980; Viera et al. 1981). Beven (1983b) provides a review of this research and how it influences runoff generation. Figure 10.4, from Bell et al. (1980), shows the relationship between measured surface soil moisture within agricultural fields to its coefficient of variation. The data shows that the variance in soil moisture is fairly constant across a wide range of soil moisture, implying that this may be due to the underlying soil properties. It must be noted that this was for homogeneous agricultural fields and larger variability would be expected for natural catchments. In addition, it would be expected that in natural catchments the variability in soil moisture would be more structured due to two important reason: (a) the spatial structure of soils, due to their formation and erosion/deposition and (b) the downslope redistribution of soil moisture due to processes described above (see section on **Runoff Mechanisms**). The first reason results in thinner, often coarser textured, soils on the upper portions of

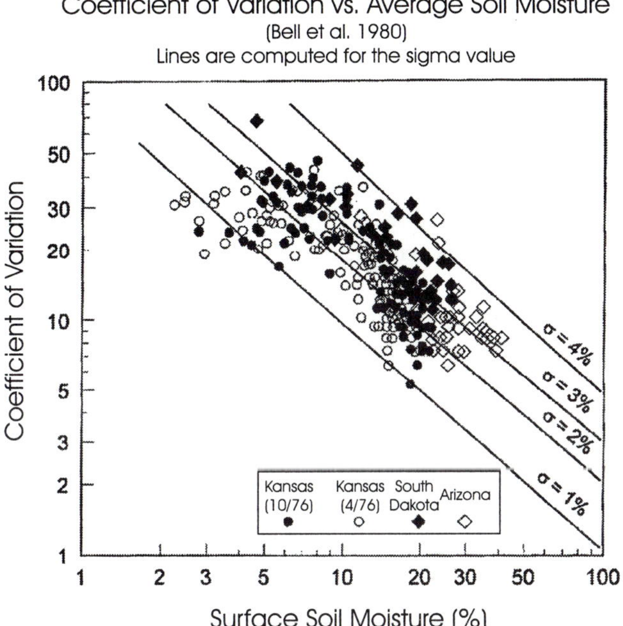

Figure 10.4 Variations in observed soil moisture collected from agricultural fields approximately 66 ha in size.

hillslopes and deeper, finer textured soils in valley bottoms. The second reason leads to higher water tables and wetter soils in valley bottoms. The end result is large spatial variability in soil moisture, with the variability quite "structured." Together, this leads to structured variability in natural vegetation and near-stream wetlands.

In summary, it can be stated that *in areas of complex topography, significant variability in soil moisture can be expected, and this variability can affect the distribution of natural vegetation, runoff generation, and hydrologically sensitive areas, resulting in significant variability in the spatial partitioning of the terrestrial water and energy fluxes (runoff, infiltration, evaporation sensible heat fluxes).*

MODELING APPROACHES

The primitive water balance equation is:

$$\Delta S = P - E - Q, \tag{10.1}$$

where ΔS is the change in soil moisture over a specified time interval; P is precipitation; E represents evapotranspiration which is the sum of evaporation from bare soil, E_S, and transpiration from vegetation, E_V; and Q is runoff which is the sum of surface or direct storm, runoff, and subsurface or base flow. The corresponding energy balance equation is

$$R_n = \lambda E + H + G \tag{10.2}$$

where R_n is net surface radiation (solar and longwave), E is the latent heat, H the sensible heat, and G the ground heat flux.

Equations 10.1 and 10.2 are not directly usable for determining the terrestrial water and energy fluxes and states; the terms (ΔS, P, E, Q, H, and G) must be parameterized in terms of hydrometeorological state variables, including the soil moisture profile, the surface, ground and near-surface air temperatures, and near-surface humidity. SVAT models can provide the necessary parameterizations.

Through the parameterization of Equations 10.1 and 10.2, the water and energy balance models (i.e., the SVAT scheme) can either explicitly account for the spatial variability in hydrological inputs, processes, and parameters (which is referred to as a spatially "distributed" model) or represent the catchment (or control volume) as being spatially homogeneous with regard to inputs, processes, and parameters — usually through averaging (which is referred to as a spatially "lumped" model.) For distributed models, spatial variability can be represented either deterministically, in which case the actual patterns of variability are used, or statistically, in which case the patterns of variability are represented statistically. Most workshop participants are familiar with the types of SVAT models used in PILPS (Henderson-Sellers and Brown 1992; Henderson-Sellers et al. 1995; Chen et al. 1997). These models are parameterized using point bio-meteorological equations applied at large grid scales, resulting in a "conceptual" representation of the terrestrial water and energy balance. Only a few models have a representation of subgrid spatial variability in soils and/or vegetation and only one or two models considering subgrid topography, lateral flows, and variability in soil moisture. As opposed to these conceptual models, there are models that numerically solve the three-dimensional Richards' equation (Richards 1931), common in groundwater modeling, which are often regarded as being the most "physically based" representation of water movement through soil. These models will be reviewed with a view towards understanding how lateral flows can be represented.

Three-dimensional Distributed Models of Surface Water Redistribution

The three-dimensional Richards' equation is often expressed as:

$$S(\psi)\frac{\partial \psi}{\partial t} = \nabla \bullet \left[K_s K_r(\psi)\nabla(\psi + z)\right], \tag{10.3}$$

where ψ, the pressure head, is the dependent variable; $S(\psi)$ is the specific moisture capacity; t, time, and z, the vertical coordinate (positive upwards). The hydraulic conductivity K is expressed as a function the conductivity at saturation, K_s, and the relative conductivity, K_r. The specific moisture capacity, hydraulic conductivity and, thus, the volumetric soil moisture, θ, which are nonlinear functions of the pressure head, are usually parameterized using relationships such as those developed in van Genuchten and Nielsen (1985) among others.

Richards' equation must be solved through numerical methods applied to finite difference or finite element approximations (e.g., see Paniconi and Wood 1993), with the set of numerical equations augmented by appropriate boundary conditions (e.g., boundary moisture fluxes of infiltrated or evaporated water.) These models are usually run at a fine grid resolution (on the order of meters) and high time resolution (on the order of tens of seconds per time step). The theoretical basis of these models and the detailed model domain has not resulted in model predictions that well match observations of moisture fluxes in the unsaturated zone. It is my belief that the poor match between model predictions and point observations is related to two causes. The first cause is the inability to measure and specify soil properties at the model grid resolution where the highly nonlinear (with respect to soil properties) unsaturated flow equations must be solved. The second cause arises from not including important, related processes like flow through soil macropores (see section on **Macropores and Redistribution of Soil Water**) and surface run-on, both of which are important at these model grid scales. Since these models rely on local gradients, the small scale variability in soil properties (on the order of one to tens of meters) can have a significant effect on model predictions.

There has been more success in utilizing these models for saturated flows, and this in fact forms the basis of current groundwater models. For saturated conditions, Richards' equation becomes a linear equation which facilitates its solution. Additionally, for many saturated flow problems, the boundary conditions are specified through pressure (head) along the boundary. While these models have appeared to give good matches for groundwater flows, when coupled to a transport equation (that predicts the movement of a constituent in the groundwater), these predictions tend to be poorer. This seems to indicate that even for saturated conditions, model-predicted local flow velocities may be in error, but that variables that are more spatially uniform (like pressure fields) are predicted better.

Conceptual SVAT Models and Water Redistribution

As stated earlier, hydrologists would classify the SVAT models used in climate models as having a "conceptual" representation of the terrestrial water balance. Only a few models consider subgrid spatial variability in soils and/or vegetation and only one or two models considering subgrid topography, lateral flows, and variability in soil moisture. The intercomparison of land surface parameterizations in climate models being conducted as part of PILPS has provided important insights in linkages between the terrestrial water and the energy balances, and in the differences among schemes in representing subsurface water drainage (Chen et al. 1997; Lettenmaier et al. 1996; Wood et al. 1998). The most recent PILPS experiment (Phase 2c) was carried out by 16 SVAT

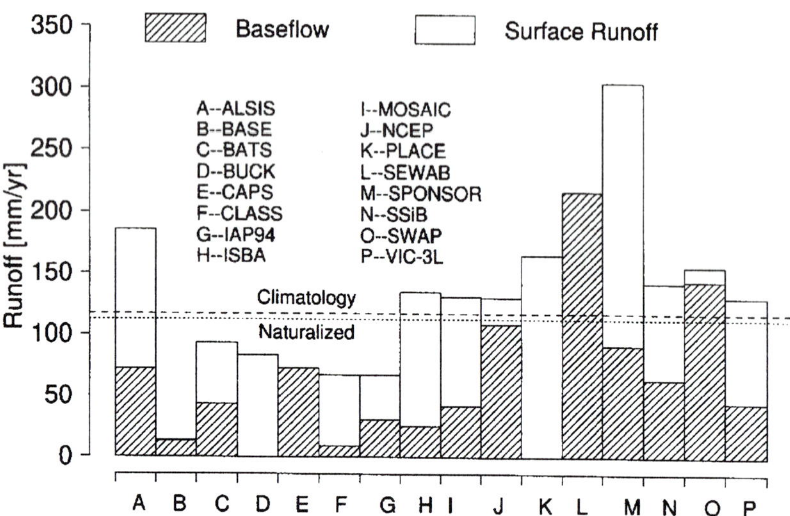

Figure 10.5 Runoff divided between surface and subsurface (baseflow) from the PILPS Phase 2c experiment (from Lohmann et al. 1998).

schemes using 10 years of hourly data from the 567,000 km^2. Arkansas–Red River Basins, divided up into 61 1° grids. Figure 10.5 shows the wide variability among schemes in representing runoff as either "surface" or "base" flow — the latter being subsurface water drainage which in these models drain through the base of the grid column but could in many aspects represent lateral flow. Getting the total volume correct is only part of the problem, the timing of the runoff is also critical to matching model predicted river discharge to observations. For those schemes with a dominance of subsurface runoff (e.g., CAPS, NCEP, SEWAB, and SWAP) the hydrograph timing will be delayed, resulting in slower recessions and flood peaks that are too small. This is illustrated in Figure 10.6 using results for two SVAT models (BATS and NCEP) from the PILPS Phase 2c Arkansas–Red River Basins experiment (Lohmann et al. 1998). The implications of this result is that model predictions of riparian wetlands, hydrologically sensitive areas and near-stream moisture conditions will be in error.

Another important aspect from the PILPS intercomparisons studies is the observation that errors in modeled runoff are reflected in compensating errors in evapotranspiration (latent heat) and thus sensible heat. This can be seen in Figure 10.7, again from the PILPS Phase2c experiment. The figure shows the basin-wide, average annual runoff and evaporation for the 16 participating models as well as observed values. It is now clear that ignoring one aspect of the water or energy cycle results in errors to other components. It is the contention of this paper that resolving the volume of runoff and the division between surface and subsurface flows is necessary for understanding the water and energy fluxes in complex landscapes.

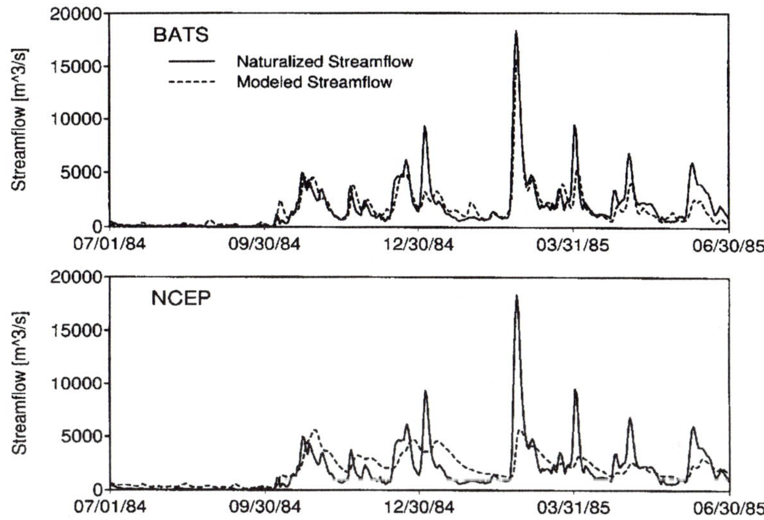

Figure 10.6 Comparison of modeled and observed daily streamflow for one year from the Arkansas–Red Basins for BATS and NCEP models. Results are part of the PILPS Phase 2c (from Lohmann et al. 1998).

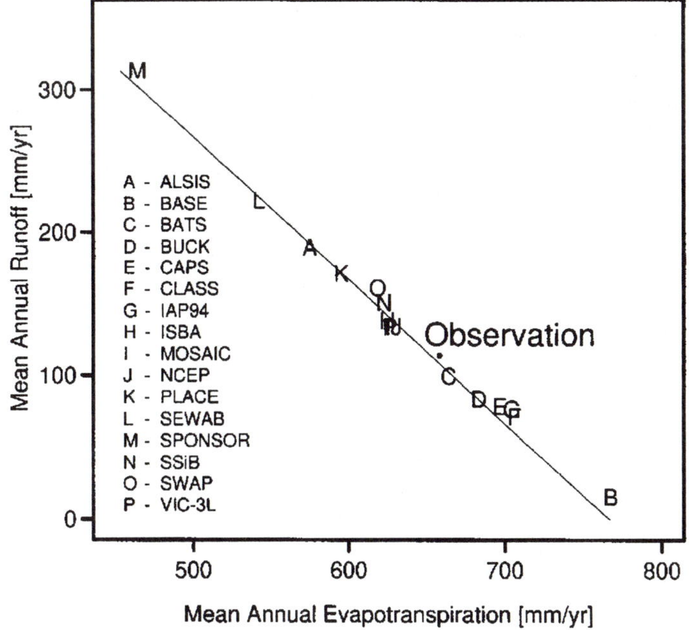

Figure 10.7 Mean annual runoff and evaporation for the Arkansas–Red Basins (from Wood et al. 1998).

This suggests that there has been an "over-focus" during recent climate field programs (like the First ISLSCP Field Experiment, FIFE; HAPEX–MOBILY, HAPEX–Sahel; and the Boreal Ecosystem Atmospheric Study, BOREAS) on vegetation energy exchanges and an "under-focus" on the terrestrial water budget which has resulted in collected data sets that are unable to resolve critical science questions related to the hydrology within complex landscapes — like the role of lateral flow and subgrid soil moisture. Further studies with a SVAT model have shown the importance of subgrid variability in precipitation and resulting subgrid soil moisture (see Liang et al. 1996). There are a wide number of SVAT models of varying complexity that do not consider spatial variability; the errors resulting from this approach are still unresolved.

There are still a wide number of models that cannot consider variability in their response at scales smaller than the catchment or grid scales. In the next section an example of a "distributed" model is presented that allows a representation of the variability in inputs (radiation and precipitation), topography, soils, and vegetation.

Topographically Based Distributed Models of Subsurface Redistribution

Parameterizations of Equation 10.2, the water balance equation, that incorporate explicitly variations in topography have been developed. One example is the parameterization developed by Beven and Kirkby (1979), TOPMODEL, which uses a spatially variable topographic index to predict the extent of saturated areas which generate surface runoff (see Figure 10.1, panel c). The model utilizes a hydrologic similarity theory in which locations within a catchment with the same topographic index respond in a similar manner to the same inputs. O'Loughlin (1981, 1986) and Moore et al. (1988) have also developed a topographic-based approach for predicting areas of saturated soils while Grayson (1990) and Grayson et al. (1992) have developed a terrain-based hydrologic model for erosion studies. With widely available digital elevation models (DEM) for topography and digital soil data bases, it is relatively straightforward to calculate the topographic index for an area.

The original concepts of TOPMODEL have been extended to consider spatial soil properties (Beven 1983a), infiltration excess mechanisms for runoff production in addition to saturation excess (Sivapalan et al. 1987; Wood et al. 1990), distributed inputs of water and energy (Famiglietti and Wood 1994; Peters-Lidard et al. 1997), and an upper thin layer whose rapid soil moisture response is dominated by local soil and vegetation characteristics (Zion 1995; Peters-Lidard et al. 1997), and a generalized theory of downslope flows that relaxes the exponential relationship and reliance on local topographic slopes (Ambroise et al. 1996). In this way the model has evolved from a rainfall–runoff (water balance) model to a land surface SVAT scheme that could be used in climate models.

One measure of the influence of topography on water and energy fluxes is to study the spatial structure of evaporation. Figure 10.8 presents the model-derived seasonally averaged evaporation over the Little Washita (OK) catchment, a 525 km^2 watershed in the southern Great Plains region of the United States. In the upper panel is the average

Figure 10.8 Model-derived seasonal evaporation over the 525 km^2. Little Washita (OK) catchment. The upper panel is summer (JJA) and the lower panel is winter (DJF). Darker areas have higher evaporation rates. Dark areas in the winter are wet areas along the river channel, farm ponds and lakes. The models was run at a 30 m resolution for 10 years.

summer season (JJA) evaporation, where the higher evaporation in the valley bottoms is clearly shown. The lower panel shows the average evaporation for the winter season (DJF). Except for a few perennial wet areas along the stream, the evaporation is quite uniform across the catchment. This is due to a combination of relatively uniform soil moisture in the upper zone during the winter and lack of vegetation transpiration, especially along the streams where the vegetation is dominated by brush and deciduous trees. From the summer results in Figure 10.8, there is a clear influence of topography and lateral flows on downslope and near-stream soil moisture and the resulting evapotranspiration.

From these modeling studies, it can be shown that lateral flows can lead to increased soil moisture and evaporation at the base of hillslopes and near-riparian areas. These modeling results are confirmed through observations of the spatial patterns of vegetation in semi-arid regions — larger water-demanding vegetation is usually found at the bottom of hillslopes and near streams. This vegetation is fed by lateral flows and has higher transpiration rates than the grasses usually found away from the streams. These model-derived evaporation patterns are similar to those reported by Holwill and

Stewart (1992) for the FIFE site based on aircraft and ground-based data (see Plates 3, 4, 7, and 8 of Holwill and Stewart 1992). Smith et al. (1992) also observed during FIFE that the valley bottom flux station (site 2) had higher latent heating over the experiment period than site 38, which was situated on a ridge top. They attribute the differences to soil moisture gradients between the valley bottom and the ridge tops. When a TOPMODEL-based SVAT was applied to FIFE, there was good agreement between site 2 and its modeled pixel, both having higher evaporation than the catchment average evaporation (see Famiglietti et al. 1992). These results, while limited, confirm the importance of lateral flows and spatial variability in soil moisture on surface water and energy balances.

ASSESSING THE IMPORTANCE OF LATERAL FLOWS AND SOIL MOISTURE VARIABILITY ON EVAPOTRANSPIRATION

To date, the necessary field experiments have not been carried out to determine definitively the importance of lateral flows and the resulting soil moisture variability on water and energy fluxes across complex landscapes. As noted above, there is ample evidence that vegetation in moderately stressed environments arranges itself along streams, which themselves are fed by lateral flows.

Wood (1997) developed an analytical model for the evaporative fraction (actual divided by potential evapotranspiration) to gain insight into the effect of spatially aggregating soil moisture. The effect of aggregation is captured in an "evaporation scaling ratio," which is the ratio of the average evaporative fraction considering soil moisture variability to the average evaporative fraction using aggregated soil moisture. Sensitivity of the evaporation scaling ratio depends on three parameters:

1. the ratio of the equilibrium soil moisture to the mean soil moisture, where the equilibrium soil moisture is defined as the soil moisture level at which the atmospheric evaporative potential and the ability of the soil-vegetation system to satisfy this potential are in equilibrium (i.e., a land–atmosphere equilibrium with regard to soil moisture, which will vary temporally),
2. the soil-vegetation sensitivity factor to soil moisture stress, which varies for different soils and vegetation types, and
3. the variability of soil moisture within a region, as measured by its coefficient of variation.

A series of sensitivity experiments was conducted by varying the soil-vegetation sensitivity factor and the soil-moisture variability. Figure 10.9 shows an example result. The analysis suggests that aggregating soil moisture levels results in evapotranspiration calculations that overestimate actual levels during periods of low atmospheric demand (early morning, late afternoon, winter periods, etc.) and underestimate evapotranspiration during periods of high demand (midday, summer periods, etc.).

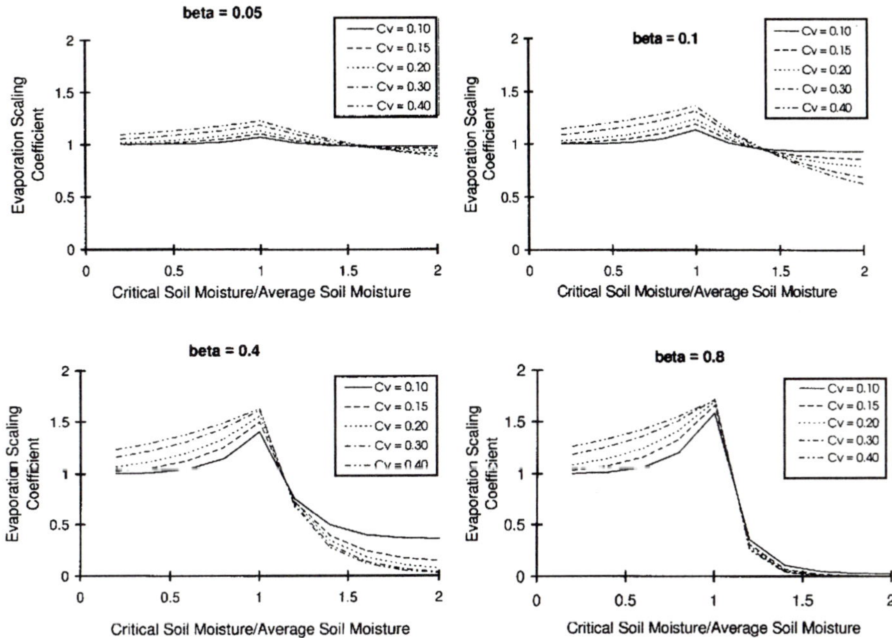

Figure 10.9 Evaporation scaling coefficient versus the ratio of critical soil moisture to areally averaged soil moisture for a soil moisture coefficient of variation C_V ranging from 0.10 to 0.40: (a) $\beta = 0.50$; (b) $\beta = 0.10$; (c) $\beta = 0.40$; (d) $\beta = 0.80$. The variable β refers to the sensitivity of the soil-vegetation system to soil moisture, with higher values being more sensitive; the "critical" soil moisture refers to the soil moisture level where the soil-vegetation system decreases evpotranspiration relative to the potential level (from Wood 1997).

From these modeling studies, it can be concluded that *lateral flows and soil moisture variability are important in complex landscapes which have significant soil moisture variability and a soil-vegetation system that is sensitive to low moisture levels.*

SUMMARY

To address the problem of the importance of lateral flows within complex landscapes it is important that the mechanisms of runoff generation within these landscapes be understood and field work observing and measuring these flows be reviewed. There is a long history of both observing these processes and modeling runoff from catchments within the hydrology community. For the Dahlem Workshop participants, the most important aspect of this work is the recognition in the hydrologic community that subsurface lateral flow plays a significant role in the spatial variation of soil moisture within a catchment, and that during inter-storm periods the subsurface redistribution of soil moisture, in conjunction with (spatially varying) soil properties, sets up spatially variable initial conditions for infiltration (during rain events) and evaporation (between events).

Currently there is debate on the degree that these recognizable physical processes can be or should be represented explicitly. It is clear that the performance of SVAT models in calculating the correct runoff volumes and timing directly affects the determination of all other water and energy flux and storage terms, i.e. the water and energy cycles are tightly coupled. This conclusion is clear from the PILPS intercomparison experiments (see Chen et al. 1997; Lettenmaier et al. 1996; Wood et al. 1997). It is less apparently obvious — but it appears to be supported by studies — that models which explicitly consider subgrid variability in its inputs and catchment (landscape) characteristics (soil, vegetation, topography) can better represent water and energy fluxes, especially at small temporal scales (Wood 1997; Sellers et al. 1995; Liang et al. 1996) and large spatial scales (Lohmann et al. 1997).

What is the minimum level of needed model complexity in the area of lateral flows? I would argue that a minimum level of model complexity is to capture statistically the subgrid spatial and temporal dynamics of soil moisture — the first-order effects of lateral flow. This allows the correct computation of runoff volumes and therefore some hope of an accurate water balance and subsequently an accurate energy balance over the long term. During periods between rain events, if the soil moisture is high or the atmospheric evaporative demand is low (early morning, late afternoon, winter), then there appears little need for accounting for spatial variability in soil moisture. This is false thinking since controls on the actual evaporation can switch between the atmosphere and the soil-vegetation system during the day, and certainly over days and seasons.

Many of the conclusions from recent climate experiments (e.g., HAPEX-MOBILY and FIFE'87) regarding scaling of land-surface processes and the need for measuring and modeling small scale hydrologic processes and subgrid soil moisture variability are simply wrong. The published conclusions in this area have been inferred from collected data that ranged from a single "golden" day to short "intensive field campaigns" which may often have atypical conditions, or are based on data collection plans that were not designed (or capable) of resolving science questions involving scaling and small scale hydrologic processes. Thus, the role and importance of processes like lateral flows and spatial variability in soil moisture are underrated.

REFERENCES

Ambroise, B., K.J. Beven, and J. Freer. 1996. Towards a generalisation of the TOPMODEL concepts: Topographic indices of hydrological similarity. *Water Res.* **32**:2135–2145.

Anderson, M.G., and T.P. Burt, eds. 1990. Process Studies in Hillslope Hydrology. Chichester: Wiley.

Bell, K.R., B.J. Blanchard, T.J. Schmugge, and M.W. Witczak. 1980. Analysis of surface moisture variations within large-field sites. *Water Res.* **16(4)**:796–810.

Beven, K.J. 1983a. Introducing spatial variability into TOPMODEL: Theory and preliminary results. Technical Report, Dept. Env. Sci., Univ. of Virginia, Charlottesville.

Beven, K.J. 1983b. Surface water hydrology — Runoff generation and basin structure. In: Contributions in Hydrology, U.S. National Report 1979–1982 to IUGG. *Rev. Geophy. Sp. Phy.* **21(3)**:721–729.

Beven, K.J. 1986. Runoff production and flood frequency in catchments of order *n*: An alternative approach. In: Scale Problems in Hydrology, ed. V.K. Gupta, I. Rodriguez-Iturbe, and E.F. Wood, pp. 107–131. Dordrecht: Reidel.

Beven, K.J., and P. Germann. 1982. Macropores and water flow in soils. *Water Res.* **18(5)**:1311–1325.

Beven, K.J., and M.J. Kirkby. 1979. A physically-based variable contributing area model of basin hydrology. *Hydrol. Sci. J.* **24(1)**:43–69.

Beven, K.J., E.F. Wood, and M. Sivapalan. 1988. On hydrological heterogeneity — Catchment morphology and catchment response. *J. Hydrol.* **100**:353–375.

Chen, T.H., et al. 1997. Cabauw experimental results from the Project for Intercomparison of Land-Surface Parameterization Schemes. *J. Clim.* **10(6)**:1194–1215.

Famiglietti, J.S., and E.F. Wood. 1994. Multi-scale modeling of spatially-variable water and energy balance processes. *Water Res.* **30(11)**:3061–3078.

Famiglietti, J.S., E.F. Wood, M. Sivapalan, and D.J. Thongs. 1992. A catchment scale water balance model for FIFE. *J. Geophys. Res.* **97(D17)**:18,997–19,008.

Freeze, R.A. 1974. Streamflow generation. *Rev. Geophys.* **12(4)**:627–647.

Germann, P.F. 1990. Macropores and hydrologic hillslope processes. In: Process Studies in Hillslope Hydrology, ed. M.G. Anderson and T.P. Burt, pp. 327–364. Chichester: Wiley.

Goodrich, D.C. 1990. Geometric simplification of a distributed rainfall runoff model over a range of basin scales. Ph.D. Diss., Dept. of Hydrology and Water Resources, Univ. of Arizona, Tucson, AZ.

Grayson, R.B. 1990. A terrain-based hydrologic model for erosion studies. Ph.D. Diss., Univ. of Melbourne, Parkville, Victoria, Australia.

Grayson, R.B., I.D. Moore, and T.A. McMahon. 1992. Physically based hydrologic modeling. 1. A terrain based model for investigative purposes. *Water Res.* **28(10)**:2639–2658.

Henderson-Sellers, A., and V.B. Brown. 1992. Project for the intercomparison of land surface parameterization schemes (PILPS): First science plan. *GEWEX Tech. Note*, IGPO Publ. Series No. 5, 53 pp.

Henderson-Sellers, A., A.J. Pitman, P.K. Love, P. Irannejad, and T.H. Chen. 1995. The project for the intercomparison of land surface parameterization schemes (PILPS): Phases 2&3. *Bull. Am. Meteorol. Soc.* **76**:489–503.

Hillel, D. 1980. Fundamentals of Soil Physics. New York: Academic.

Hjelmfelt, A.T., and R.E. Burwell. 1984. Spatial variability of runoff. *J. Irrig. Drain. Eng.* **110(1)**:46–54.

Holwill, C.J., and J.B. Stewart. 1992. Spatial variability of evaporation derived from aircraft and ground-based data. *J. Geophys. Res.* **97(D17)**:18,673–18,680.

Horton, R.E. 1945. Erosional development of streams and their drainage basins: Hydrophysical approach to quantitative morphology. *Bull. Geol. Soc. Am.* **56**:275–370.

Kirkby, M.J., Ed. 1978. Hillslope Hydrology. Chichester: Wiley.

Lettenmaier, D.P., D. Lohmann, E.F. Wood, and X. Liang. 1996. PILPS–2c Draft Workshop Report: Report of a Workshop Held at Princeton University, Oct. 28–31. Internet http://earth.princeton.edu.

Liang, X., D.P. Lettenmaier, and E.F.Wood. 1996. A one-dimensional statistical-dynamic representation of subgrid spatial variability of precipitation in the two-layer VIC model. *J. Geophys. Res.* **101(D16)**:21,403–21,422.

Lohmann, D., et al. 1998. The Project for Intercomparison of Land-Surface Parameterization Schemes (PILPS) Phase-2c Red-Arkansas River Basin Experiment. 3. Spatial and temporal analysis of water balance fluxes. *J. Glob. Plan. Change* **19**:161–179.

McCord, J.T., D.B. Stephens, and J.L.Wilson. 1991. Hysteresis and state-dependent anistropy in modeling unsaturated hillslope hydrologic processes. *Water Res.* **27**:1501–1518.

Moore, I.D., E.M. O'Loughlin, and G.J. Burch. 1988. A contour-based topographic model for hydrological and ecological applications. *Earth Surf. Proc. Landforms* **13**:305–320.

Nielsen, D.R., J. Biggar, and K. Erh. 1973. Spatial variability of field measured soil-water properties. *Hilgardia* **42**:215–260.

O'Loughlin, E.M. 1981. Saturation regions in catchments and their relations to soil and topographic properties. *J. Hydrol.* **53**:229–246.

O'Loughlin, E.M. 1986. Prediction of surface saturation zones in natural catchments by topographic analysis. *Water Res.* **22**:794–804.

Paniconi, C., and E.F. Wood. 1993. A detailed model for simulation of catchment scale subsurface hydrologic processes. *Water Res.* **29(6)**:1601–1620.

Peters-Lidard, C., M. Zion, and E.F. Wood. 1997. A soil-vegetation-atmosphere transfer scheme for modeling spatially variable water and energy balance processes. *J. Geophys. Res.* **102(D2)**:4303–4324.

Philip, J.R. 1969. Theory of infiltration. *Adv. Hydrosci.* **5**:215–290.

Richards, L.A. 1931. Capillary conduction of liquids through porous mediums. *Physics* **1**:318–333.

Rodriguez-Iturbe, I., and A. Rinaldo. 1997. Fractal River Basins: Chance and Self-organization. Cambridge: Cambridge Univ. Press.

Sellers, P.J., et al. 1995. Effect of spatial variability in topography, vegetation cover and soil moisture on area-averaged surface fluxes: A case study using FIFE 1989 data. *J. Geophys. Res.* **100(D21)**:25,607–25,629.

Sivapalan, M., K.J. Beven, and E.F. Wood. 1987. On hydrological similarity. 2. A scaled model of storm runoff production. *Water Res.* **23**:2266–2278.

Sklash, M.G. 1990. Environmental isotope studies of storm and snowmelt runoff generation. In: Process Studies in Hillslope Hydrology, ed. M.G. Anderson and T.P. Burt, pp. 401–435. Chichester: Wiley.

Sklash, M.G., R.N. Favolden, and P. Fritz. 1976. A conceptual model of watershed response to rainfall, developed through the use of oxygen-18 as a natural tracer. *Can. J. Earth Sci.* **13**:271–283.

Smith, E.A., W.L. Crosson, and B.D. Tanner. 1992. Estimation of surface heat and moisture fluxes over a prairie grassland. 1. *In situ* energy budget measurements incorporating a cooled mirror dew point hygrometer. *J. Geophys. Res.* **97(D17)**:18,557–18,582.

Stahler, A.N. 1957. Quantitative analysis of watershed geomorphology. *EOS Trans., Am. Geophys. Union* **38**:921–920.

Stewart, M.K., and J.J. McDonnell. 1991. Modeling base flow soil water residence times from deuterium concentrations. *Water Res.* **27(10)**:2681–2695.

van Genuchten, M.T., and D.R. Nielson. 1985. On describing and predicting the hydraulic properties of unsaturated soils. *Ann. Geophys.* **3(5)**:615–628.

Viera, S.R., D.R.Nielsen, and J. Biggar. 1981. Spatial variability of field measured infiltration rates. *Soil Sci. Soc. Am. J.* **45**:1040–1048.

Willgoose, G.R., R.L. Bras, and I. Rodriguez-Iturbe. 1992. The relationship between catchment and hillslope properties: Implications of a catchment evolution model. *Geomorphology* **5**:21–38.

Wood, E.F. 1997. Effects of soil moisture aggregation on surface evaporative fluxes. *J. Hydrol.* **190**:397–412.

Wood, E.F., M. Sivapalan, and K. Beven. 1990. Similarity and scale in catchment storm response. *Rev. Geophys.* **28(1)**:1–18.

Wood, E.F., M. Sivapalan, K. Beven, and L. Band. 1988. Effects of spatial variability and scale with implications to hydrologic modeling. *J. Hydrol.* **102**:29–47.

Wood, E.F., et al. 1998. The Project for Intercomparison of Land-Surface Parametrization Schemes (PILPS) Phase-2(c) Red–Arkansas River Experiment. 1. Experiment description and summary intercomparisons. *J. Glob. Plan. Change* **19**:115–135.

Yoshida, H., M. Hashino, T. Tamura, and K. Muraoka. 1995. Formation process of streamwater chemistry in small forested mountrain basin. *J. Hydrosci. Hydr. Eng.* **13(2)**:83–97.

Zion, M.S. 1995. Use of operational satellite and radiosonde data to estimate radiation forcings for hydrologic models. Master of Science Thesis, Dept. Civil Eng. and Oper. Res., Princeton Univ., Princeton, NJ.

11

The Importance of Root Distributions for Hydrology, Biogeochemistry, and Ecosystem Functioning

R.B. JACKSON
Dept. of Botany, Duke University, Durham, NC 27708, U.S.A.

ABSTRACT

Rooting depth is a strong determinant of many hydrological and biogeochemical processes. This chapter examines the consequences of root distributions, particularly deep roots, for carbon and water fluxes. Although more than 90% of root biomass is usually in the top meter of soil, there are dozens of plant species that grow roots more than 10 m deep and at least half a dozen with roots below 50 m. The exchange of deep- and shallow-rooted species can alter the lateral movement of water, the depth of the water table, and conditions of salinization and waterlogging. In the Amazon, ignoring roots below 2 m would have underestimated evapotranspiration by up to 60%, and ignoring the soil below 1 m would have omitted a carbon pool larger than either the vegetation or the top meter of soil. In forests of the Carolina Piedmont, at least 8 m of the soil profile bears the biogeochemical imprint of biological activity. Correlational evidence there and in desert systems indicates that rooting depth may control the distribution of microbes and soil fauna to at least 12 m depth. To integrate the effects of deep soil processes (and to predict when the deep soil may be unimportant), global data bases of depth to bedrock, soil texture, water holding capacity, waterlogged areas, and maximum rooting depth would be useful for ecosystem and global models and for testing hypotheses. This chapter highlights numerous unanswered questions, including whether deep roots increase ecosystem productivity through hydraulic lift and enhanced deep percolation of water, and how rooting depth affects whether an ecosystem is relatively open or closed.

INTRODUCTION

Plants are important mediators of the transfer of energy and materials between the soil and the atmosphere (Vitousek et al. 1986; Field et al. 1995). They also play a key role in global hydrological cycles and in the biogeochemical transfer of nutrients and

Integrating Hydrology, Ecosystem Dynamics, and Biogeochemistry in Complex Landscapes
Edited by J.D. Tenhunen and P. Kabat

weathering of parent material. Plant resources in the environment are taken up primarily through roots and leaves; leaves capture carbon and energy, while roots take up water and nutrients. Despite this simplistic characterization, few plant processes are neatly partitioned into above- or belowground categories. Water is taken up by roots but is lost by leaves, and hormonal signaling between roots and shoots is common. Carbon and energy are taken up by leaves, but roots are a prominent, sometimes the prominent, sink for carbon acquired in terrestrial net primary productivity (e.g., Nadelhoffer and Raich 1992; Jackson et al. 1997).

Where roots are found in the soil is an important component of ecosystem functioning, as is the position of leaves. Leaf area distribution with canopy height is fairly well known for many ecosystems, though overcoming such uncertainties as variable phenology and leaf angles remains challenging. Estimates of root abundance and their distribution with depth in the soil are considerably poorer. In general, more than 90% of root biomass is in the top meter of soil (Jackson et al. 1996). Despite the majority of roots being in this upper layer, deep roots play an important role in many ecosystems (Stone and Kalisz 1991; Nepstad et al. 1994; Richter and Markewitz 1995; Canadell et al. 1996). Productivity in water-limited systems can be tightly linked to the availability of deep soil water and its uptake by roots (Rawitscher 1948; Lewis and Burgy 1964; Greenwood 1992). Rooting depth and deep-soil resources are important for determining the boundary between evergreen and deciduous vegetation (e.g., Neilson 1995). Rooting depth also interacts with plant nutrition to affect stomatal conductance, global transpiration, and regional hydrology (e.g., Schulze et al. 1994; Field et al. 1995).

This chapter addresses the importance of roots for ecosystem functioning, hydrology, and biogeochemistry in complex landscapes. It emphasizes the consequences of roots deep in the soil and their role in global carbon and water fluxes. The chapter begins with a discussion of root distributions and their relationship to soil carbon globally. It continues with an analysis of global patterns of maximum rooting depth, including which ecosystems and plant functional types have been shown to grow roots deep in the soil. The consequences of deep roots for carbon and water fluxes are then examined in three systems: the Amazon, the southeastern pine forests of the U.S., and the deserts of North America. A more general discussion follows, including how global data bases may be used a priori to predict regions and climatic regimes where deep roots should be important. The chapter ends with some unanswered questions (there are many) and with recommendations for future research.

DISTRIBUTION OF ROOTS AND ORGANIC CARBON IN SOIL

Global Root Distributions

The International Biological Programme (IBP), established by the International Council of Scientific Unions in 1964, strove to understand "the biological basis of productivity and human welfare." It made great strides in improving estimates of above- and belowground plant biomass and primary productivity (e.g., Lieth and Whittaker

1975). Despite these improvements, however, knowledge of root distributions remains relatively poor. Recent analyses made further progress for several systems, including grasslands and forests (Sun et al. 1997; Vogt et al. 1996; Cairns et al. 1997).

To examine root distributions globally and their relationships to environmental variables, Jackson et al. (1996, 1997) constructed a global data base of climate, soil, and root attributes with depth in the soil. Root data from more than 300 studies were collected and separated by terrestrial biome and plant functional type. The data were then fitted to an asymptotic model of vertical root distribution:

$$Y = 1 - \beta^{d} \tag{11.1}$$

where Y is the cumulative root fraction (a proportion between 0 and 1) from the soil surface to depth d (in cm) and β is the fitted extinction coefficient (Gale and Grigal 1987). β is the only parameter estimated in the model and provides a simple numerical index of rooting distributions. High values of β (e.g., 0.98) correspond to a greater proportion of roots at depth and low β values (e.g., 0.90) imply a greater proportion near the soil surface. The mathematical model is easily transformed to a negative exponential or other commonly used mathematical equations.

Global root distributions with depth showed clear patterns for terrestrial biomes (Figure 11.1). Tundra, boreal forests, and temperate grasslands had the shallowest root profiles on average, with β values of 0.913, 0.943, and 0.943, respectively (Figure 11.1; Table 11.1). Deserts, temperate coniferous forests, and savannas had some of the deepest root distributions ($0.970\ \beta \leq \beta \leq 0.980$). Tundra typically had 60% of roots in the upper 10 cm of soil while deserts had approximately 20% of roots in the same depth increment. Of the approximately 50 studies that sampled roots to at least a meter depth (those studies used to fit curves for each biome), only fifteen or so measured root distributions below 2 m (Jackson et al. 1996). In consequence, some of the estimated distributions may be biased to shallow depths.

The data base was also used to examine patterns of global root biomass and annual belowground net primary production (NPP) (Table 11.2). Average root biomass on an area basis ranged from 4–5 kg m^{-2} for most forest systems to < 1 kg m^{-2} for deserts and crops. When root biomass estimates were combined with the area covered by each biome, total root biomass was estimated as approximately 290×10^{15} g globally (or approximately 140×10^{15} g C). The global biomass of fine roots (< 2 mm diameter) was approximately 80×10^{15} g, containing carbon equivalent to 5% of the atmospheric pool. If one assumes conservatively that fine roots turn over on average once per year, then 20×10^{15} g C of net primary production cycles through fine roots annually, approximately one-third of total NPP for plants globally (Jackson et al. 1997).

Significant quantities of nitrogen and other important elements also cycle through fine roots annually. Based on more than 50 studies, the average C:N:P ratio of fine roots was 450:11:1 globally (Jackson et al. 1997; Gordon and Jackson unpublished). Coarser roots ($2 \leq x \leq 5$ mm), less likely to turn over each year, had a significantly higher C:N:P ratio of 850:11:1. Based on the assumption that retranslocation of

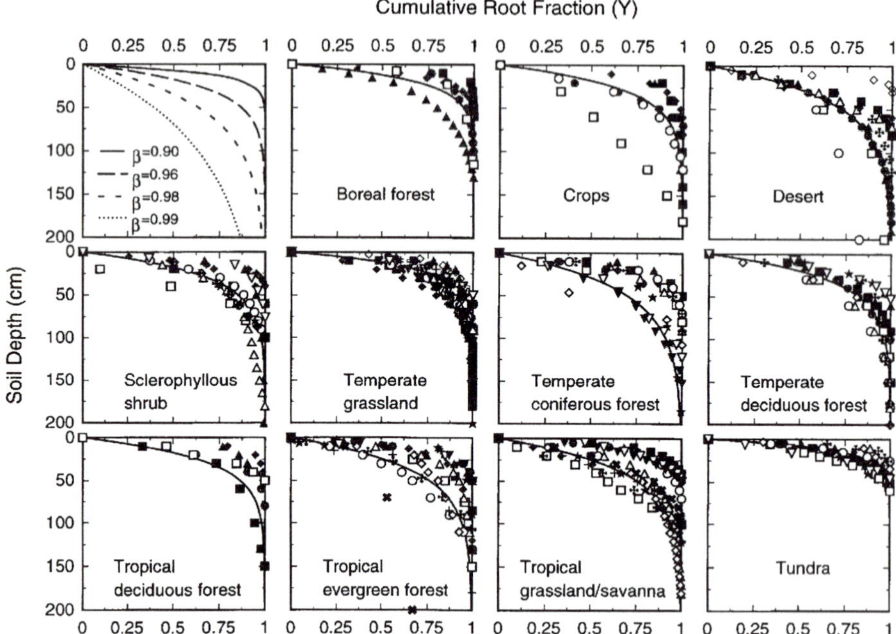

Figure 11.1 Cumulative root distributions (cumulative proportion) as a function of soil depth for eleven terrestrial biomes and for the theoretical model of Gale and Grigal. The curve in each panel is the least squares fit of β for all studies with data to at least 1 m depth in the soil (data from shallower studies are shown in the figure but not included in the calculations of β). Specific values of β are in Table 11.1 and the key to the symbols in each panel can be found in Table 2 of Jackson et al. (1996). The crop data presented are simply examples from a few studies.

nutrients from fine roots is minimal (e.g., Aerts 1990), approximately 40×10^{12} g of P cycle through fine roots annually. Similarly, about 0.5×10^{15} g of N cycle through fine roots each year, as much as one-third of the standing pool of nitrogen in terrestrial plants (Schlesinger 1997).

Analyses of biome data are useful for broad analyses at regional and global scales. In making predictions at finer scales or across complex landscapes, biome-based data may be too coarse in resolution. Plant functional types provide one tool for helping to scale ecosystem processes in complex environments (Körner 1994). In general, woody species tend to be more deeply rooted than grasses and herbs (e.g., Walter 1954). Based on data for species in > 20 ecosystems, β values were 0.957 for grasses and 0.972 and 0.980 for trees and shrubs, respectively. Cumulative root biomass in the top 30 cm of soil varied from 45% for an average shrub to almost 75% for a typical grass. Shrubs had 87% of their root biomass in the top meter of soil while grasses had almost 99% of root biomass in the top meter. These estimates for total root biomass are slightly deeper than those in Jackson et al. (1996) because only profiles to ≈ 2 m soil depth or greater were used in the revised analysis.

Table 11.1 β values (r^2) for biomes and the proportion of total root biomass in the upper 30 cm for total root biomass and fine root biomass (< 2 mm). β is defined in the model of Gale and Grigal (1987) in the form $Y = 1 - \beta^d$, where Y is the cumulative root fraction with depth (a proportion between 0 and 1), d is soil depth (in cm), and β is the fitted parameter. β is a simple numerical index of rooting distribution; larger values imply more deeply rooted profiles. Data from Jackson et al. (1996, 1997).

	Total Root Biomass			Fine Root Biomass		
Biome	β	(r^2)	% root biomass in upper 30 cm	β	(r^2)	% root biomass in upper 30 cm
Boreal forest	0.943	(0.89)	83	0.943	(0.89)	83
Desert	0.975	(0.95)	53	0.970	(0.99)	60
Sclerophyllous shrubs	0.964	(0.89)	67	0.950	(0.84)	79
Temperate coniferous forest	0.976	(0.93)	52	0.980	(0.96)	45
Temperate deciduous forest	0.966	(0.97)	65	0.967	(0.96)	63
Temperate grassland	0.943	(0.88)	83	0.943	(0.88)	83
Tropical deciduous forest	0.961	(0.99)	70	0.982	(0.99)	42
Tropical evergreen forest	0.962	(0.89)	69	0.972	(0.91)	57
Tropical grassland/ savanna	0.972	(0.95)	57	0.972	(0.97)	57
Tundra	0.914	(0.91)	93	0.909	(0.90)	94

There are a number of uncertainties in averaging data across such broad categories as biomes or plant functional types. Seasonal and spatial dynamics are masked by pooling information within and across sites, as are climatic and edaphic factors. Fine and coarse topographic variation, as well as other sources of small-scale variability, can be important to individual plants (e.g., Snaydon 1962; Burke et al. 1997). To take seasonal dynamics into account, the above estimates of root distributions can be used in models with the most appropriate root phenology that researchers describe for a particular site. Bearing such caveats in mind, three goals for the estimates are (a) to uncover broad patterns and test hypotheses among biomes and plant functional types in the field, (b) to improve the representation of belowground processes in ecosystem and global models, and (c) to provide a benchmark for improvement as additional data become available.

Table 11.2 Global land area (10^6 km^2), total root biomass per unit area of soil (kg m^{-2}), total root biomass (10^9 Mg), total fine root biomass (10^9 Mg), and live fine root biomass (10^9 Mg) for the global classification scheme of Whittaker (1975). The estimates are based on data in Jackson et al. (1996, 1997).

Biome	Land area (10^6 km^2)	Total root biomass (kg m^{-2})	Total root biomass (10^9 Mg)	Total fine root biomass (10^9 Mg)	Live fine root biomass (10^9 Mg)
Tropical Rainforest	17.0	4.9	83	9.7	5.7
Tropical Seasonal Forest	7.5	4.1	31	4.3	2.1
Temperate Evergreen Forest	5.0	4.4	22	4.1	2.5
Temperate Deciduous Forest	7.0	4.2	29	5.6	3.1
Boreal Forest	12.0	2.9	35	7.2	2.8
Woodland and Shrubland	8.5	4.8	41	4.4	2.4
Savanna	15.0	1.4	21	14.9	7.7
Temperate Grassland	9.0	1.4	14	13.6	8.5
Tundra/Alpine	8.0	1.2	10	7.7	2.7
Desert	18.0	0.8	6.6	4.9	2.3
Cultivated	14.0	0.2	2.1	2.1	1.1
Totals	**121**		**292**	**78.2**	**40.8**

Global Comparisons of Soil Carbon and Root Biomass with Depth

Soils contain the largest terrestrial pool of organic carbon in the world. Approximately 1500×10^{15} g (Pg) of organic C are found in the upper meter of soil and another 800 Pg are found in inorganic forms, primarily as carbonates (e.g., Schlesinger 1977; Post et al. 1982; Eswaran et al. 1993). Soil organic C was estimated as 700 Pg in the upper 30 cm, 1500 Pg in the upper meter, and 2400 Pg in the upper 2 m by Batjes (1996). The 900 Pg of organic C at 1–2 m soil depth highlights the biogeochemical importance of relatively deep soil layers and suggests that soil organic C below 2 m may be substantial (e.g., Nepstad et al. 1994). Such C in the deep soil is usually ignored in estimates of global C pools (e.g., IPCC 1992).

A recent analysis by Batjes (1996) compiled global estimates of C and N in the upper 2 m of soil. The analysis was based on 4350 soil profiles stratified by 26 FAO-UNESCO soil units. These data provide an interesting comparison with the global root distributions discussed above. On average, soil organic carbon (SOC) in the upper 2 m of soil is distributed much more deeply than are plant roots (Figure 11.2); soil N and other nutrients are distributed more deeply still. Globally, almost 40% of total SOC in the top 2 m is between 1 and 2 m depth; for plants < 10% of root biomass on average is distributed in the same depth increment (Jackson et al. 1996). Even for the most deeply rooted functional type, shrubs, almost 90% of biomass is in the upper meter of soil

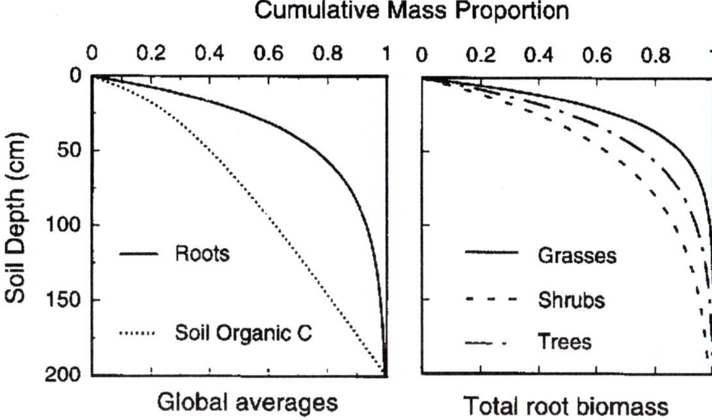

Figure 11.2 Left panel: The average global distribution of soil organic carbon to 2 m (Batjes 1996) and average root biomass for soil profiles of ≈ 2 m soil depth or greater (Jackson et al. 1996). Right panel: Average distributions for total root biomass of grasses, shrubs, and trees based on soil profiles of ≈ 2 m soil depth or more (see Table 11.1). From Jobaggy and Jackson (unpublished data).

(Figure 11.2). Couple these relatively shallow root distributions with surface inputs from aboveground litter, and the discrepancy between C inputs to the soil and the distribution of SOC is even more obvious. Potential mechanisms for this discrepancy include the underestimation of root biomass at depth, slower decomposition at depth in the soil, leaching of soil C to deeper layers, and the build-up of the soil profile over time. Understanding this difference between carbon inputs to the soil and the long-term distribution of SOC with depth is critical for understanding C cycling locally and C sequestration globally.

Maximum Rooting Depth by Biome and Plant Functional Type

The β values described above are one way to describe root abundance with depth in the soil. One of the reasons for generating such values is to improve the representation of belowground processes in models (Jackson et al. 1996; 1997). In using β values for models (or any representation of roots, for that matter), questions of root functioning quickly become apparent. Are all roots active for water and nutrient uptake? Does uptake correlate linearly with total root biomass, fine root biomass, or neither? With length or surface area? Under what circumstances are a small number of roots disproportionately important for resource uptake?

Deep roots provide an interesting example because they are usually a small fraction of total root biomass but can be extremely important for resource uptake. Gregory et al. (1978) examined water uptake by winter wheat from different depths in the soil and its correlation with root biomass. In a dry summer period, almost half of the water taken up by the plants was estimated to come from below 60 cm depth, where only 20% of root biomass was present. Only 3% of the root system below 1 m supplied approximately 20% of water uptake. The authors emphasized the importance of a few deep roots during dry periods for the maintenance of growth.

Maximum rooting depth is inherently difficult to measure. Sampling deep soils is problematic, and there is no unequivocal way to determine when the "maximum" rooting depth has been found. Despite these difficulties, there have been several recent attempts to describe broad patterns of maximum rooting depth for biomes and plant functional types. Stone and Kalisz (1991) studied maximum rooting depth in the literature for almost 100 genera of trees and documented tree roots to 60 m for *Juniperus* and *Eucalyptus* spp. More recently, Canadell et al. (1996) analyzed patterns for more than 250 plant species classified by biome and plant functional type (grasses, shrubs, and trees). Based on functional types, the average maximum rooting depth for herbaceous plants, including crops, was 2–2.5 m (Figure 11.3). Average maximum rooting depth of woody species was considerably deeper, 5 m and 7 m on average for trees and shrubs, respectively (Figure 11.3). Such differences between woody and herbaceous plants are one reason the two-layer model of water partitioning proposed for savannas has been widely adopted (Walter 1954).

At least 50 plant species have been shown to grow roots more than 5 m deep in the soil, and at least eight species grow roots to more than 40 m depth (Stone and Kalisz 1991; Canadell et al. 1996). Arid and semi-arid systems contain some of the most deeply rooted plants in the world (e.g., Cannon 1960; Phillips 1963). Examples include such species as *Alhagi maurorum* (15 m), *Retama raetam* (20 m), *Tamarix aphylla* (20 m), and *Prosopis farcta* (15 m) in Israel, *Juniperus* and *Prosopis* spp in the U.S. (61 and 53 m, respectively), and *Acacia erioloba* (60 m) and *Boscia albitrunca* (68 m) in the Kalahari. Of the 255 woody and herbaceous species examined by Canadell et al. (1996), 80% had roots at least 2 m deep and roughly 10% had roots below 10 m. Globally, species rooting to > 10 m have been found in at least four biomes: deserts, sclerophyllous shrublands/woodlands, tropical evergreen forest, and savannas (Figure 11.3). Based on the classification scheme of Whittaker (1975), such systems cover approximately 60 million km^2 and 40% of terrestrial land area.

IMPORTANCE OF DEEP ROOTS: EXAMPLES FROM THE AMAZON AND NORTH AMERICA

Why plants should bother growing roots to 10, 50, or even 100 m is debatable, but the most likely explanation is water uptake. Water deep in the soil may be an important resource for deeply rooted plants throughout the year, or it may represent insurance during droughts. Water uptake from deep in the soil can be a strong determinant of the type of vegetation that grows in an area and the ability of that vegetation to maintain a green canopy. This may be especially true for savannas and tropical deciduous forests, vegetation intermediate between grasslands and evergreen forests. The role of deep roots in

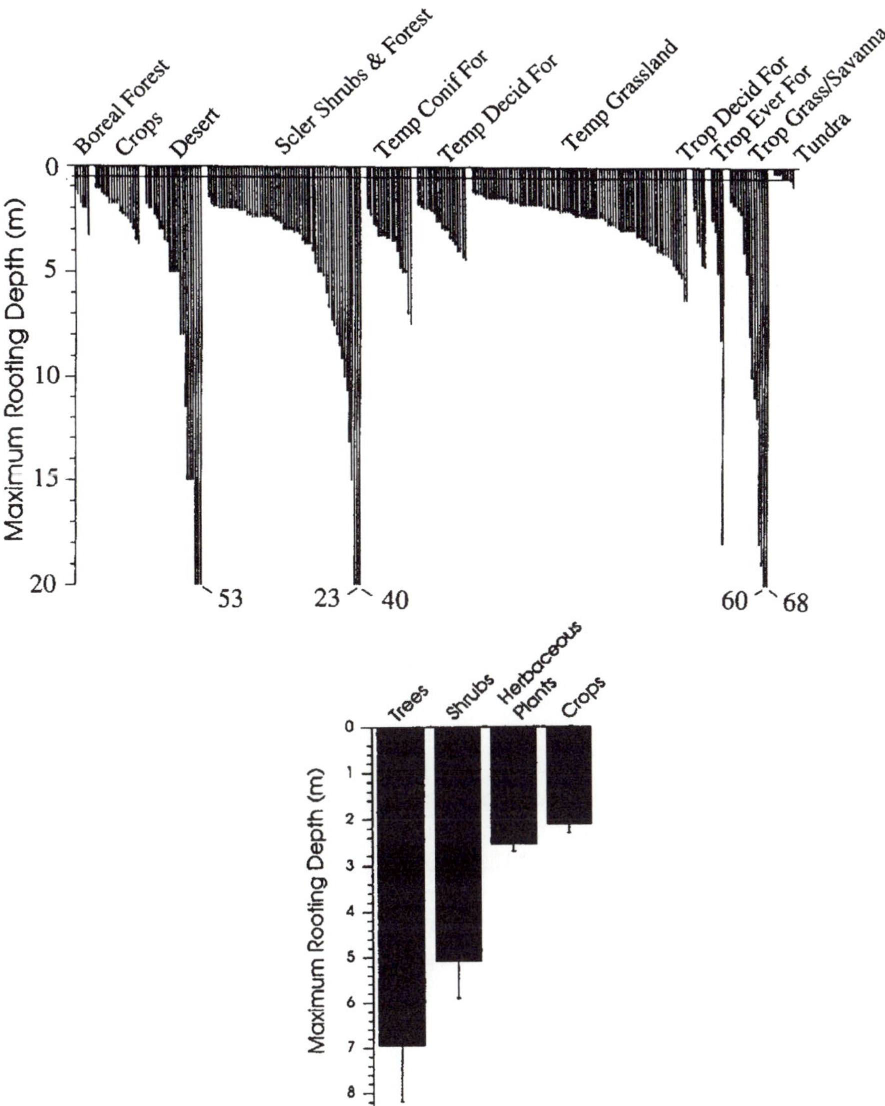

Figure 11.3 Upper panel: Maximum reported rooting depth for individual species grouped by terrestrial biome. Lower panel: Average maximum rooting depth by plant functional type (mean ± se; n = 82, 71, and 85 species for trees, shrubs, and herbs, respectively). From Canadell et al. (1996).

defining the transition between evergreen and deciduous vegetation is especially important for determining the ecological consequences of climate change.

As discussed earlier, roots are one of the most important sources of carbon inputs into soil. More than twice the carbon is held in the soil than in either the atmosphere or terrestrial vegetation (Schimel 1995). Based on these and other factors, the importance of deep roots for water and carbon cycling will now be examined for three systems: the Amazon, pine forests of the southeastern U.S.A., and the deserts of North America.

The Amazon

Over a century ago, Eugen Warming spent three years in the 1860s studying the floristics of the Cerrado region of Brazil. Campos Cerrados vegetation in this savanna system has trees < 8 m tall and the region has a pronounced dry period from May to August. Warming (1892) noted that leaf-out and sprouting in many of the trees occurred at the end of the dry season, often prior to the first rains of the wet season. He inferred that the plants were using water stored in the soil to initiate growth since "new" water was unavailable in some years.

Fifty years later, Rawitscher (1948) documented the presence of tree roots (most likely *Andira humilis*, of the Fabaceae) and standing water at 18 m depth in the same region of Brazil. Because Rawitscher drilled a series of wells at the highest points of a rolling landscape, there was little likelihood that lateral movement supplied the water; more likely, excess summer rains filled the soil profile *in situ*. Rawitscher compared the stomatal behavior of relatively deep-rooted plants (mostly trees and shrubs) with more shallowly rooted grasses and herbs in the system. Deep-rooted species maintained higher transpiration values than did shallow species. His transpiration experiments were crude by today's standards (and problematic), but his conclusion that the presence of deep roots allowed many of the woody species to maintain their foliage during the seasonal dry period was almost certainly correct.

More recently, Nepstad et al. (1994) studied forests and adjacent pastures of southern and eastern Amazonia using field experiments and satellite imagery. They estimated that half of the closed forests in the 2 million km^2 area depended on deep roots for maintaining a green canopy. The phenology also changed in the pasture system that was more shallowly rooted. Leaf area in the degraded pasture decreased by 68% during the dry season but by only 16% in the adjacent forest. In total, more than three quarters of the water transpired in each system during the dry period came from below 2 m in the soil. Estimates and modeling analyses that ignored roots below 2 m would have underestimated evapotranspiration by more than 60%.

Deep roots were also important for carbon storage in Amazon soils. There was more carbon in the soil below one meter than in the uppermost meter of soil or in aboveground vegetation (Nepstad et al. 1994). Just as importantly, analysis of the ^{14}C signature of the soil indicated that up to one-sixth of deep-soil carbon turned over on annual or decadal timescales. Given the importance of the soil for terrestrial carbon budgets (Schimel et al. 1995), land-cover change and altered plant species composition may decrease carbon inputs to the deep soil and reduce carbon storage in ways beyond the obvious effects of biomass burning.

The importance of deep-rooting for water fluxes in the Campos Cerrados rests in part on Rawitscher's observation that the top 2.5 m or so of soil dried out sufficiently during the dry season to supply little water. This observation was echoed by Nepstad et al. (1994) for the wetter Amazon forest in which they worked. To address when deep roots and deep water availability are likely to be important, one criterion of obvious importance is dry-down of the surface soil layers. Dry-down of the upper 2 m of soil, or some other plausible cut-off, might be one way to screen systems where deep resources should be important. A second method would be to estimate annual transpiration for systems globally and to ask which of them are likely to have sufficient water supplied from the upper 2 m of soil. Such an analysis should include soil texture, the amount and timing of precipitation in the region, temperature, and leaf area. There are certainly areas in the Amazon where deep soil processes are unimportant (and where waterlogging may limit roots to shallow depths), but the studies just described highlight the need for critical evaluation and further research

Forests of the Southeastern United States

A second example of the importance of deep soil processes is provided by the pine forests of the southeastern United States. Richter and Markewitz (1995) studied a secondary *Pinus taeda* ecosystem at the Calhoun Experimental Forest in the Carolina Piedmont. Soils at the site were Ultisols, with B horizons typically acidic, low in fertility, and clayey. Ultisols dominate 20 million ha of the southeastern U.S. and are one of the most common soil types in the tropics. At the Calhoun Experimental Forest, depth to bedrock was > 8 m and pine roots were observed to more than 4 m depth in the soil. Pronounced seasonality in the depletion of soil water indicated a hydrologic rooting depth of > 3 m. Microbial activity was also prevalent in the deep soil. Bacteria, which depend on plant inputs and leached substrates for growth, were still abundant at 8 m soil depth; densities were 1.8×10^7 cells g^{-1} soil at 8 m compared to 1.2×10^8 cells g^{-1} soil in the top meter of soil. Concentrations of CO_2 in the soil reflect the timing and amount of CO_2 produced and its vertical and lateral diffusion in soil. Soil CO_2 concentrations at the site also indicated root and microbial activity deep in the soil profile. CO_2 concentrations were strongly seasonal at depth (Figure 11.4). The data indicated production of CO_2 by root respiration and microbes through at least the upper 4 m of the soil profile and a strong seasonality of biological activity. Overall, the authors conclude that at least 8 m of the soil profile bear the imprint of biological activity and its effects on biogeochemical processes.

North American Deserts

As discussed earlier, deserts contain many of the deepest rooted plants in the world, including at least four species with roots more than 50 m deep in the soil. Deserts of North America provide a third system in which the presence of deep roots has been documented and some consequences suggested for hydrology, biogeochemistry, and

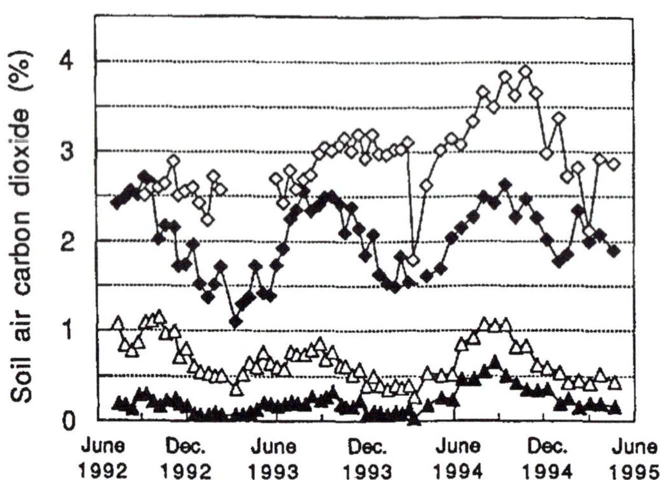

Figure 11.4 Seasonality of soil CO_2 concentrations as illustrated by mean CO_2 concentrations from four soil depths in the Calhoun Experimental Forest (Richter and Markewitz 1995). ▲ – 0.15 m; Δ – 0.6 m; ♦ – 1.75 m; ◊ – 4.0 m

ecosystem functioning. Freckman and Virginia (1989) studied grasslands and shrublands in the Chihuahuan Desert of southern New Mexico. In the late 1800s, shrublands increased from approximately 5% of the region to 50%, and black grama grasslands declined from 90% to 25%. Coincident with, and perhaps causing the change, was a 10-fold increase in cattle grazing. Black grama (*Bouteloua eriopoda*) is a relatively shallow rooted species whose pattern of photosynthesis is tightly coupled to water availability in the upper meter of soil. The two dominant shrubs in the region, mesquite (*Prosopis glandulosa*) and creosote (*Larrea tridentata*) are more deeply rooted, and their patterns of photosynthesis tend to be less coupled with shallow water availability (Schlesinger et al. 1990). Freckman and Virginia (1989) examined root biomass and soil nematode densities to more than 10 m depth in this system. Roots and nematodes were found to at least 12 m in the mesquite playa. Creosote and grassland sites tended to be more shallowly rooted, with maximum rooting depths of 3.5 to 5 m. Caliche layers were apparently not significant barriers to roots or nematodes at any of the sites. Interestingly, nematode densities were fairly well correlated with rooting depth across the five locations. The maximum depth of nematode occurrence was always within a meter or two of the maximum depth of roots. This was also the case for microarthropod fauna in general (Silva et al. 1989). Additional circumstantial evidence that the distribution of soil microbes and fauna might be coupled to rooting depth can be found in the example for southeastern forests discussed in the previous section (Richter and Markewitz 1995).

That there may be a general, global relationship between the distribution of roots with depth and the distribution of soil organisms is an important hypothesis that needs to be tested. If such a pattern exists, it also needs to be shown that roots "control" the frequency and depth to which soil organisms occur. A positive correlation could be driven by some covarying soil property, such as a barrier to growth, or other environmental or biological factors. A causal relationship, or even a correlational one, has important implications for carbon and nutrient cycling and the diversity and distribution of soil organisms.

USING GLOBAL DATA BASES FOR EXAMINING THE IMPORTANCE OF ROOTING DEPTH

For deep roots to be important for hydrology, biogeochemistry, and ecosystem functioning, the vegetation must be able to grow roots deep in the soil and the soil must be penetrable and hold resources at depth. In the Brazilian example discussed earlier, bedrock is generally more than 20 m below the soil surface. Such depths are not unusual globally. Depth to bedrock can be 50 m in the Piedmont of the eastern United States (Richter and Markewitz 1995), 20 m on the Malaysian peninsula (Eswaran and Bin 1978), and 100 m in Hong Kong (Ruxton and Berry 1957). Historical floodplains such as the Rio Grande Valley of central New Mexico hold hundreds of meters of alluvial material over bedrock. Based on such "soil" depths, two clear needs for improving hydrological and biogeochemical models are apparent. The first is to understand when such deep soil is important for resource uptake by plants (coupled with an equally important goal of understanding where shallow soil conditions and waterlogging limit root growth). The second is to chart where deep soil exists globally. Both needs would be addressed in part by a global map of soil depth to bedrock. To my knowledge no such map exists (Casper and Jackson 1997). The closest analog is a global map of soil profile thickness constructed by Webb et al. (1993) using data from the FAO/UNESCO Soil Map of the World (1971–1981). Part of this dataset is available on the ISLSCP CD-ROM produced by NASA (Sellers et al. 1995). The analysis by Webb et al. is based on horizon texture and thickness for > 100 soil types divided into nine continental regions: Africa, South-Central Asia, North-Central Asia, Southeast Asia, Australia/South Asia, Europe, Mexico/Central America, South America, and North America. It includes unconsolidated material of the C horizon in its definition of soil.

Plate 11.1 presents the FAO/UNESCO data of soil profile thickness in two forms: eight depth increments between 0 and 5 m (upper panel) and two depth increments indicating regions with soil profile depths < or ≥ 3 m (lower panel). Most of the dataset has soil profiles shallower than 3 m. For the examples discussed earlier, the Amazon rainforest has a soil profile thickness of approximately 2.7 m in the FAO dataset and the southwestern deserts of the United States have a soil profile thickness of approximately 2 m. Such estimates are in sharp contrast to the actual rooting depth of 10 m or more known from both systems (Rawitscher 1948; Freckman and Virginia 1989). Discrepancies between known rooting depth in natural systems and soil profile depths in the analysis of Webb et al. (1993) are understandable for several reasons. First, the FAO data base is comprised primarily of data from agricultural soils. Second, soil depth isn't measured at each location; rather representative soil profiles are taken and extrapolated for a given region (the same limitation we all face). Third, soil scientists tend to view what constitutes "soil" fairly conservatively compared to ecologists, hydrologists, and biogeochemists (Richter and Markewitz 1995). The Soil Survey Staff (1992) of the U.S. Department of Agriculture recommends that the lower limit of soil be considered 2 m for the sake of convenience and consistency. Webb et al. (1993) recognize the depth limitations of their data: "In many cases, the soil profile thicknesses represent minimum possible values because profile descriptions do not always extend to subsurface bedrock."

In addition to soil physical characteristics, Webb et al. (1993) also estimate three water-holding parameters globally: 1) potential water storage in the soil profile, 2) potential water storage in the rooting zone, and 3) potential water storage based on soil texture. Their estimates of potential water storage globally use data for water storage capacity in different soil textures. Estimates range from a low of 4% water storage by volume for sandy soils to 60% for highly organic soils (Table 11.3). The determination of rooting depth in the analysis is based on the eight simplified vegetation types of Matthews (1983) as prescribed for the GISS GCM (Table 11.3). The deepest rooting depth in the model is 2 m for woodlands and deciduous and evergreen forests. Maximum rooting depth for rainforests is 0.8 m. Potential storage of total water in the rooting zone of the Amazon as estimated by Webb et al. (1993) is approximately 450 mm; Nepstad et al. (1994) estimated *available* soil water as approximately twice that amount if soil layers 2–8 m deep were included.

The analysis of Webb et al. (1993), accompanying FAO/UNESCO and ISLSCP data, and more recent DIS, WISE, and USDA data bases (e.g., Batjes 1996) provide an important foundation for examining the importance of rooting depth and deep soil processes. However, existing data bases are insufficient for estimating soil profile depth and plant rooting depth globally. An improved data base should be constructed to combine our best estimates of depth to bedrock, soil texture, water holding capacity, and maximum rooting depth globally. Additional data for areas where waterlogging limits rooting depth would also be useful. Such a data base could greatly improve ecosystem and global models and would generate numerous hypotheses testable in the field. Improving such data bases provides an opportunity for hydrologists, soil scientists, and ecologists to combine different approaches for addressing carbon and water fluxes globally.

WHEN SHOULD DEEP ROOTS BE IMPORTANT?

Throughout this chapter I have made suggestions for when and where the distribution of roots and deep roots in particular might be important. This final section organizes those suggestions and highlights some of the many unanswered questions that remain.

As discussed earlier, several conditions must be met if deep roots are to be functionally important. The most obvious is that the vegetation must be capable of growing roots deep in the soil. Depending on one's definition of "deep," this may exclude many plants, especially those whose root systems are adventitious. Grasses (monocots in

Table 11.3 Spatially averaged maximum rooting depth (m) of eight major vegetation regions as prescribed for the GISS GCM and the relative amount of available water for different soil textures (Webb et al. 1993).

Vegetation type	Maximum rooting depth (m)
Desert	0.005
Tundra	0.1
Grassland	1.1
Shrub	1.5
Woodland	2.0
Deciduous	2.0
Evergreen	2.0
Rainforest	0.8

Soil type	Storage capacity (% volume)
Sand	4
Loamy sand	8
Sandy loam	14
Sandy clay loam	14
Clay loam	14
Sandy clay	14
Silty clay	14
Clay	14
Silty clay loam	17
Loam	17
Silt loam	20
Silt	20
Organic soil	60

general?) and some herbaceous species fall into this category, though there are clear examples of "herbs" with deep roots. Alfalfa at 40 m depth is one example (Meinzer 1927). The second condition is that the soil must be penetrable to both roots and resources. Impediments to root growth and water percolation, such as permafrost and waterlogging, will clearly limit the importance of deep roots. The third condition is that the soil must hold resources at depth. Passing through 50 m of bedrock is fruitless if roots do not access a limiting resource. In principle, each of these conditions is fairly simple and testable with a suite of approaches, including field experiments and modeling (e.g., Kleidon and Heimann 1998). A combination of such approaches from geology and ecology could address each of these conditions globally.

A discussion of when deep roots should be important must also consider which resources are most likely to be available at depth. The most prevalent deep resource is probably water. Water can accumulate at depth in a number of ways, including the downward percolation of recent rainfall (as in the system of Rawitscher [1948]), the buildup of unused water over many years, or the aggregation of water by lateral flow across landscapes (Greenwood 1992). Vegetation change today often involves the exchange of woody and herbaceous vegetation (e.g., deforestation, afforestation, and shrub encroachement). A summary of 94 catchment studies showed that for each 10% change in the cover of conifers and eucalypts there was a 40 mm change in water yield for the catchment; analgous changes in hardwood and scrub communities were 25 and 100 mm, respectively (Bosch and Hewlett 1982). Rooting depth is clearly just one of many important factors associated with such changes, but it appears to be a dominant factor in a number of studies (e.g., Pierce et al. 1993; Scott and Lesch 1997).

Building on the three conditions described in this section, how could the importance of such deep water be addressed? To integrate its importance requires:

1. Predicting the climate regimes, soil profiles, and vegetation types that would allow water to accumulate at depth
2. Understanding the conditions under which subsurface lateral flow of water allows it to build up in the soil
3. Describing the biomes and plant functional types that would have roots sufficiently deep in the soil to transpire such water.

Regional and global maps of systems where subsurface water is likely to accumulate, either as standing water or held in the soil profile, would be useful for predictions of hydrology and ecosystem functioning. Some uncertainties in the availability and importance of deep water are described below.

It is important to remember that maximum rooting depth is only one of a host of factors important for hydrology, biogeochemistry, and ecosystem functioning. In the Chihuahuan desert system of Freckman and Virginia (1989), grassland and creosote sites had a shallower root system than the deeply rooted mesquite playa, but their total root biomass was almost twice as large. Grasslands in general maintain some of the highest belowground productivities with some of the shallowest root systems in the world (Jackson et al. 1997). Defining which systems have an inherently shallow zone of biogeochemical activity would be a good first step to answering the question posed above. Such biomes as tundra, boreal forest, wetlands, and some grasslands may fall into this category. Ruling out the presence of deep roots across broad regional zones would be useful from both experimental and modeling perspectives.

UNANSWERED QUESTIONS AND SUGGESTIONS FOR RESEARCH

With the above discussion as background, there are many unanswered questions that remain. I end by discussing some of those questions and suggesting a few ways to help answer them.

Scaling and Variability in Complex Landscapes

Much of this chapter has focused on relatively broad scales, particularly analyses for biomes and plant functional types. Complex landscapes (a complex term with myriad definitions) can be a mosaic of landscape units and a mixture of plant functional types. While the use of remotely sensed data provides an important tool for estimating landscape complexity, the consequences of that complexity for water and nutrient fluxes remain challenging. There may be important nonlinearities at the boundary of landscape units, complicating the averaging of processes across units. To understand such processes, ecosystem experiments within landscape units must be combined with techniques that scale across units, including stable isotope analyses, eddie flux measurements, and the use of remotely sensed data. Understanding landscape complexity and its consequences remains a scientific challenge, one that will become increasingly important with human population growth and further landscape fragmentation.

Positive Feedbacks between the Occurrence of Deep Roots and Deep Water?

Macropores have long been recognized as important by hydrologists for understanding water movement in the soil. The presence of macropores is one reason why traditional bucket models may often be inaccurate. The occurrence of a deep-rooted woody plant, or the invasion of such a plant into a formerly shallow-rooted community, might increase the likelihood of water percolating deep into the soil. Water may run down the stem and taproot of the plant, follow any of the existing roots into the soil, or flow through channels where roots have died. If soil fauna occur deeper in the soil when deep roots are present, then they could also increase the flow of water deep into the soil. These and other mechanisms suggest there may be positive feedbacks between the presence of deep roots and the availability of deep water.

Hydraulic Lift

Hydraulic lift is another potentially important role for deep roots (Breazeale 1930; Richards and Caldwell 1987). Hydraulic lift is the passive movement of water through roots from relatively wet, deeper soil layers to relatively dry, shallower ones (though the reverse may also occur), and it has now been observed in more than 25 plant species (Caldwell et al. 1998). Hydraulic lift acts to bridge resource separation when deep- and shallow-rooted plants are present in an ecosystem. Deep water can flow into shallow layers and become available to the plant "lifting" the water and to its neighbors. Although plants often compete for water, the coupling of deep- and shallow-rooted species may therefore enhance overall ecosystem productivity at times. By "irrigating" the upper soil, decomposition and nutrient diffusion may also be enhanced, further increasing ecosystem productivity by mobilizing nutrients in a surface soil that might otherwise be dry. Such techniques as soil psychrometry, stable isotope analyses, and modeling are being used to examine the importance of hydraulic lift (e.g., Ehleringer and Dawson 1992).

Nutrient Availability

Nutrient availability in the soil is generally greatest in upper soil layers. In contrast, water may be most available at depth and, at times, absent from surface layers. Consequently, plants likely face situations where water and nutrients are separated spatially and temporally. Are deep and shallow roots equally capable of taking up water and nutrients? Do deep roots show the same kind of plasticity as their shallow counterparts? Under what conditions are nutrients available at depth?

Modeling Resources in the Soil/Plant System

Modeling water and nutrient uptake is even more difficult than modeling carbon and energy uptake by leaves. Although sun angle changes with time, it does so in a predictable fashion and there is rarely more light available at the bottom of a canopy than at the top. In contrast, water availability may be greatest at 5 m depth, greatest in surface layers, or equally available in both depending on the time of year and the pattern of precipitation. Another difficulty is knowing the proportion of root biomass that is active in resource uptake. Shallow roots both take up resources and provide the infrastructure that supports deeper roots. Coarse roots can be a large proportion of biomass but be inconsequential for resource uptake.

A scientist modeling water uptake might have good data for root biomass or length at different soil depths (perhaps even good fine-root data). Given such data, how should water uptake from different soil layers be modeled? One possibility is a proportional weighting that combines root abundance and water availability. An alternative scheme might be to use the shallowest water first and then have the plant take up water progressively deeper in the soil as needed. A third scheme might simply take water from the soil with the most positive water potential, regardless of depth. Issues of root biomass to functioning, and temporal and spatial variation in resource availability (e.g., Jackson et al. 1990), demonstrate the complexity of integrating hydrology, ecosystem dynamics, and biogeochemistry in complex landscapes.

Barriers to Root Growth

There are a number of barriers that can impede root growth and the movement of water and nutrients. Permafrost and bedrock are two of the most dramatic. Caliche in arid systems and waterlogging are two others. Global maps showing where such barriers to root growth occur regionally would be very useful.

Caution is needed in interpreting barriers to root growth. With the exception of permafrost, there is probably no barrier through which roots do not sometimes pass. Even bedrock can be passable through cracks and fissures (e.g., Lewis and Burgy 1964). The Edwards Plateau, a region in central Texas covering more than 100,000 km^2, provides an interesting example. The Plateau is named for the outcrops of early Cretaceous limestone that dominate the area and is characterized by thin, limestone-derived soils (often < 30 cm in depth). Immediately to the east of the Plateau are deep, rich Blackland Prairie soils formerly dominated by tussock grasses. The transition from the deep, loamy soils of the Prairie to the shallow, limestone soils of the Plateau is immediate, defined by the 500 km-long Balcones Escarpment. This striking contrast in vegetation is clear evidence that factors other than climate control regional vegetation. The presettlement vegetation of the Edwards Plateau was originally a live-oak savanna, but it is currently dominated by cedar (*Juniperus ashei*), oak (*Quercus* spp.), and mesquite

(*Prosopis glandulosa*) trees. Since the average precipitation is 800–900 mm, it seems unlikely that a 30-cm soil would be able to support such trees. It is likely (and there is evidence) that the trees are tapping water through cracks in the bedrock. The shallow soils of the Edwards Plateau have a strong effect on vegetation for more than 100,000 km^2, but the bedrock itself may not completely limit root growth.

Open vs. Closed Ecosystems

The extent to which matter and energy is retained in an ecosystem determines whether that system is relatively "open" or "closed." There are many practical consequences to the distinction, with nitrate contamination of groundwater as just one example. In principle, rooting depth can play an important role in how open or closed an ecosystem might be. Soil water in a deep-rooted ecosystem must percolate deeper into the soil to move beyond the rooting zone of the system. The same is true for nitrate and other nutrients. Increased microbial and soil faunal abundance at depth in ecosystems with deep roots may also contribute to making a system more closed.

Issues for Discussion

In summary, the following ten issues are highlighted for discussion:

1. What is the best way to estimate a "hydrologically active" rooting depth?
2. Does landscape complexity aboveground necessarily imply the belowground system is complex? Can systems be complex belowground but appear homogenous from above?
3. How representative of forests are the measurements of Rawitscher (1948), Nepstad et al. (1994), and others documenting roots below 10 m? What is the importance of such roots for biogeochemical cycling, water fluxes, and the maintenance of leaf area? In which systems can the importance of deep roots be excluded?
4. What is the importance of rooting depth in determining whether an ecosystem is relatively open or closed?
5. Does the presence of deep roots make it more likely that water percolates deep in the soil?
6. What are the global patterns of depth to bedrock? In what areas do shallow bedrock and other impediments to root growth determine climax vegetation?
7. Is the distribution, abundance, or diversity of microbes and soil fauna positively correlated with rooting depth?
8. Are deep roots only important for water uptake? If not, under what conditions are nutrients available deep in the soil?
9. How prevalent is hydraulic lift and what is its biogeochemical and hydrological importance?
10. In which systems are the phenologies of above- and belowground growth tightly coupled?

ACKNOWLEDGMENTS

I wish to thank members of a workshop on root distributions and global models at the National Center for Ecological Analysis and Synthesis (University of California at Santa Barbara). This research was supported by the National Sciencc Foundation (DEB 97–33333), the Department of Energy's National Institute for Global Environmental Change (TUL–038–95/98), the Inter-American Institute for Global Change Research, and the Andrew W. Mellon Foundation. It is a contribution to the Global Change and Terrestrial Ecosystems (GCTE) Core Project of the International Geosphere Biosphere Programme (IGBP). L.J. Anderson, W.A. Hoffmann, and numerous Dahlem participants provided helpful comments on the manuscript.

REFERENCES

Aerts, R. 1990. Nutrient use efficiency in evergreen and deciduous species from heathlands. *Oecologia* **84**:391–397.

Batjes, N.H. 1996. Total carbon and nitrogen in the soils of the world. *Eur. J. Soil Sci.* **47**:151–163.

Bosch, J.M., and J.D. Hewlett. 1982. A review of catchment experiments to determine the effect of vegetation change on water yield and evapotranspiration. *J. Hydrol.* **55**:3–23.

Breazeale, J.F. 1930. Maintenance of moisture-equilibrium and nutrition of plants at and below the wilting percentage. *Ariz. Agr. Exp. Sta. Tech. Bull.* **29**:137–177.

Burke, I.C., W.K. Lauenroth, and W.J. Parton. 1997. Regional and temporal variation in net primary production and nitrogen mineralization in grasslands. *Ecology* **78**:1330–1340.

Cairns, M.A., S. Brown, E.H. Helmer, and G.A. Baumgardner. 1997. Root biomass allocation in the world's upland forests. *Oecologia* **111**:1–11.

Caldwell, M.M., T.E. Dawson, and J.H. Richards. 1998. Hydraulic lift: Consequences of water efflux from the roots of plants. *Oecologia* **113**:151–161.

Canadell, J., R.B. Jackson, J.R. Ehleringer, H.A. Mooney, O.E. Sala, and E.-D. Schulze. 1996. Maximum rooting depth for vegetation types at the global scale. *Oecologia* **108**:583–595.

Cannon, H.L. 1960. The development of botanical methods of prospecting for uranium on the Colorado Plateau. U.S. Geological Survey Bulletin 1085–A, 50 pp.

Casper, B.B., and R.B. Jackson. 1997. Plant competition underground. *Ann. Rev. Ecol. Syst.* **28**:545–570.

Ehleringer, J.R., and T.E. Dawson. 1992. Water uptake by plants: Perspectives from stable isotope composition. *Plant Cell Env.* **15**:1073–1082.

Eswaran, H., and W.C. Bin. 1978. A study of deep weathering profile on granite in peninsular Malaysia. I. Physiochemical and micromorphological properties. *Soil Sci. Soc. Am. J.* **42**:144–149.

Eswaran, H., E. Van den Berg, and P. Reich. 1993. Organic carbon in soils of the world. *Soil Sci. Soc. Am. J.* **57**:192–194.

FAO-UNESCO. 1971–1981. Soil Map of the World, 1:5,000,000. Vols. II–X. Paris: UNESCO.

Field, C.B., R.B. Jackson, and H.A. Mooney. 1995. Stomatal responses to increased CO_2: Implications from the plant to the global scale. *Plant Cell Env.* **18**:1214–1225.

Freckman, D.W., and R.A. Virginia. 1989. Plant-feeding nematodes in deep-rooting desert ecosystems. *Ecology* **70**:1665–1678.

Gale, M.R., and D.F. Grigal. 1987. Vertical root distributions of northern tree species in relation to successional status. *Can. J. For. Res.* **17**:829–834.

Greenwood, E.A.N. 1992. Deforestation, revegetation, water balance, and climate: An optimistic path through the plausible, impracticable and the controversial. *Adv. Bioclim.* **1**:89–154.

Gregory, P.J., M. McGowan, and P.V. Biscoe. 1978. Water relations of winter wheat. 2. Soil water relations. *J. Agr. Sci., Camb.* **91**:103–116.

IPCC (Intergovernmental Panel on Climate Change). 1992. Climate Change 1992: The Supplementary Report to the IPCC Scientific Assessment, ed. J.T. Houghton, B.A. Callander, and S.K. Varney. Cambridge: Cambridge Univ. Press.

Jackson, R.B., J. Canadell , J.R. Ehleringer, H.A. Mooney, O.E. Sala, and E.-D. Schulze. 1996. A global analysis of root distributions for terrestrial biomes. *Oecologia* **108**:389–411.

Jackson, R.B., J.H. Manwaring, and M.M. Caldwell. 1990. Rapid physiological adjustment of roots to localized soil enrichment. *Nature* **344**:58–60.

Jackson, R.B., H.A. Mooney, and E.-D. Schulze. 1997. A global budget for fine root biomass, surface area, and nutrient contents. *Proc. Natl. Acad. Sci. USA* **94**:7362–7366.

Kleidon, A., and M. Heimann. 1998. A method of determining rooting depth from a terrestrial biosphere model and its impacts on the global water and carbon cycle. *Glob. Change Biol.* **4**:275–286.

Körner, C. 1994. Scaling from species to vegetation: The usefulness of functional groups. In: Biodiversity and Ecosystem Function, ed. E.-D. Schulze and H.A. Mooney, pp. 117–140. Berlin: Springer.

Lewis, D.C., and R.H. Burgy. 1964. The relationship between oak tree roots and groundwater in fractured rock as determined by tritium tracing. *J. Geophys. Res.* **69**:2579–2588.

Lieth, H., and R.W. Whittaker. 1975. Primary Productivity of the Biosphere. Berlin: Springer.

Matthews, E. 1983. Global vegetation and land use: New high-resolution data bases for climate studies. *J. Clim. Appl. Meteorol.* **22**:474–487.

Meinzer, O.E. 1927. Plants as indicators of ground water. U.S. Geological Survey, Water Supply Paper 577, 95 pp.

Nadelhoffer, K.J., and J.W. Raich. 1992. Fine root production estimates and belowground carbon allocation in forest ecosystems. *Ecology* **73**:1139–1147.

Neilson, R.P. 1995. A model for predicting continental-scale vegetation distribution and water balance. *Ecol. Appl.* **5**:362–385.

Nepstad, D.C., C.R. de Carvalho, E.A. Davidson, P.H. Jipp, P.A. Lefebvre, G.H. Negreiros, E.D. da Silva, T.A. Stone, S.E. Trumbore, and S. Vieira. 1994. The role of deep roots in the hydrological and carbon cycles of Amazonian forests and pastures. *Nature* **372**:666–669.

Phillips, W.S. 1963. Depth of roots in soil. *Ecology* **44**:424.

Pierce, L.L., J. Walker, T.I. Downling, T.R. McVicar, T.J. Hatton, S.W. Running, and J.C. Coughlan. 1993. Ecohydrological changes in the Murray-Darling basin. III. A simulation of regional hydrological changes. *J. Appl. Ecol.* **30**:283–294.

Post, W.M., W.R. Emanuel, P.J. Zinke, and A.G. Stangenberger. 1982. Soil carbon pools and world life zones. *Nature* **298**:156–159.

Rawitscher, F. 1948. The water economy of the vegetation of the "Campos Cerrados" in southern Brazil. *J. Ecol.* **36**:237–268.

Richards, J.H., and M.M. Caldwell. 1987. Hydraulic lift: Substantial nocturnal water transport between soil layers by *Artemisia tridentata* roots. *Oecologia* **73**:486–489.

Richter, D.D., and D. Markewitz. 1995. How deep is soil? *BioScience* **45**:600–609.

Ruxton, B.P., and L. Berry. 1957. Weathering of granite and associated erosional features in Hong Kong. *Geol. Soc. Am. Bull.* **8**:1263–1292.

Schimel, D.S. 1995. Terrestrial ecosystems and the carbon cycle. *Glob. Change Biol.* **1**:77–91.

Schlesinger, W.H. 1977. Carbon balance in terrestrial detritus. *Ann. Rev. Ecol. Syst.* **8**:51–81.

Schlesinger, W.H. 1997. Biogeochemistry: An Analysis of Global Change. San Diego: Academic.

Schlesinger, W.H., J.F. Reynolds, G.L. Cunningham, L.F. Huenneke, W.M. Jarrell, R.A. Virginia, and W.G. Whitford. 1990. Biological feedbacks in global desertification. *Science* **247**:1043–1048.

Schulze, E.-D., F.M. Kelliher, C. Körner, J. Lloyd, and R. Leuning. 1994. Relationships among maximum stomatal conductance, ecosystem surface conductance, carbon assimilation rate, and plant nitrogen nutrition: A global ecology scaling exercise. *Ann. Rev. Ecol. Syst.* **25**:629–660.

Scott, D.F., and W. Lesch. 1997. Streamflow responses to afforestation with *Eucalyptus Grandis* and *Pinus Patula* and to felling in the Mokobulaan experimental catchments, South Africa. *J. Hydrol.* **199**:360–377.

Sellers, P.J., B.W. Meeson, F.G. Hall, G. Asrar, R.E. Murphy, R.A. Schiffer, F.P. Bretherton, R.E. Dickinson, R.G. Ellingson, C.B. Field, K.F. Huemmrich, C.O. Justice, J.M. Melack, N.T. Roulet, D.S. Schimel, and P.D. Try. 1995. Remote sensing of the land surface for studies of global change: Models-algorithms-experiments. *Remote Sens. Env.* **51**:3–26.

Silva, S., W.G. Whitford, W.M. Jarrell, and R.A. Virginia. 1989. The microarthropod fauna associated with a deep rooted legume, *Prosopis glandulosa*, in the Chihuahuan desert. *Biol. Fert. Soils* **7**:330–335.

Snaydon, R.W. 1962. Micro-distribution of *Trifolium repens* L. and its relation to soil factors. *J. Ecol.* **50**:133–143.

Stone, E., and P.J. Kalisz. 1991. On the maximum extent of tree roots. *For. Ecol. Manag.* **46**:59–102.

Sun, G., D.P. Coffin, and W.K. Lauenroth. 1997. Comparison of root distributions of species in North American grasslands using GIS. *J. Veg. Sci.* **8**:587–596.

Vitousek, P.M., P.R. Ehrlich, A.H. Ehrlich, and P.A. Matson. 1986. Human appropriation of the products of photosynthesis. *BioScience* **36**:368–373.

Vogt, K., D.J. Vogt, P.A. Palmiotto, P. Boon, J. O'Hara, and H. Asbjornsen. 1996. Review of root dynamics in forest ecosystems grouped by climate, climatic forest type and species. *Plant Soil* **187**:159–219.

Walter, H. 1954. Die Verbuschung, eine Erscheinung der subtropischen Savannengebiete, und ihre ökologischen Ursachen. *Vegetatio* **5/6**:6–10.

Warming, E. 1892. Lagoa Santa. Et Bidrag til den biologiske Plantegeografi. K. Kanske videns K. 6, Räkke VI. (as cited in Rawitscher 1948).

Webb, R.S., C.S. Rosenzweig, and E.R. Levine. 1993. Specifying land surface characteristics in general circulation models: Soil profile data set and derived water-holding capacities. *Glob. Biogeochem. Cyc.* **7**:97–108.

Whittaker, R.H. 1975. Communities and Ecosystems. London: Macmillan.

12

Regional-scale Export of C, N, P, and Sediment: What River Data Tell Us about Key Controlling Variables

N. CARACO and J. COLE
Institute of Ecosystem Studies, Box AB, Millbrook, NY 12545, U.S.A.

ABSTRACT

One way in which terrestrial and aquatic ecosystems are linked is by the downstream transfer of materials. At the regional scale, humans have potentially large impacts on this linkage by altering transfers of nitrogen, phosphorus, carbon, and sediment. Large river studies allow for regional-scale quantification of this impact and testing of conceptual models. A review of the literature and comparative analyses of riverine data sets reveal a strong signal of human impact for some components, but not others. For the export of both NO_3 and PO_4, humans may be by far the most important control of terrestrial transfer and this human control can be described by very simple models which explain up to 90% of the variation in export. For dissolved organic C (DOC) there is little detectable impact of humans on total export. At the regional scale DOC was predicted largely from water runoff (Q) alone ($r^2 = 0.68$). The C:N ratio of exported dissolved organic matter (DOM) is, on the other hand, strongly influenced by humans. The export of both dissolved organic N (DON) and sediment show intermediate responses. For sediments, the human impact is detectable at the regional scale, but this export is also strongly influenced by physical features of the watershed.

The extremely strong impact of humans on NO_3 and PO_4 export and the extremely weak effect of humans on DOC export are surprising and suggest some revision of existing conceptual models is needed. The continued study of materials export in rivers presents a means to both validate revised models and directly observe responses to changes in disturbance regime or management practices at the regional scale.

INTRODUCTION

Aquatic systems integrate information from the watershed and store this information in both the water column and sediments. Because of this integration, aquatic systems

Integrating Hydrology, Ecosystem Dynamics, and Biogeochemistry in Complex Landscapes
Edited by J.D. Tenhunen and P. Kabat

are studied not only to gain information about the aquatic system itself, but also to gain information on the surrounding terrestrial system. This information has included changes in terrestrial export of toxins, nutrients, and hydrogen ions (Sulliban et al. 1990; Turner and Rabalais 1994).

In addition to the studies of storage within aquatic systems, the flux of materials through aquatic systems can give important information about terrestrial processing and losses. Flux through aquatic systems can best be used to gain information about the surrounding watershed if the bulk of water and material input, in fact, comes from the watershed. Running waters (rivers and streams), unlike some lakes and coastal waters, have very high watershed to surface area ratios and receive essentially all input from the surrounding watershed. Additionally, the high watershed/surface area ratio means that rivers and streams have short residence times. This feature is of interest as, for water chemistry to be used as the indicator of terrestrial inputs, the aquatic processing must either be small or well known. For P, NO_3, and dissolved organic C (DOC) it has been shown that net uptake within lakes is correlated to hydrologic residence time (Ahlgren et al. 1988; Dillon and Molot 1990, 1997). Further, although not yet tested for P and DOC, river retention for N is within the general range expected based on lake models (Howarth et al. 1996). That is, rivers have comparatively low retention of dissolved inorganic nitrogen (DIN) (Figure 12.1) and, based on lower DOC than DIN retention in lakes, they might additionally be expected to have low processing of DOC.

Because of their high watershed to surface area ratio and related short residence time, running waters are particularly useful to study terrestrial transfers. Not surprisingly, there is, in fact, an extensive literature that has used running waters to study terrestrial processing and transfers which has contributed greatly to understanding of evapotranspiration, weathering, and nutrient cycling in terrestrial systems (e.g., Likens et al. 1977). In most of these studies, small streams with watersheds much less than 0.1 km^2 are used. Less often used are large rivers, which tend to have huge, regional-scale watersheds. However, some studies on regional to global transfer of sediment, organic carbon, and N and P have all been done in the past twenty years. These studies have obvious value for testing our understanding of the important controls of regional-scale terrestrial export.

Here we review what river studies tell us about the important controls of N, P, C, and sediment transfer at the regional scale. The transfer of all of these materials is thought to be influenced by both human activity in the watershed and a variety of natural processes. In this study, we rely on cross-system studies of large rivers with differing human impact to test if human activities appear to exert a major control on export. Further, we compare human control with control by simple hydrologic and geologic characteristics. This review is primarily based on information from major rivers with very large-scale watersheds (10,000 to 7,000,000 km^2). These landscapes are all complex and contain, in varying proportions, forests, wetlands, agricultural lands, and urban areas. For NO_3 we also test if simple models developed at this very large scale can be used to make predictions for complex landscapes at smaller scales.

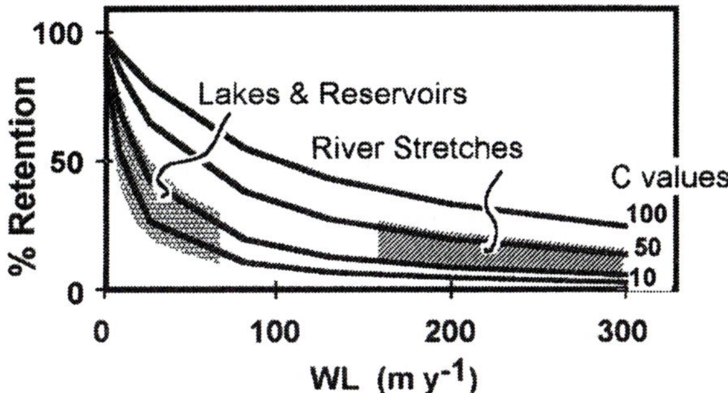

Figure 12.1 Modeled percent retention of NO_3 in aquatic systems versus water load (WL) to the aquatic systems. For each of these values retention (R) is modeled as $R = 100 \times C/(C + WL)$, where C is a constant that describes loss from the water column in m y^{-1}. For rivers this ratio is generally high and rivers, therefore, have high water loads as compared to lakes or estuaries. These high water loads should result in lower in-system retention. After Howarth et al. (1996).

ANALYSIS OF TERRESTRIAL EXPORT

Data

In addition to reviewing results from a variety of published studies on large rivers, we test some of these ideas for data from forty rivers with global distribution. This river data set has already been used to test human control of NO_3 and PO_4 export (Cole et al. 1993; Caraco 1995; Caraco and Cole 1999). NO_3 export is available for the full data set (Cole et al. 1993), PO_4 for most of the rivers (Caraco 1995). For the purpose of this chapter we have compiled DOC, sediment, and dissolved organic N (DON) export for a subset of these systems (Schlesinger and Melack 1981; Meybeck 1982; Milliman and Meade 1983; Esser and Kohlmaier 1991). Additionally, in regard to watershed attributes, runoff, watershed area, maximum elevation, land use (crop, pasture, etc.), human population, urbanization, N and P fertilizer use, and atmospheric deposition of NO_y are available for the full data set (Cole et al. 1993; Caraco 1995).

For NO_3 export at the small scale we used data for tributaries to the Hudson River, U.S.A. (Phillips and Hanchar 1996). The 19 sub-watersheds of these tributaries ranged in size from 150 to 20,000 km^2 and vary from completely forested to mixed forest, agriculture, and urban area. The human population density in the watershed varies from 3 to 100 individuals km^{-2} (Phillips and Hanchar 1996). In addition to export of NO_3 from the sub-watersheds, estimates of N inputs from fertilizer, atmospheric deposition, and sewage loads are available (Phillips and Hanchar 1996).

Model of Export

Esser and Kohlmaier (1991) described a simple conceptual model of DOC export that included human influenced and "natural" features of the watershed. A modified

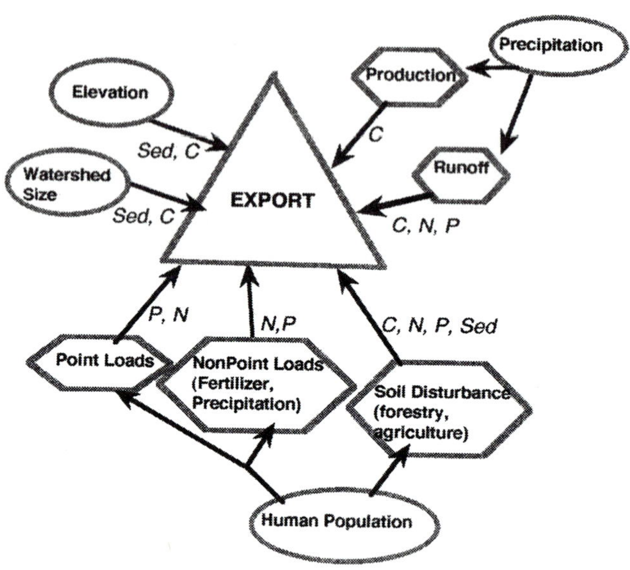

Figure 12.2 Simple conceptual model of key variables controlling C, N, P, and sediment (Sed) export from terrestrial systems. Key controlling variables are listed in circles and related proximate variables are in hexagons.

version of this model serves as a basis for our river analysis. This model includes the major pathways by which C, N, P, and sediment (Sed) export are thought to be controlled (Figure 12.2). For sediments, elevation or slope in the watershed are viewed as major natural controls of release (Ludwig et al. 1996; Milliman and Meade 1983) while for dissolved C, N, and P runoff is viewed as a major control (Meybeck 1982). Additionally, significant human impact on all these components is thought to occur with different processes dominating for different components (Figure 12.2). For some components of this model, tests against regional-scale export in rivers have not yet been made.

Large River Results

Sediment Transport

In their classic review of sediment export from nearly 50 rivers, Milliman and Meade (1983) demonstrated that there was nearly a 10,000-fold variation in area weighted sediment export from different rivers (expressed in tons km^{-2} watershed [WS] y^{-1}). Although rivers were used in this study primarily as an integrative measure of sediment export to the ocean, some analysis on terrestrial controls of export were examined. This limited analysis showed that size of river was inversely related to areal export from the watershed (Figure 12.3). This inverse relationship is likely driven by progressive retention in downstream areas of sediment eroded in steeper headwater

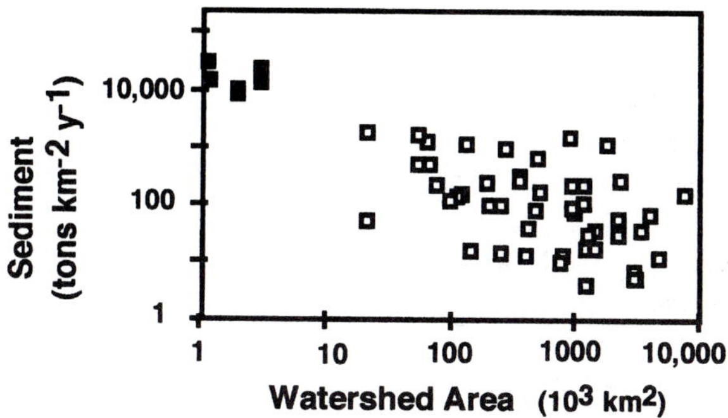

Figure 12.3 Sediment discharge against watershed size (drainage area) in thousands of km^2 (from Milliman and Meade 1983). For all data these variables are marginally related ($r^2 = 0.2$). For large watersheds (10,000 km^2) open squares are used; filled squares are those with small watersheds. Without the 10 small systems, there is little relationship between sediment export and drainage area ($r^2 = 0.05$).

areas (Milliman and Meade 1983). At any rate, the actual predictive power of the relationship is rather low ($r^2 = 0.2$) and for large rivers (10,000–6,000,000 km^2 WS) there is almost no relationship between watershed size and the 3-order of magnitude variation in export ($r^2 = 0.05$; Figure 12.3).

A more recent paper looking at sediment export in rivers found that the key features driving sediment export are slope in the watershed (closely related to elevation difference), runoff, and rainfall variability over the year (Ludwig et al. 1996). This same study did not find a relationship between sediment export and human impacts as indicated by total human population density in the watershed or the percent that was cultivated land.

For a subset of Milliman's data (n = 22), we found that the impact of human population and a multiple regression using rural population density and watershed elevation as independent variables explained 70% of the variation in sediment export. Thus, our limited data analysis suggests that human impact on sediment export may be discernable at the regional scale. It is likely that the important activities that link sediment export and human population, rural population in particular, are agriculture on steep terrain and erosion associated with forest clearing for fuel and construction.

DOC Export

For most terrestrial systems, several percent of the primary production is exported as organic carbon in rivers, and DOC is generally the major form of organic export in most systems (Meybeck 1982; Esser and Kohlmaier 1991). Organic export in river systems is thought to be impacted by precipitation inputs and related changes in primary production and runoff (Figure 12.2). Furthermore, humans can increase export through deforestation and agriculture, which lead to loss of soil organic C (Figure 12.2).

There have been several studies examining regional-scale organic C exports. Schlesinger and Melack (1981) looked at organic C export in a large variety of aquatic systems and found that export was roughly related to water discharge. Esser and Kohlmaier (1991) studied this same process for DOC and POC in large rivers only and found a very good relationship between export and areal water discharge (Q). This relationship was stronger than any of the other relationships with variables that are related to discharge, including precipitation and net primary production. For DOC the relationship was fit with the linear equation

$$\text{DOC (g/m}^3\text{)} = 6.4\ \text{Q (m/y)}, \tag{12.1}$$

and Q explained about 80% of DOC export. Similarly, for our data set we found

$$\text{DOC (g/m}^3\text{)} = 4.9\ \text{Q (m/y)} + 0.17, \tag{12.2}$$

with Q explaining almost 70% of the variation in DOC export (Figure 12.4).

Q is in general related to primary production within terrestrial systems so DOC export also relates to primary production. However, DOC export appears to relate better to Q than it does to primary production (Esser and Kohlmaier 1991; Ludwig et al. 1996). Thus, the data suggest either that leaching of organic material is limited by

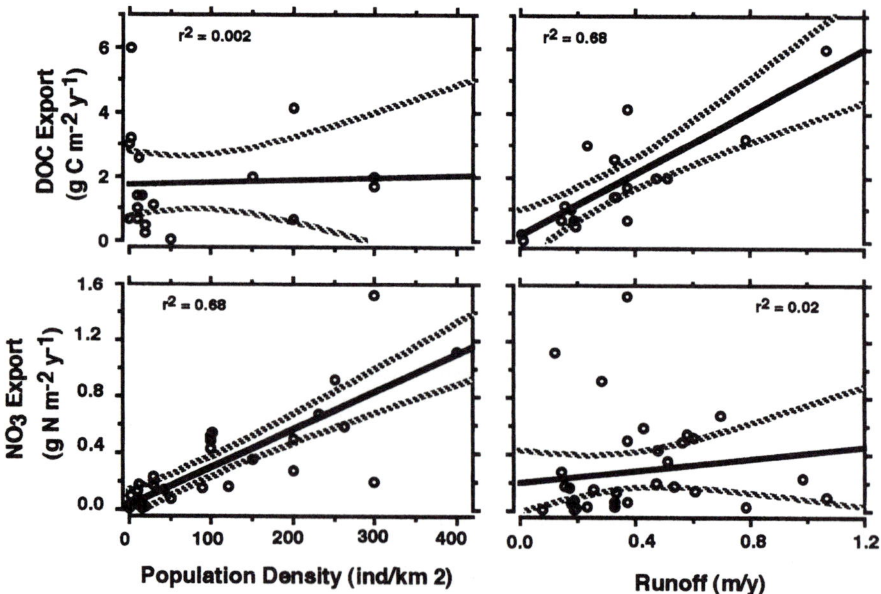

Figure 12.4 DOC (upper) and NO_3 (lower) export in large rivers against population density (left) and watershed runoff (Q, right). The relationship between human population and NO_3 export and that between DOC export and Q are both highly significant ($p < 0.01$). The relationships between DOC and human population and NO_3 and Q are not significant ($p > 0.05$).

water throughflow or that DOC, through some equilibrium physical processes, comes to a value of between 4 and 7 g C m^3 in most rivers.

For both our data set and that of Esser and Kohlmaier (1991), the only variable related to DOC export, that was not in turn related to water discharge, was elevation in the watershed. In neither data set was human impact on DOC export apparent. Esser and Kohlmaier (1991) did not find a positive relationship between agricultural area or increase in agricultural area in the past 30 years and DOC export. In fact, change in agricultural area was actually slightly negatively related to DOC export. Another human impact, forest clearing, was not tested against river export in the analysis of Esser and Kohlmaier (1991), and we know of no study that has formally tested the relationship between forest clearing and organic C export at the regional scale. Lastly, in regards to human impact, our data set showed no relationship between DOC export and any of a number of indicators of human influence, including human population density in the watershed (Figure 12.4), agricultural or crop area, or fertilizer use in the watershed.

One possible reason that human activity does not show any clear relationship to DOC export is that a large but variable fraction of the DOC that enters rivers is metabolized. High within-river metabolism seems in opposition to models of retention developed for lakes (Dillon and Molot 1997). If these lake models hold for rivers it would suggest that, considering the relatively short residence time of water in most rivers (Howarth et al. 1996; Vorosmarty 1997), only about 3%–10% of DOC would be retained within rivers by metabolism. This model developed for lakes may not, however, be applicable to rivers, and rivers may metabolize far more DOC than is expected from lake models.

In agreement with the possibility of relatively high DOC metabolism within rivers is the observation of extremely high CO_2 concentrations in many of these systems (Kempe et al. 1982; Raymond et al. 1997; Cole unpubl. data). These high CO_2 concentrations suggest that CO_2 flux from rivers is in the range of 50 to 500 g C m^{-2} (river) y^{-1} or about 0.5 to 5 g C m^{-2} (Ws) y^{-1}. This CO_2-implied metabolism is co-equal to DOC export from the average river (ca. 2 g C m^{-2} [WS] y^{-1}) and suggests metabolism of near 50% of material entering the river. Determining whether large amounts of organic C that enter rivers are metabolized will be key in the future to linking models of organic C export from watersheds to regional-scale DOC export from large rivers.

NO_3 and Organic N

The two major forms of N being exported from rivers are NO_3 and dissolved organic N, and these two forms combined generally comprise more than 70% of the total N in rivers (Meybeck 1982). NO_3 is the most frequently measured form of N in aquatic systems due to ease of measurement, clear links to eutrophication, and health impacts (Peierls et al. 1991). The export of NO_3 has, therefore, been subject to greater analysis than have other forms of N. Studies on total N or organic N export have been fewer (see, however, Howarth et al. 1996).

While regional studies of NO_3 export suggest that water discharge might relate to export in pristine systems (Meybeck 1982), human activities are clearly the most important control of NO_3 export (Meybeck 1982; Cole et al. 1993; Jordan and Weller 1996; Caraco and Cole 1998). In a study of rivers with global distribution it was found that human population density in the watershed explained about 50% of the variation in NO_3 export from rivers (Peierls et al. 1991; Cole et al. 1993). Further, for these rivers, a model that included consideration of human impacts from sewage loads and precipitation and fertilizer inputs explained 89% of the variation in export (Figure 12.5, linear regression; Caraco and Cole 1998). Similarly, for regions of the United States, total anthropogenic inputs of N were closely related to variation in NO_3 export (Jordan and Weller 1996).

DON export from large rivers might be expected to behave like NO_3 export and show strong responses to human activity, or like DOC export and show no observable response. There is little literature addressing this topic. Our analysis of a limited data set suggests, however, that the answer is intermediate. That is, there does appear to be some indication of a human impact on DON export, but it is less severe than is the response of NO_3 export to human activities. For example, over a 1000-fold range in population density, NO_3 export in rivers on average increases by 100-fold (Cole et al. 1993; from 0.01 to 1 g N m^{-2} y^{-1}). Over the same population range, DON export increased by only about 3-fold (Caraco and Cole 1999, from 0.1 to 0.4 g N m^{-2} y^{-1}). The lower human impact on DON as compared to NO_3 means that humans change the ratio of NO_3:DON export from aquatic systems. While NO_3:DON export ratios appear to be near 0.1:1 in relatively pristine rivers, they are as high as 3:1 in some heavily impacted rivers of Asia, North America, and Europe.

The impact of humans on DON but not DOC suggests that human activity leads to a decrease in C:N ratios of dissolved organic matter (DOM) exported in rivers. Indeed we found some evidence for this in that C:N ratio of DOM is related to NO_3 concentration in river water (Figure 12.6); NO_3 is in turn strongly impacted by humans (Figures 12.4 and 12.5). At low NO_3 concentration, river C:N ratios in riverine DOM are near 70:1, at higher NO_3 values, C:N ratios are as low as 10:1. The mechanism for the increased N content of organic material export could be heterotrophic immobilization of DIN within aquatic systems (Caraco et al. 1998) or could be due to differences in material export from the terrestrial system (Currie et al. 1996). In either case, if higher DIN leads to lower C:N ratios of exported DOC, it implies that human activity may have increased the quality of organic material leaving terrestrial systems. If this is true, any human impact leading to higher DOC export may be masked by higher metabolism of this material in downstream aquatic systems.

PO_4 Export and N:P Ratios

For PO_4 export in rivers, some studies suggest that a major control of export in rivers is chemical equilibrium processes within the river system (Fox 1993). Other studies suggest that human control dominates, similar to NO_3. At the global scale, human

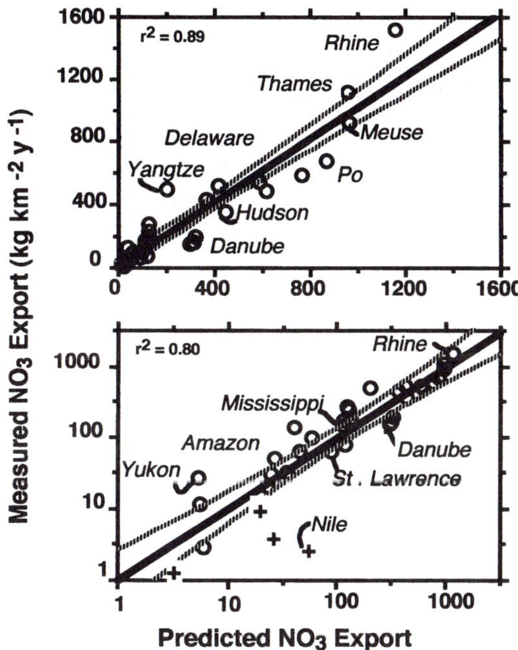

Figure 12.5 Measured NO_3 export for 35 rivers against that predicted from a simple model of human impact on N load. Both linear (top) and log–log relationships are shown. The points with the + signs are those rivers in dry regions with very low Q from the watershed. From Caraco and Cole (1999).

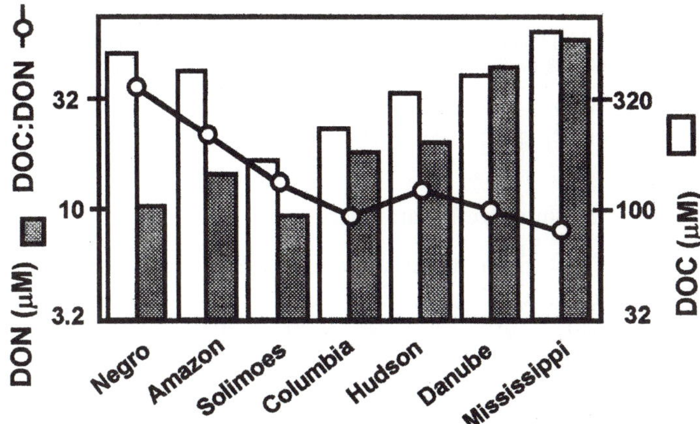

Figure 12.6 DOC, DON, and DOC:DON ratio in 7 rivers with nitrate concentration ranging from 2–100 µM. Rivers are ordered by nitrate concentration (low to high). While DOC and nitrate are not correlated ($r^2 = 0.05$), DON is positively correlated to nitrate ($r^2 = 0.78$) and DOC:DON is negatively related ($r^2 = 0.77$). Data from Meybeck (1982) and Esser and Kohlmaier (1991).

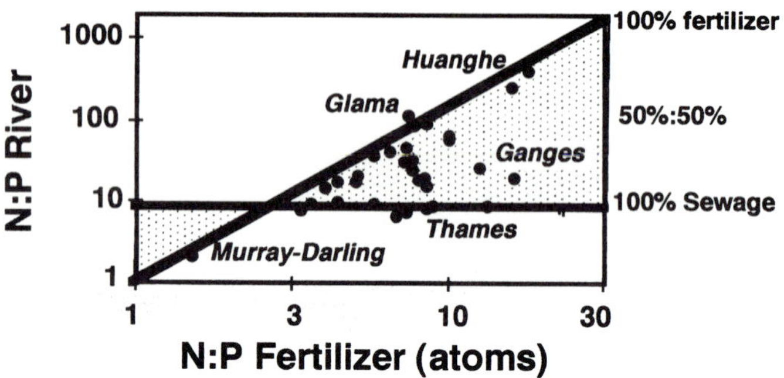

Figure 12.7 NO_3 to PO_4 ratios being exported from rivers against N to P of fertilizer use in the watershed. The two lines that enclose most of the data are meant to represent agricultural runoff (changing against fertilizer N:P) and a point source load (constant with fertilizer load). Fertilizer N:P alone explains 40% of the variation in N:P of export, while a multiple regression that includes fertilizer N:P and sewage input explains 50% of the variation in N:P or export. Both regressions are significant ($p < 0.01$). After Caraco (1995).

population density explains about 45% of the variation in export while models that consider point load and fertilizer load to the watershed explain 72% of the variation in export (Caraco 1995).

In addition to having large impacts on N and P export, an impact of humans may be to change the ratio of N:P export. While some studies suggest that this impact may be a unidirectional decrease (Billen et al. 1991; Howarth et al. 1995), one study done on rivers with global distribution suggests a more complicated effect of humans (Caraco 1995). That is, on a global scale the NO_3:PO_4 ratio of export is related positively to fertilizer N:P (Figure 12.7) but negatively to urban population density (a measure of sewage loading). These different impacts mean that in some areas, like China, with high fertilizer N but low fertilizer P use and a generally rural population, humans may have decreased N:P ratios being exported to coastal waters in rivers.

Export from Intermediate-sized Watersheds

Measured export of NO_3 from the tributaries of the Hudson River varied from 150 to 700 kg N km^{-2} y^{-1} (0.15 to 0.7 g N m^{-2} y^{-1}). Runoff of water did not vary greatly between these watersheds; however, population density in the watershed varied between 3 and 100 individuals per km^{-2}. As for the large river data, the variation in human population density related well to variation in NO_3 export, explaining 69% of the variation in export. Furthermore, a simple model that was used successfully to explain NO_3 export at the large regional scale (Figure 12.5) explained well the variation in NO_3 export at the sub-watershed scale (Figure 12.8). The model explained 74% of the variation in river export and gave close to a 1:1 prediction between modeled and measured export. Thus, at this intermediate spatial scale (150 to 20,000 km^2), as for the very

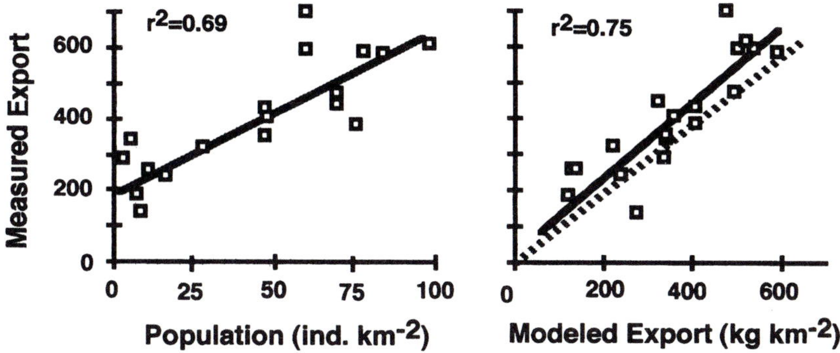

Figure 12.8 NO_3 export from 19 subbasins of the Hudson River vs. human population density in the watershed (Left Panel) and predicted export based on a simple model that considers only N loads to the watershed from fertilizer and precipitation and sewage inputs of N to running waters (Right Panel, Caraco and Cole 1999).

large scale, NO_3 export from complex landscapes may in some cases be described relatively well from simple models of human activity. More studies at this scale are needed to determine the generality of this finding.

Integrating River Studies with Models of Export

Rivers are an important source of N, P, and C to many coastal waters and the export of these materials ultimately drives burial of materials in the deep sea and the long-term gas flux from oceans (Meybeck 1982; Cole et al. 1993; Caraco 1995; Howarth et al. 1995). The N and P input to coastal waters is of great interest as increased nutrient input to coastal waters can increase productivity of coastal waters and at times cause unwanted eutrophication (Cole et al. 1993; Howarth et al. 1996). The negative impacts of eutrophication include algal blooms that at times can be toxic, anoxic bottom waters, and related fish and shellfish death. The long-term (hundreds of thousands to hundreds of millions of years) control of gas flux from the ocean is of interest as the oceans play a major role in net global gas exchange and, therefore, are key in controlling atmospheric oxygen and CO_2 concentrations (Holland 1990).

Because of the importance of rivers to coastal and some whole ocean processes, the export of materials from rivers was studied first from the point of view of summed inputs (Meybeck 1982; Milliman and Meade 1983). More recently a major focus of large river work has shifted to the source of materials to rivers and human impacts on terrestrial transfers. Studies on large rivers that may be particularly valuable in this regard are those that link models with actual observations of river export.

Such studies have already revealed interesting aspects about terrestrial export at the very large scale. For example, for some compounds, very simple models (and in the extreme single variables) can describe much of the variation in export (Esser and Kohlmaier 1991; Cole et al. 1993). For example, for DOC, water runoff (Q) in itself describes much of the variation in export, while for NO_3, human population density describes much of the export (Figure 12.4) with many of the other variables thought to influence export showing no explanatory power (Figures 12.2 and 12.4). Further, data from sub-watersheds of the Hudson River suggest that at intermediate scales, control of export may also be relatively simply described (Figure 12.8). That single variables explain much of the variation in export at large and intermediate scales is surprising. First, even simple conceptual models of export include many points of control for N and C export (Figure 12.2). Second, studies at the very small scale often reveal much more complexity. Variables like wetland area, soil thickness, and vegetation type appear to all contribute substantially to the large variation at this scale. Studies on running waters ranging in size from streams to large rivers will be useful in examining at which scale complexity increases.

Another aspect revealed by river studies is the variability of human impact on regional export. NO_3 export, for example, is dominated by human activity while DOC shows no detectable impact (Figure 12.4). For both of these compounds, conceptual models of terrestrial export and processing could suggest similar impacts (Figure 12.2). Considering N, although humans have increased new inputs of N to terrestrial systems, to date, natural inputs are still co-equal to human inputs (Vitousek and Matson 1994). Thus, humans are not completely swamping natural inputs. Further, before being exported, new inputs of N go through much processing in soils, groundwaters, riparian areas, wetlands, and in river systems. Nevertheless, recent models that have considered only new human inputs and have ignored natural N loading by fixation and complex processing of N have been successful in predicting NO_3 and total N export at regional scales (Cole et al. 1993; Howarth et al. 1996; Jordan and Weller 1996; Caraco and Cole 1999).

As for N export, humans are expected to have multiple impacts on organic C. These impacts include forest clearing and agricultural disturbance of soils (Esser and Kohlmaier 1991). For organic C, however, this impact is not dominant or even readily detectable at the regional scale. The difference between the response of N (or at least NO_3) and organic C could be due to the fact that more of the C is processed in the downstream path after human impact (Esser and Kohlmaier 1991). This prediction is in opposition to predictions of retention generated for lakes (Figure 12.1); there could, however, be a difference in processing of organic C in lakes and rivers (or at least human impacted rivers). This is possible in that it seems that DOC in heavily human impacted rivers, with high NO_3, have lower C:N ratios than those in more pristine systems. Lower C:N ratios could be related to higher processing rates (Figure 12.6).

For N export, another interesting aspect revealed from the river studies is the differential impact of humans on NO_3 and organic N. That is, much of human impact is on NO_3 export with a far smaller impact on organic N export. While this might be explained by the fact that NO_3 is very mobile, this explanation does not seem entirely consistent with the fact that DON export dominates in many pristine systems (Meybeck 1982). The pattern could be explained by the fact that some of the human N input enters as NO_3 (from atmospheric sources or NH_4NO_3 fertilizers) and simply passes through the system without being substantially retained. Studies, suggest, however, that at the present most (about 80%) of the human input of N is retained (Jordan and Weller 1996; Howarth et al. 1996; Caraco and Cole 1999). In addition, some recent isotope (^{18}O) studies have shown that much of the NO_3 input to terrestrial systems is processed to organic matter in terrestrial systems before being later exported as NO_3 (Burns and Kendell, pers. comm.).

Saturation of N uptake in the terrestrial system is another possible explanation for differential export of N in organic forms from pristine systems and NO_3 in human impacted systems. That is, NO_3 export occurs only after uptake potential of plants is saturated. This saturation is unlikely to occur from the self-regulated process of N_2 fixation. Organic N on the other hand could originate from leakage of long-term accumulated N pools (Vitousek and Matson 1994), some of which may be unavailable to plant uptake and have little to do with saturation of plants. At any rate, if saturation explains NO_3 export, one would expect an accelerating release of N with increased N export to terrestrial systems. While some cross-system studies of NO_3 export have shown this acceleration (Jordan and Weller 1996), others have not demonstrated an obvious acceleration (Caraco and Cole 1999). More work at the regional scale is needed to demonstrate the nature of N saturation.

In conclusion, river studies can reveal controls of regional-scale element processing and export and are especially useful when compared to expectations of conceptual models (Figure 12.2). Considering the importance of regional-scale processing, there are relatively few studies of this kind. Advances could be made by extending analysis of river data to a broader number of systems so that minor controlling variables can be more easily seen, looking across different scales to examine how controls change from one scale to another, and examining carefully the export of a greater number of compounds. Studies to date suggest that investigations of organic N, DOC quality, and inorganic C (both DIC and CO_2) in rivers could be particularly useful.

REFERENCES

Ahlgren, I., T. Frisk, and L. Kamp-Nielsen. 1988. Empirical and theoretical models of phosphorus loading, retention and concentration vs. lake trophic state. *Hydrobiologia* **170**:285–303.

Billen, G., C. Lancelot, and M. Meybeck. 1991. N, P, and Si retention along the aquatic continuum from land to ocean. In: Ocean Margin Processes in Global Change, ed. R.F.C. Mantoura, J.M. Martin, and R. Wollast, pp. 19–44. Dahlem Workshop Report PC 9. Chichester: Wiley.

Caraco, N.F. 1995. Influence of humans on phosphorus transfers to aquatic systems: A regional scale study using large rivers. In: Phosphorus in the Global Environment: Transfers, Cycles and Management, ed. H. Tiessen, pp. 235–244. Chichester: Wiley.

Caraco, N.F., and J.J. Cole. 1999. Human impact on nitrate export: An analysis using major world rivers. *Ambio* **28(2)**, in press.

Caraco, N.F., G. Lampman, J.J. Cole, K.E. Limburg, M.L. Pace, and D. Fischer. 1998. Microbial assimilation of DIN in a nitrogen rich estuary: Implications for food quality and isotope studies. *Mar. Ecol. Prog. Ser.* **167**:59–71.

Clair, T.A., T.L. Pollock, and J.M. Ehrman. 1994. Exports of carbon and nitrogen from river basins in Canada's Atlantic Provinces. *Glob. Biogeochem. Cyc.* **8**:441–450.

Cole, J.J., B.L. Peierls, N.F. Caraco, and M.L. Pace. 1993. Nitrogen loading of rivers as a human-driven process. In: Humans as Components of Ecosystems: The Ecology of Subtle Human Effects and Populated Areas, ed. M.J. McDonnell and S.T.A. Pickett, pp. 141–157. Berlin: Springer.

Currie, W.S., J.D. Aber, W.H. McDowell, R.D. Boone, and A.H. Magill. 1996. Vertical transport of dissolved organic C and N under long-term N amendments in pine and hardwood forests. *Biogeochemistry* **35**:471–505.

Dillon, P.J., and L.A. Molot. 1990. The role of ammonium and nitrate retention in the acidification of lakes and forested catchments. *Biogeochemistry* **11**:23–43.

Dillon, P.J., and L.A. Molot. 1997. Dissolved organic and inorganic carbon mass balances in central Ontario lakes. *Biogeochemisty* **36**:29–42.

Esser, G., and G.H. Kohlmaier. 1991. Modelling terrestrial sources of nitrogen, phosphorus, sulphur and organic carbon to rivers. In: Biogeochemistry of Major World Rivers, ed. E.T. Degens, S. Kempe, and J.E. Richey, pp. 297–322. Chichester. Wiley.

Fox, L.E. 1993. The chemistry of aquatic phosphate: Inorganic processes in rivers. *Hydrobiologia* **253**:1–16.

Holland, H.D. 1990. Origins of breathable air. *Nature* **347**:17.

Howarth, R.W., G. Billen, D. Swaney, et al. 1996. Regional nitrogen budgets and riverine N & P fluxes for the drainages to the North Atlantic Ocean: Natural and human influences. *Biogeochemistry* **35**:75–139.

Howarth, R.W., H. Jensen, R. Marino, and H. Postma. 1995. Transport and processing of phosphorus in near-shore and oceanic waters. In: Phosphorus in the Global Environment: Transfers, Cycles and Management, ed. H. Tiessen, pp. 323–346. Chichester: Wiley.

Jordan, T.E., and D.E. Weller. 1996. Human contributions to terrestrial nitrogen flux. *BioScience* **46**:655–664.

Kempe, S. 1982. Compilation of carbon and nutrient discharge from major world rivers. *Mitt. Geol.Pal. Inst. Univ. Hamburg* **58**:29–32.

Likens, G.E., F.H. Bormann, R.S. Pierce, J.S. Eaton, and N.M. Johnson. 1977. Biogeochemistry of a Forested Ecosystem. New York: Springer.

Ludwig, W., J.-L. Probst, and S. Kempe. 1996. Predicting the oceanic input of organic carbon by continental erosion. *Glob. Biogeochem. Cyc.* **10**:23–41.

Meybeck, M. 1982. Carbon, nitrogen, and phosphorus transport by world rivers. *Am. J. Sci.* **282**:401–450.

Milliman, J.D., and R.H. Meade. 1983. World-wide delivery of river sediment to the oceans. *J. Geol.* **91**:1–21.

Peierls, B.L., N.F. Caraco, M.L. Pace, and J.J. Cole. 1991. Human influence on river nitrogen. *Nature* **350**:386–387.

Phillips, P.J., and D.W. Hanchar. 1996. Water-quality assessment of the Hudson River Basin in New York and adjacent states — Analysis of available nutrient, pesticide, volatile organic compound, and suspended-sediment transport. U.S. Geological Survey, Water-Resources Investigations Report 96–4065.

Raymond, P.A., N.F. Caraco, and J.J. Cole. 1997. CO_2 concentration and atmospheric flux in the Hudson River. *Estuaries* **20**:381–390.

Schlesinger, W.H., and J.M. Melack. 1981. Transport of organic carbon in the world's rivers. *Tellus* **33**:172–187.

Sulliban, T.J., D.F. Charles, J.P. Smol, et al. 1990. Quantification of changes in lakewater chemistry in response to acidic deposition. *Nature* **345**:54–58.

Turner, R.E., and N.N. Rabalais. 1994. Coastal eutrophication near the Mississippi river delta. *Nature* **368**:619–621.

Vitousek, P.M., and P.A. Matson. 1994. Agriculture, the global nitrogen cycle, and trace gas flux. In: Biogeochemistry of Global Change, ed. R.S. Oremland, pp. 193–208. New York: Chapman and Hall.

Vorosmarty, C.J., K.P. Sharma, B.M. Fekete, A.H. Copeland, J. Marble, and J.A. Lough. 1997. The storage and aging of continental runoff in large reservoir systems of the world. *Ambio* **26**:210–219.

Standing, left to right:
Nina Caraco, Rob Jackson, Herbert Lang, Eric Wood, Lucas Menzel
Seated, left to right:
Carol Johnston, George Hornberger, Otto Richter

13

Group Report: How Is Ecosystem Function Affected by Hydrological Lateral Flows in Complex Landscapes?

L. MENZEL and O. RICHTER, Rapporteurs

N. CARACO, G.M. HORNBERGER, R.B. JACKSON, C.A. JOHNSTON, H. LANG, and E.F. WOOD

INTRODUCTION

Complexity, like beauty, is in the eye of the beholder. A landscape may be either "simple" or "complex" depending on what aspects of the lateral fluxes of water, chemicals, and energy are considered. Nevertheless, some guidance can be given for making the distinction. In this report we discuss the minimum spatial scale of what is considered to be a "landscape" and possibilities to describe complexity on such a scale, together with the question of how to capture complexity in models.

Lateral flow in complex landscapes redistributes water, organic matter, and minerals and can lead to complex, interrelated patterns of soil moisture, vegetation, and biogeochemistry in terrestrial landscapes. Lateral flow into open water aquatic systems causes considerable losses of nutrients, as described below. Groundwater recharge largely depends on soil heterogeneity and related subsurface lateral flows. Understanding ecosystem dynamics is crucially related to the monitoring of such below-ground variations, including horizontal and vertical root distributions (see section on **LATERAL FLOW AND SUBSURFACE PROCESSES**).

In terms of linking measurements and modeling in complex landscapes, much remains to be learned. In particular, subgrid-scale processes must be parameterized in models. Both purely model-based (e.g., sensitivity) studies and comprehensive field observations are needed to develop and test the parameterizations. Consequently, we address possibilities to capture essential processes of lateral flow at different scales into geostatistical approaches and emphasize the need for interdisciplinary field

Integrating Hydrology, Ecosystem Dynamics, and Biogeochemistry in Complex Landscapes
Edited by J.D. Tenhunen and P. Kabat

studies on a regional scale. Finally, we list our recommendations to assess the importance of lateral flows better within future research work frameworks.

LANDSCAPE COMPLEXITY

To define exactly what is meant by complex landscapes, typical attributes of "complexity" and their possible measurement have to be recognized. Further, if a landscape is regarded as complex, the relation of these attributes with heterogeneity in general or the sensitivity of the considered system in hydrologic and ecosystem response should be identified. Finally, the question arises of how landscape complexity (i.e., complexity on a regional scale) can be measured and integrated into appropriate models. In this context it is also of interest whether this integration is in fact necessary to describe complex systems beyond the local scale.

What Is a Complex Landscape?

Complexity can occur at all scales, from microbial environments to the globe. There was discussion as to the minimum spatial scale that is relevant to the problems addressed at the workshop; this scale is approximately 100 to 1000 m (the hillslope scale), which is also consistent with the scale of VIS/IR remote sensing (Thematic Mapper and AVHRR) and extends up to catchment and regional scales. At the landscape scale, landscape complexity is imparted by variations in the spatial distribution of geomorphologic features, biotic features, and disturbances, including human land uses. Most landscapes encompass a mixture of these three variables.

1. *Geomorphology* is the template upon which biota and disturbances are superimposed. It is the most enduring, least alterable source of landscape complexity. Geomorphologic complexity includes variation in bedrock, landforms, soils, elevation (mountainous vs. level terrain), and degree of dissection.
2. *Biotic complexity* is imparted by the variety of life forms that populate a landscape. Biotic complexity is thought of primarily in terms of vegetation distribution; however, the distribution of microbial biota can also influence key ecosystem functions, such as soil carbon accumulation, nutrient mineralization, and fluxes of methane and N_2O to the atmosphere.
3. *Disturbances* further influence landscape complexity, primarily by setting back biotic succession, but also by altering hydrology. Disturbances such as fire and windthrow of trees act directly upon vegetation, whereas floods and drought alter hydrology, often with additional feedbacks to vegetation. Humans are major agents of disturbance, adding to the complexity of landscapes by altering biota (agriculture, logging, urban development) and hydrology (water withdrawals, land irrigation). Geomorphic complexity is altered only by extreme disturbances that rapidly move massive amounts of matter, such as volcanic eruptions, earthquakes, and landslides.

These sources of landscape complexity are dependent upon each other. Geomorphic complexity is linked to biotic complexity via controls on soil moisture distribution and nutrient supply. Geomorphic and biotic complexity are linked to the spread of disturbances such as fire (see Noble, this volume).

Topography is a forcing function that controls lateral hydrologic flows (surface and subsurface) and results in the formation of streams and riparian wetland areas (see Wood, this volume). Over long time scales, these flows result in soil erosion and redistribution such that finer-grained soils are often found at the bottom of hillslopes and near streams. These distributions of soil and water manifest themselves in the distribution of vegetation, especially in water-limited environments. The influence of the underlying geomorphologic template on the vegetation distribution is a result of nutrient and water supply limitations; however, subsurface adaptations allow plants to overcome some of these limitations.

A question that remains open is to what extent a complex landscape (or complexity in general) should also be defined by climate. For instance, when humidity increases (increase in mean annual precipitation), does complexity decrease? In this case vertical subsurface fluxes would take on an increasingly dominant role, and horizontal soil water redistribution would have less influence on vegetation patterns. As an example, in humid alpine regions, the influence of climate (e.g., differences in radiation fluxes, in temperature, in wind, and in soil moisture through inclination and various exposures) plays a dominant role in the distribution of vegetation.

Finally, we need to determine whether it is helpful to define a "measure" of complexity to characterize a landscape as complex and how important the related lateral fluxes are. Value of the measure could be composed of a combined recognition of natural or anthropogenic "driving forces," such as climate, topography, and disturbance.

What Are Attributes of Complex Landscapes and How Are They Measured?

The description of attributes of landscape complexity and their measures remained unresolved during the workshop. Complexity may be viewed from several perspectives, as illustrated by a catchment divided into two nested scales. Figure 13.1a illustrates a catchment with a river network overlain by a fine-scale grid or "pixel" map. Complexity is related inextricably to measures of heterogeneity in the landscape; it is such heterogeneities — in atmospheric inputs, in elevation, in soil properties, etc. — that give rise to lateral hydrological flows and related variations in soil moisture, in vegetation cover, and in biogeochemical fluxes. If the scale of the heterogeneity is very small relative to the scale of the landscape unit being considered (the pixel), then the variations over the landscape can be treated statistically and the system can be considered as homogeneous (i.e., "simple"). If the scale of the heterogeneity is relatively large (say about 10% the size of the pixel), then the landscape will be hydrologically "complex" and spatial patterns in landscape heterogeneity are likely to be important. We felt that a landscape may be either simple or complex depending on what aspects of the lateral fluxes of water, chemicals, and energy are considered.

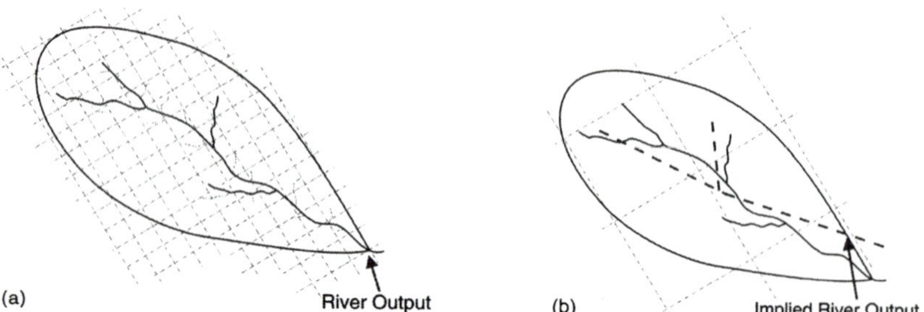

Figure 13.1 Illustrative catchment with a river network overlain by a fine (a) and coarse grid (b).

Across a watershed transect or in adjacent plots within one large grid, a process may be complex if the variance (σ^2) of nearby estimates is high. Complexity in this form is equated with high local variability (what we might term Type 1 complexity). A second form of complexity (Type 2) arises when measurements or predictions at fine scales fail to aggregate to the actual value at a larger scale. In our illustration, this arises if estimates from the fine-scale grids fail to give the appropriate large grid value (see Figure 13.1b). The local variance can be small, even nonexistent, but our predictive ability is nonetheless poor. Type 2 complexity arises when the hydrologic, ecological, and/or biogeochemical processes are nonlinear with respect to the heterogeneous landscape-dependent variables — like soil moisture. Ignoring lateral transport across grid boundaries is one mechanism that would contribute to Type 2 complexity. In contrast, local variance may be high in Type 1 complexity, but the underlying processes may be linear (or weakly nonlinear) to allow aggregation, or we may have appropriate methods or models for aggregating or predicting the specific values. For example, adjacent plots may be north- and south-facing slopes, respectively. The measurements vary at fine scales, but we can correctly predict the large-scale outcome. Both types of complexity may be important in natural systems.

How Do We Develop Models or Methods to Measure This Complexity?

One perspective of flows within a landscape can be given by the storage equation with only horizontal fluxes:

$$\frac{dS}{dt} = -D_h F_h + P - kS \,, \tag{13.1}$$

where dS/dt is a change of storage in a landscape state variable (soil moisture, chemical, biota, etc.); F_h is the horizontal flux related to S; $D_h F_h$ is the divergence of the horizontal flux F_h; P and $-kS$ are source/sink terms. The equation is valid across all scales. For specific scales (pixel sizes), it is usually parameterized by hydrologic and biogeochemical process equations. Possibly a landscape is complex if the divergence $D_h F_h$ is large. This concept needs to be compared, however, to the Type 1 and Type 2 complexity discussed earlier.

LATERAL FLOW AND ITS IMPACT ON COMPLEX LANDSCAPES

Lateral flow redistributes moisture within landscapes and ultimately results in water, organic matter, and mineral loss from the terrestrial system to adjoining aquatic systems. These aquatic systems include wetlands, lakes, streams, rivers, and the coastal ocean. At a regional scale, lateral flow of rivers is an important output of water and nutrients. Groundwater is often an important link between the terrestrial system and the open water aquatic system.

Small-scale redistribution of water leads to variations in soil moisture often impacting the vegetation and biogeochemistry of the terrestrial landscape. Wetter areas can have very different vegetation types and cover than drier locations, and have both greater primary production and respiration. Additionally, wet areas can act as hotspots of trace gas flux. For example, CH_4 flux from a landscape will occur from only the wettest areas (Naiman et al. 1991; Bridgham et al. 1995).

The redistribution of moisture by water within the terrestrial gradients is driven by a number of system components. For soil moisture, geomorphologic and topographic gradients and plant growth are likely key features controlling redistribution of moisture. Interestingly, however, some correlative studies between topography and surface soil moisture have been conducted by Famiglietti and Wood (1991), who found good correlations, and Troch et al. (1993), who found weak correlations. These studies were done during the summer; results are likely different during wetter seasons. During winter months, blowing of snow can also lead to redistribution of moisture (Johnston 1993). This redistribution is impacted by air flow patterns and topographic features in the landscape.

At the regional scale, lateral flow has important biogeochemical implications. Runoff removes, on average, about 30% of the water that falls on terrestrial systems as rain and snow (Berner and Berner 1987). However, this amount varies from trace amounts to nearly 100% in different landscapes and at different times within a given landscape. Variation in water runoff is driven not only by precipitation inputs but also by temperature, slope, vegetation cover, and soil type (Ludwig et al. 1996). At the scale of large river basins (watershed area ~ $9 \cdot 10^3$ to $6 \cdot 10^6$ km^2), 82% of the between region variation in average annual runoff is explained by precipitation, temperature, and watershed slope alone (Ludwig et al. 1996). On smaller space or time scales, more complicated models including vegetation, rooting depth, and soil type are needed to predict runoff (e.g., Borman et al. 1974; Jackson, this volume).

Water runoff carries with it organic carbon and mineral nutrients. These losses are ultimately driven by new inputs to the system by weathering, microbial fixation (for N), and human inputs derived from fossil fuel combustion and fertilizer production. For organic carbon, the major input is autotrophic carbon fixation (net primary production, NPP). However, the amount of water runoff plays a key role in determining the amount of weathering and NPP that occurs and can impact greatly the retention of new inputs to the system (Ludwig et al. 1996). For example, at an average annual runoff of 1000 mm per year, 40% of the new inputs of N from fertilizer and NO_x

deposition are on average exported. On the other hand, in systems with 100 mm average annual runoff, only 6.3% of these inputs, on average, leave as dissolved nutrients (Caraco and Cole 1999). Similarly runoff amount strongly influences organic carbon export from terrestrial systems and at a regional scale may, in fact, be the most important variable controlling dissolved organic carbon (DOC) export (Ludwig et al. 1996). This relationship implies that, where runoff is greater, a terrestrial system may be more likely to be net autotrophic. On the other hand, much of the organic material lost from terrestrial systems as DOC may be ultimately respired in adjoining aquatic systems, causing these systems to be strongly heterotrophic (Cole et al. 1994). In a landscape study in arctic tundra, Oberbauer et al. (1996) raised the question of whether such effects can explain the outgassing of CO_2 (net negative balance) observed on tussock tundra slopes. They emphasized the need to couple hydrology and terrestrial ecosystem models better, in order to examine fully dependencies of carbon balance on vertical versus lateral flows.

Runoff of water, minerals, and organic matter is somewhat difficult to quantify when it occurs as groundwater flows. Runoff in streams and rivers is easier to quantify, and a large number of stream and river gauging stations are available worldwide for this purpose. Humans can change residence times in waters dramatically by both changing river morphometry and lateral flow in the landscape. These changes may have been most severe in dry climates. In general, human activity has probably resulted in increased within-river retention of nutrients since residence time has been increased.

LATERAL FLOW AND SUBSURFACE PROCESSES

Lateral hydrological flows occur along the surface of landscapes and in the subsurface. Such flows both affect and are affected by ecosystem properties and are important from several perspectives. They present as well an opportunity to monitor and understand ecosystem processes. Gauge measurements and chemical monitoring of streams and rivers are an obvious example appropriate to a broad range of scales (Caraco and Cole, this volume). Subsurface measurements are more difficult, but such tools as groundwater-level monitoring and lysimeters (e.g., point measurements with ceramic cup lysimeters) or tracer studies can help estimate the movement of elements in the soil and in the aquifers (Menzel and Demuth 1993; Johnston et al. 1995). Lateral flow at scales from local to regional represents an underutilized means for monitoring ecosystem processes.

Lateral hydrological flows also interact strongly to influence ecosystem functioning. There are important direct effects on biogeochemical cycles, such as when nutrients and organic matter are transported in water. Eutrophication and erosion are two aboveground examples of such important processes. Lateral hydrological flows also redistribute nutrients and organic matter in the soil. The chemical composition of tropical and highly leached soils reflects the long-term consequences of soil water movement. A second way that lateral flow affects ecosystem processes and biogeochemical cycling is through its "indirect" effects. Many ecosystem processes, particularly microbial ones, depend on adequate water availability. Trace gas emissions and decomposition fall into this category. For trace gas emissions, minor changes in topography can mean the difference between a sink or a source at a particular location.

The interaction between river water and groundwater in riparian zones causes specific conditions of lateral flows of water and matter. In estuaries, particular problems of seawater intrusion into terrestrial freshwater ecosystems occur.

Subsurface heterogeneity presents one of the most difficult challenges to understanding the ecosystem consequences of subsurface lateral flow. Macropores, such as those found around roots, earthworm channels, and animal burrows, are examples of how such pores form and why water flow, predicted solely from soil texture data, is almost always inaccurate at the landscape scale. Changes in the depth to and the topography of bedrock, caliche layers, and other horizon changes also contribute to heterogeneity. The inability to characterize subsurface heterogeneity leads to great uncertainty in data interpretation, in upscaling from point measurements, and in modeling at landscape scales. The emphasis placed on predicting how contaminants move to and then within groundwater aquifers has driven the development of new tracer and geophysical techniques for inferring hydrological heterogeneity. Some of them (e.g., ground penetrating radar, seismic refraction, tomography) may be useful to define subsurface heterogeneity across landscape units. Acquiring such data sets might also be used to determine the importance of heterogeneity to landscape processes and could lead to insights of how smaller-scale processes may be parameterized within landscape models.

Because the soil is so heterogeneous at fine scales (e.g., Jackson and Caldwell 1993), it is unlikely that we will ever generate deterministic models of soil water movement through the soil at broad spatial scales. On a regional scale the application of statistics to predict average values in soil moisture as well as the usage of remotely sensed data or new soil data bases (see final section) will help to model some consequences of lateral flow.

Vegetation across the landscape also interacts with subsurface lateral flows. Introducing deep-rooted species into a previously shallow-rooted ecosystem increases the zone in which water and nutrients can be recycled (Jackson, this volume). Water is taken up by plants even in deep soil layers, resulting in a reduced groundwater recharge. In conversions of grasslands to shrublands, this depth can be considerable (from a few to tens of meters). In the Amazon, converting deep-rooted trees to pasture land does the reverse: the zone of root uptake decreases from tens of meters to just a few. The presence of deep roots also increases the likelihood that macropore flow is important.

FIELD EXPERIMENTS

Parameterization of models including lateral flow and related processes requires observational data bases, which are rarely available on a regional scale (100 m up to 1 km or even larger). In this context, a new series of intensive field studies is recommended. The consideration of most recent theories about the variability of soil physical properties and their prediction (see Appendix) will help to plan new observation networks. Instrumental improvements concerning design, accuracy, or frequency of measurement should be taken into account. For example, the development of the time domain reflectometry (TDR) has allowed soil moisture to be monitored over larger units without using single sensors or the application of geophysical techniques to examine subsurface topography.

The impact of lateral flow on complex landscapes is affected by a variety of mutually coupled processes. For instance, nitrogen mineralization and related fluxes in the topsoil layer are highly correlated to variations in soil moisture. Therefore, interdisciplinary field experiments should be carried out. One type of experiment needed to identify the importance of lateral flow is the creative exploration of hydro-ecological models, e.g., the improvement of topography-related measurements to identify ecosystem functions in various types of landscapes. Such studies might be linked to field measurements at a site where airborne laser altimetry or air photogrammetry surveys, including GPS-based (global positioning system) ground control, could be used to generate a very detailed description of topography. Watershed level simulation studies using high-resolution data sets of topography and soils show significant differences in the timing of peak runoff, soil moisture variability, evapotranspiration, and NPP when they are operated with and without the consideration of lateral flow. Another possibility for integrating the knowledge of various disciplines is the coupling of remotely sensed data with topographic characteristics and belowground measurements. For instance, soil hydraulic properties or variations in soil depth on a hillslope could be reasonably represented with such an approach.

For a variety of complex landscapes, different observation networks are needed. In mountainous regions, the spatial density of measurements is required to be higher than in flat areas and should include the assessment of vertical gradients. For example, less is known about airflow patterns over complex terrain, although this can lead to significant differences in precipitation, snow depth, evapotranspiration, or NPP.

CONCLUSIONS AND RECOMMENDATIONS

Our discussions led us to conclude that for landscapes at the scale of hundreds to thousands of square kilometers, lateral hydrological flows are likely to be important for, and in some cases linked closely to, ecosystem processes. Furthermore, in previous field campaigns, inadequate attention has been given to such flows. Scientific advances will require creative approaches to linking mathematical models, flux measurements, and areal information from remote sensing. New studies should focus on integrated approaches and should include the various aspects of lateral flow; ongoing studies should, to the extent possible, incorporate tests of models and hypotheses regarding linkages of hydrological flows and ecosystem functions. To this end, we propose the following recommendations for future research.

Need for Continuous Measurements

Measurements of water fluxes, sediments, and dissolved materials in streams and rivers give an integrated result for lateral hydrological and geochemical flows from complex landscapes. Often these results provide answers to important questions in and of themselves. The data are also extremely important for placing bounds on (testing) the closure of element budgets derived from small-scale measurements and modeling. Thus, it is critical that a network of gauging stations be maintained (and expanded as appropriate) where river flow and concentrations of dissolved and suspended material are routinely measured.

Links with Remote Sensing and Tracer Techniques

Methods must be developed to assess the importance of lateral flows within complex landscapes better. This assessment must include the development of observational techniques at scales from hundreds to thousands of meters. It is our belief that this assessment can be done through a combination of remote sensing, isotopic and dye tracer methods, and direct and inverse hydrological modeling. The combination of direct observations of lateral flows at these scales and indirect observations may be the only feasible method at the landscape scale.

Data Bases on a Landscape Scale

Although scientists were once limited primarily to the FAO/UNESCO global soils data base, a number of more detailed data bases are now available. These include, for example, IGBP/DIS, the WISE data base, a new U.S. data base of more than 24,000 soil cores (USDA 1997), and a similar soil network for Brazil. These new or revised data are still based on point measurements and will consequently be of limited value in predicting lateral flow directly. They will, however, be useful to test hypotheses on the consequences of lateral flow for ecosystem functioning at broad spatial scales. For example, the depth distributions of soil organic matter and soil nutrients can be compared for different plant functional types and climatic conditions. The biogeochemical consequences of altered rooting depth of different vegetation types can be compared at similar climatic conditions (e.g., aridland shrubs and grasses). These data bases are not likely to help us predict lateral flow across a 100 m hillslope, but they will help generate and test hypotheses at landscape, regional, and global scales.

Linkages at Landscape Scales

We now have a number of tools that can measure and predict ecosystem and hydrologic processes at large scales. Most of these tools, by necessity, focus on aboveground phenomena. By combining these tools, we may be able to infer some consequences of lateral flow for ecosystems. For example, remote sensing can now provide a fairly robust estimate of leaf area index in many systems, particularly system phenology. Such estimates, combined with eddy flux measurements and climate data, may be used to test the sensitivity and accuracy of models. If the rooting depth represented in the model is one meter, has there been sufficient rainfall recently to maintain the observed activity of the plants? Over the course of the growing season, is there sufficient water holding capacity in the layer to maintain observed plant activity? Based on the amount and pattern of rainfall observed, is lateral flow necessary to obtain sufficient water availability at a given point? Although there are practical and mathematical limitations for doing so, inverting models may also be a way to derive rooting depth and soil water for ecosystem functioning.

New Modeling, Upscaling, and Downscaling Methods

Deterministic models have to be incorporated into stochastic models of soil heterogeneity (model parameters derived from soil properties). Stochastic models of soil heterogeneity should nest stochastic point processes into Boolean processes (stochastic geometries). Upscaling can then be achieved by (stochastic) integration. Water transport models in the unsaturated zone have to be completed with dynamic root growth models.

Research on Model–Data Linkages

One method for calculating lateral flows in the subsurface or to infer soil hydraulic properties from observations is to use the Darcy–Richards equation. Effective use of this method for lateral flows at the landscape scale will require: (a) establishment of pedo-transfer functions for model parameters governing lateral flows; (b) establishment of statistical parameters for soil hydraulic properties, e.g., variance and mean and shape of probability density function of the saturated hydraulic conductivity; and (c) development of efficient mathematical/numerical methods for the solution of ill-posed inversion problems.

Soft Computing and Knowledge Engineering

We recommend that new methods of soft computing be applied. These would include: (a) application of newly developed methods of soft computing such as fuzzy logic, neuronal networks, and artifical intelligence (AI) approaches for the derivation of model parameters from qualitative and/or incomplete soil information (extension of pedo-transfer function to the case where no regression equation can be set up); (b) modeling of uncertainties in data by data fusion (combining incoherent data of different sources and different degrees of reliability by means of AI, fuzzy approaches, possibility theory); and (c) modeling uncertain processes by means of soft computing (e.g., fuzzy expert systems). Vulnerability of ecosystems by lateral surface/subsurface transport of pesticides provides an example.

Importance of High-resolution Topography

Patterns and rates of hydrological lateral flows depend very strongly on topography. Ecosystem processes (as indicated by vegetation cover, for example) may likewise be conditioned by topography. New instrumentation (laser altimeters and aerophotogrammetry, including GPS) can be used to obtain very high-resolution topography over complex landscapes. Studies should be undertaken to obtain very high-resolution topography over one or more catchments to test the sensitivity of hydrological and linked hydrological—ecological models to different levels of resolution of topography to determine the need for better representations of the land surface.

Need for Combination of Catchment Studies and Flux Measurements

We recommend installing stream gauge stations at selected Euroflux and Ameriflux sites. Such data could be used to check water fluxes estimated from eddy flux measurements. Elemental and organic measurements might also be used to help close biogeochemical budgets.

REFERENCES

Berner, E.K., and R.A. Berner. 1987. The Global Water Cycle. London: Prentice Hall.

Borman, F.H., G.E. Likens, T.G. Siccama, R.S. Pierce, and J.S. Eaton. 1974. The export of nutrients and recovery of stable conditions following deforestation at Hubbard Brook. *Ecol. Monogr.* **44**:255–277.

Bridgham, S.D., C.A. Johnston, J. Pastor, and K. Updegraff. 1995. Potential feedbacks of northern wetlands on climate change. *Bioscience* **45**:262–274.

Caraco, N.F., and J.J. Cole. 1999. Human impact on nitrate export: An analysis using major world rivers. *Ambio* **28(2)**, in press.

Cole, J.J., N.F. Caraco, G.W. Kling, and T.K. Kratz. 1994. Carbon dioxide supersaturation in the surface waters of lakes. *Science* **265**:1568–1570.

Famiglietti, J.S., and E.F. Wood. 1991. Comparison of passive microwave and model derived estimates for soil moisture fields. In: 5th Intl. Colloq. on Physical Measurements and Signatures in Remote Sensing, Courcheval, Jan. 14–18, 1991, pp. 321–326. ESA SP–319. Noordwijk: European Space Agency.

Jackson, R.B., and M.M. Caldwell. 1993. Geostatistical patterns of soil heterogeneity around individual perennial plants. *J. Ecol.* **81**:683–692.

Johnston, C.A. 1993. Material fluxes across wetland ecotones in northern landscapes. *Ecol. Appl.* **3**:424–440.

Johnston, C.A., G. Pinay, C. Arens, and R.J. Naiman. 1995. Influence of soil properties on the biogeochemistry of a beaver meadow hydrosequence. *Soil Sci. Soc. Am. J.* **59**:1789–1799.

Ludwig, W., S. Kempe, and J.-L. Probst. 1996. Predicting the oceanic input of organic carbon by continental erosion. *Glob. Biogeochem. Cyc.* **10**:23–41.

Menzel, L., and N. Demuth. 1993. Tracerhydrologische Untersuchungen am Lysimeter Rietholzbach. Vol. 52 of *Berichte und Skripten*. Zurich: Dept. of Geography ETH.

Naiman, R.J., T. Manning, and C.A. Johnston. 1991. Beaver population fluctuations and tropospheric methane emissions in boreal wetlands. *Biogeochemistry* **12**:1–15.

Oberbauer, S.F., W. Cheng, C.T. Gillespie, B. Ostendorf, A. Sala, R. Gebauer, R.A. Virginia, and J.D. Tenhunen. 1996. Landscape patterns of carbon dioxide exchange in tundra ecosystems. In: Landscape Function and Disturbance in Arctic Tundra, ed. J.F. Reynolds and J.D. Tenhunen, pp. 223–256. Ecological Studies Series 120. Berlin: Springer.

Troch, P.A., M. Mancini, C. Paniconi, and E.F. Wood. 1993. Evaluation of a distributed catchment scale water balance model. *Water Res.* **29(6)**:1805–1817.

USDA (U.S. Dept. of Agriculture). 1997. Soil survey laboratory characterization data (CD-ROM). Lincoln, NE: National Soil Survey Center, Natural Resources Conservation Service.

APPENDIX: THEORETICAL APPROACHES

Basic Models

To capture the essential processes of lateral flow in a mathematical model, physically based hydrological models for subsurface and overland flow have to be coupled with spatial stochastic models for the distribution of soil parameters and vegetation cover, including root distribution models. In addition, a digital elevation model is required.

Classic mathematical models for subsurface flow are based on the combination of Darcy's law and mass conservation, yielding Richard's equation for the unsaturated zone. Overland flow is frequently modeled by the Bussinesque equation, where the flow is primarily influenced by the slope and the surface roughness, which is in turn related to vegetation cover. Each of these equations is derived from basic physical laws and function well at the laboratory scale under homogeneous conditions.

In contrast, a real landscape is composed of regular and nonregular patterns with respect to soil properties, vegetation cover, and other variables. The challenge is to imbed hydrological models into such a complex landscape context at different scales. Sir Ernest Rutherford once said, "If you need statistics in your experiments, you ought to have done better experiments." In contrast to this statement, however, we are in a situation where we have to model statistical variation explicitly by the use of spatial stochastic theory.

Geostatistical Models

Geostatistics provides a theoretical basis for coping with complexity at different scales. Consider any hydrologically important soil parameter, such as the hydraulic conductivity. In geostatistics, K is considered to be a spatial random field.

Let us assume that $\{Z(\vec{x}), x \in D\}$ is a spatially stochastic process over the domain D, which is of at least second-order stationarity. $Z(\vec{x})$ is often found to be composed of a sum of spatial random processes at different spatial scales. Consider, for example, a process composed of two spatial processes overlaid by a pure Gaussian process:

$$Z(\vec{x}) = \mu + \omega(\vec{x}) + \eta(\vec{x}) + \varepsilon(\vec{x}) \qquad \text{(A13.1)}$$

where μ is the general mean; $\omega(\vec{x})$is the spatial process with small characteristic length (with respect to grid size); $\eta(\vec{x})$ is the spatial process with large characteristic length(with respect to grid size); and $\varepsilon(\vec{x})$: pure Gaussian process. The process is then characterized by the form of the variogram function:

- The correlation length characterizes the pertinent scale.
- The sill is a measure of the overall variability.
- A nugget value different from zero indicates that there might be a further process nested at a smaller scale.
- The mathematical form of the variogram, especially the slope at the origin, indicates whether the manifold of a realization is smooth (slope ~ zero) or rough (slope >> 0). Therefore, the form of the variogram, the slope at the origin, and the sill constitute a convenient measure of complexity with respect to *one* variable.

This concept can be extended to a situation in which is a vector composed of several soil parameters.

Upscaling and Downscaling in Geostatistical Models

Upscaling and downscaling are performed by stochastic integration. Each geostatistical measure, such as the variogram function or Krige predictors, can be referred to any scale, ranging from a point to blocks of any size. This classical geostatistical model can be embedded into Boolean models for random geometries (Cressie 1993). Patterns are simulated by a Boolean process; to each Boolean realization, a random point process is allocated.

Coupling Geostatistical and Hydrological Models

Once variogram or correlogram functions have been established, it is possible to simulate the stochastic process by the turning bands method and to solve the boundary value problem over the realization of a random field. If one is repeating the simulation, it is possible to arrive at a *representative parameter set* for the hydrological model. In this way, representative parameter sets can be derived for each soil/vegetation type (Richter and Diekkrüger 1997). It should be kept in mind that the identification of the underlying spatial processes demands a huge experimental effort.

Upscaling Models

The application of models at any scale requires the parameterization of the specific process equations and this parameterization may vary with scale. There are a number of mathematical approaches to scaling, including geostatistics, self-similarity theory, and so forth. These approaches require certain mathematical assumptions. For example, kriging requires that the spatially stochastic process be at least intrinsically stationary.

What May Be the Form of Large-scale (Macroscale) Models?

An unresolved question is whether the parameterization of the model equation at different scales has the same form but scale-dependent parameters, or whether the underlying process equations vary with landscape scale and complexity. It is usually assumed that the form of the parameterizations does not change with scale or complexity, but this needs to be challenged. In particular, we need measurements (e.g., sensible heat) at scales consistent with the application of the balance model across complex (heterogeneous) landscapes. Currently, measurements are made at point (local) scales by flux towers.

Statistical Closure Models

Another perspective is that the balance equation can be divided into two components: a resolvable "macroscale" balance equation and an unresolvable "subgrid" equation. The former represents the mean behavior while the latter, a closure model, represents statistically subgrid processes (fluxes) that influence the mean behavior. This approach is very common in turbulence but is untested in hydrological and ecosystem process models to account for landscape heterogeneity and complexity. In climate and weather prediction models, only the "macroscale" model is parameterized and included.

Estimation of this subgrid equation can be conceptualized by computing the spatial (or temporal) spectrum and parameterizing the higher frequency fluctuations, i.e., developing a statistical closure model. If high-frequency fluctuations have a spectral separation from the lower frequencies, then this demonstrates that small-scale processes are filtered out of the large-scale (or mean) behavior, i.e., a separation of scales. Initial work in this area shows that the spectral parameterization is time-dependent for a fixed landscape. More research is required to develop general parameterizations that take into account both landscape characteristics (vegetation, soil, and topographical patterns) and moisture states.

Alternatively, such closure models could be developed or hypothesized from theory. The balance equation can be rewritten into a mean term and a fluctuation around the mean. The fluctuation terms (correlation terms) can also be parameterized, and this parameterization potentially can be tested through field experiments. This approach appears rather daunting given the wide range of potential conditions. It might be worth exploring whether laboratory experiments could be used as substitute for *in situ* field experiments. It is important that both approaches be explored if progress is to be made.

Homogenization

Another way of upscaling is performed by homogenization. The following example, which illustrates a case where the mathematical model structure is invariant under scale transitions, is taken from Bourgeat (1997). The stationary monophasic transport in a porous medium is described by the combination of mass conservation

$$\nabla \cdot q^{(e)}(x) = Q(x), \quad x \in D, \tag{A13.2}$$

and Darcy's Law

$$q^{(e)}(x) = -K(x, y)\nabla\Psi^{(e)}(x), \tag{A13.3}$$

where $q^{(e)}$ is the Darcy flow; $\Psi^{(e)}$ is the hydrodynamic potential; and K is the conductivity tensor. The superscripts e and o refer to the local scale and global scale, respectively; x denotes the global scale variable; and $y = x/\varepsilon$ denotes the local-scale variable.

When $\varepsilon \to 0$, one obtains the equation system at the global scale:

$$\begin{aligned} &\nabla \cdot q^{(o)} = Q(x), \\ &q^{(o)}(x) = -\tilde{K}(x)\nabla\Psi^{(o)}(x), \end{aligned} \tag{A13.4}$$

where

$$\tilde{K}(x) = \frac{1}{\left|\varepsilon^{-1}D\right|} \iiint_{y \in \varepsilon^{-1}D} K(x, y)C(x, y)dy \tag{A13.5}$$

where $C(x,y)$ denotes the local tortuosity tensor.

Rigorous mathematical analyses of this kind are important from the purely scientific point of view. They give us some confidence that we are using one and the same model at different scales. On the other hand, it is not feasible in practice to relate the tensor field $C(x,y)$ to field data at the landscape level. A further drawback is that the theory only works for transitions between two scales in the stationary case. Here, more theoretical work is still to be done. For practical purposes, the methods pointed out in the previous section seem to be more feasible.

It should be kept in mind that in other situations new model types can emerge at a higher level, e.g., a transition from Navier–Stokes equation to Darcy equation, when changing from microscopic to macroscopic scale.

Coupling Root Growth Models with Water Transport Models

So far, only the physical aspects of water flow have been considered. In the unsaturated zone, root distribution and water extraction by roots have a major influence on the water balance. This means specifying the term $Q(x)$ in the above mass conservation equation.

To study the interaction between slope, soil heterogeneity, and root growth, it is proposed to solve the following two-dimensional boundary value problem over a slope of about 100 m length (see Figure A13.1):

$$\begin{aligned} \frac{\partial\theta}{\partial t} &= -\nabla \cdot q - Q(x, z) \\ q &= -K\nabla\Psi \\ Q(x, z) &= \rho(x, z, t)\ f(\Delta\Psi_{rs}) \end{aligned} \tag{A13.6}$$

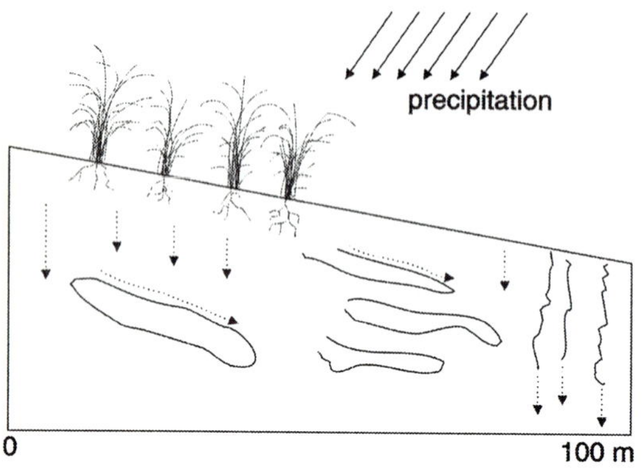

Figure A13.1 Layout of a proposed boundary value problem including lateral flow, soil heterogeneity, and root growth. The heterogeneity is caused by clay lenses and by macropores.

plus initial and boundary conditions, where $\rho(x,z,t)$ denotes the time-dependent root density distribution, f denotes the water extraction rates as a function of the potential difference between root and soil, and θ denotes the volumetric water content. $\rho(x,z,t)$ can either be described by simple geometrical models or obtained by the solution of a dynamic root growth model.

Coping with Uncertainties

Pedo-transfer Functions

Pedo-transfer functions map soil physical parameters into model equations, e.g., into the parameters of the Van Genuchten water retention curve. Most pedo-transfer functions are based on multiple regression approaches (Rawls and Brakensiek 1985; Rawls et al. 1985; Tietje and Tapkenhinrichs 1993).

There are two promising lines of future development. The first concerns the extension of pedo-transfer functions to statistical properties of the parameters. Consider as an example the saturated hydraulic water conductivity K_S. It is well known that K_S is lognormally distributed even in homogenous soils such as a Loess soil. K_S is the decisive parameter determining the partitioning of precipitation between infiltration and overland flow. If one derives a model for a unit of land using the mean value, one gets the partitioning completely wrong. Thus, it is important to infer information not only on the mean parameter values but also on their statistical distribution.

The other line of development concerns the application of methods of soft computing, such as fuzzy logic, neural networks, and possibility theory to infer model parameters from qualitative soil data, which are not amenable to serve as inputs into regression equations.

Uncertainties in Models

Most hydrological models in operation are deterministic partial or ordinary differential equations. If the statistical distributions of model parameters are known, confidence limits of model predictions can be constructed by Monte Carlo methods (stratified random sampling, Latin hypercube sampling). This is the classical approach. It tacitly assumes precise knowledge of the underlying processes (differential equations!) and the statistical distribution of parameters.

However, due to incomplete data and also incomplete process knowledge, this is not always the case. Here, it might be promising to resort to methods of soft computing to be able to model the processes. Fuzzy approaches have recently been developed for the prediction of the vulnerability of aquifers with respect to pesticide leaching (Freixenet 1997) and for the prediction of kinetic properties of pesticides in heterogeneous soils (Richter et al. 1996).

New Theoretical Approaches

Modeling coupled atmosphere–hydrological–ecological systems is a challenge. At the landscape scale, the traditional approaches favored to date in hydrology — those based on classical continuum mechanics — may not work. One challenge faced by those seeking to model hydro-ecosystems at landscape scales is to derive quantitative hypotheses that may lead to macroscale theories. An example of this has been put forward by Beven and Kirkby (1979). They proposed that hydrological functioning of catchments was conditioned very strongly by topography as represented by a simple topographic index and derivable from digital elevation model data. Models based on this concept do not represent point processes but capture many important hydrological patterns. Deriving quantitative hypotheses (and models) that match the scale of landscape measurements (e.g., using remote sensing) is an activity that deserves greater attention.

References

Beven, K.J., and M.J. Kirkby. 1979. A physically-based variable contributing area model of basin hydrology. *Hydrol. Sci. J.* **24**:43–69.

Bourgeat, A. 1997. Quelques problèmes de changement d'échelle pour la modélisation des écoulements souterrains complexes. In: Tendances nouvelles en modélisation pour l'environnement. Paris: Elsevier.

Cressie, N.A.C. 1993. Statistics for Spatial Data. New York: Wiley.

Freixenet, C. 1997. Estimation des imprécisions dans la modélisation du devenir des produits phytosanitaires dans les sols: Une méthode fondée sur la logique floue. Ph. D. diss., Laboratoire d'hydraulique de France. Grenoble.

Rawls, W.J., and D.L. Brakensiek. 1985. Prediction of soil water properties for hydrological modelling. In: Proc. Symp. of Watershed Management in the Eighties, pp. 239–299. Denver.

Rawls, W.J., D.L. Brakensiek, and K.E. Saxton. 1985. Estimation of soil water properties. *Trans. ASAE* **108**:1316–1320.

Richter, O., and B. Diekkrüger. 1997. Translating environmental xenobiotic fate models across scales. *Hydrol. Earth Sys. Sci.* **4**:895–904.

Richter, O., B. Diekkrüger, and P. Nörtersheuser. 1996. Environmental Fate Modelling of Pesticides. Weinheim: Wiley-VCH.

Tietje, O., and M. Tapkenhinrichs. 1993. Evaluation of pedo transfer functions. *Soil Sci. Soc. Am. J.* **57**:1088–1095.

Plate 1.1 Simulated water use and CO_2 exchange by the landscape at the ECOMONT intensive study site at Monte Bondone in Southern Tyrolia, considering topography and distribution of six landscape functional units. A: Photograph of the study area showing typical distribution of grassland types in the valley and coniferous forest on the mountain slopes. Numbers indicate the equivalent locations shown on the adjacent vegetation map. B: Vegetation map indicating the distribution of six LFUs. C: Map of estimated daily local transpiration for full canopy development and before mowing with typical July weather conditions for the site. Fluxes vary by a factor of ca. 2 as a result of radiation differences and differences in LAI development in response to fertilization regimes. D: Map of estimated CO_2 exchange at 4 p.m. with typical July weather conditions at the site. With lower sun angles at this time of day, the largest differences in CO_2 exchange occur as a result of radiation distribution gradients at intensities below saturation for net photosynthesis.

Plate 4.1 Spatial determination of plant parameters of corn with AVIRIS for July 5, 1991 (Bach et al. 1995).

Plate 4.2 Comparison of measured and modeled soil-moisture distribution in part of the Ammer watershed S. Germany (Rombach and Mauser 1997).

Plate 6.1 Model output cloud and water vapor mixing ratio fields in the Texas-Oklahoma panhandle region of the United States at 21 Greenwich Mean Time on May 15, 1991. The clouds are depicted by white surfaces with $q_c = 0.01$ g/kg, with the sun illuminating the clouds from the west. The vapor mixing ratio in the planetary boundary layer is depicted with $q_v = 8$ g/kg. Areas formed by the intersection of clouds or the vapor field with lateral boundaries are flat surfaces, and visible ground implies $q_v<8$ g/kg. The vertical axis is height and the backplanes are the north and east sides of the grid domain. (a) used the current landscape of the region while (b) used the natural (shortgrass) landscape. Otherwise, the model runs were identical (from Pielke et al. 1997).

Plate 6.2 Accumulated precipitation (mm), at 6 p.m., in the five simulated domains illustrated in Figure 6.3 (adapted from Avissar and Liu 1996).

Plate 11.1 Upper panel: Soil profile thickness as determined from the FAO/UNESCO Soils Map of the World. Lower panel: The same data masked to show those locations where soil profile thickness is >3 m depth. Note that soil profile depth for regions with no information is arbitrarily set to 3.6 m. From Webb et al. (1993) as incorporated into the ISLSCP CD-ROM (Sellers et al. 1995).

A.
1
2
3
Monte Bondone study site of ECOMONT
B.
Landscape Functional Units
Beech
Spruce
Fallow
Meadow
Pasture
Bog
C.
0
2
4
July
Transpiration
mm d-1
D.
26
0
-60
CO2 Exchange
in July 4 p.m.
(mmol m-2 h-1)

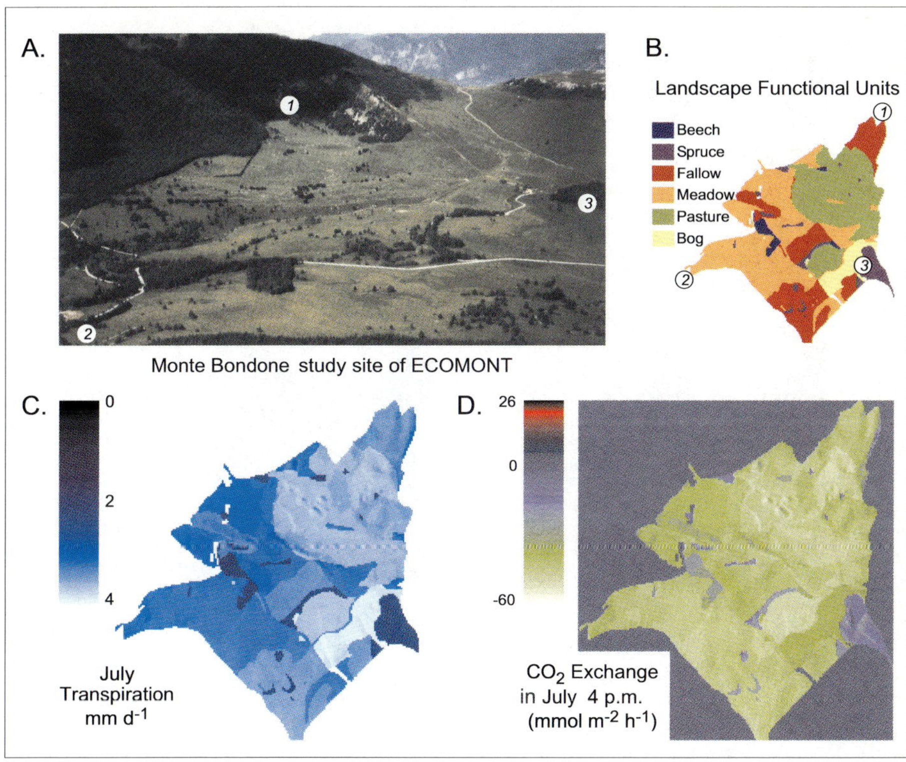

Plate 1.1

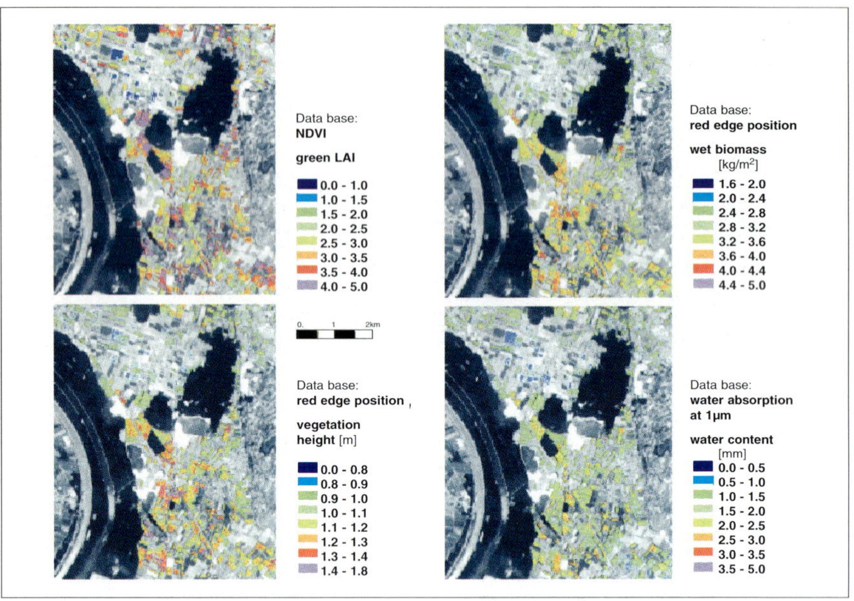

Plate 4.1

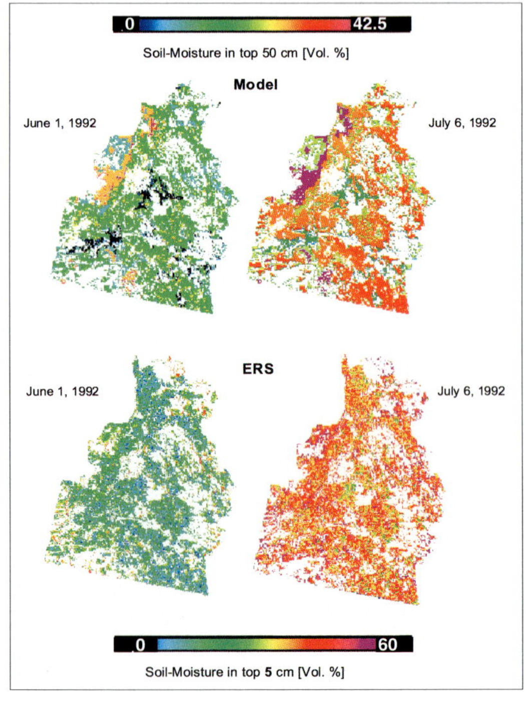

Plate 4.2

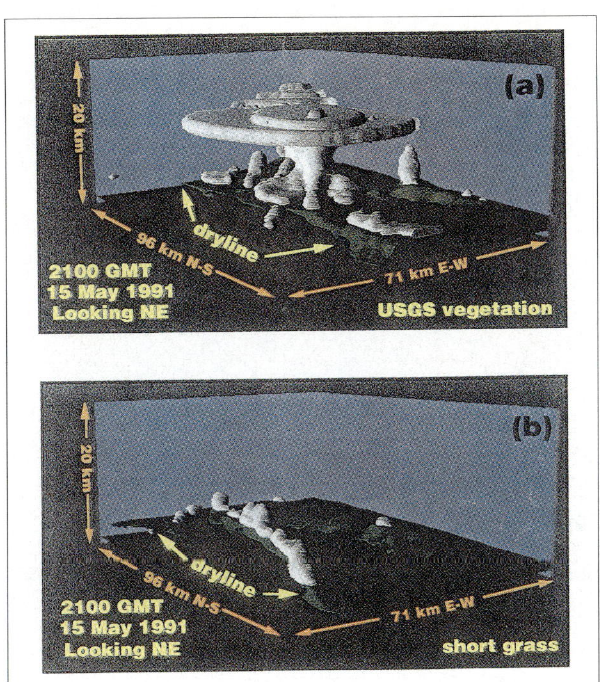

Plate 6.1

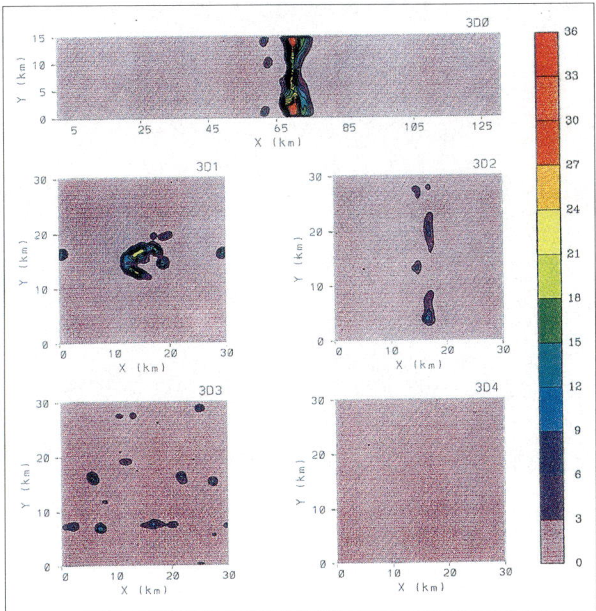

Plate 6.2

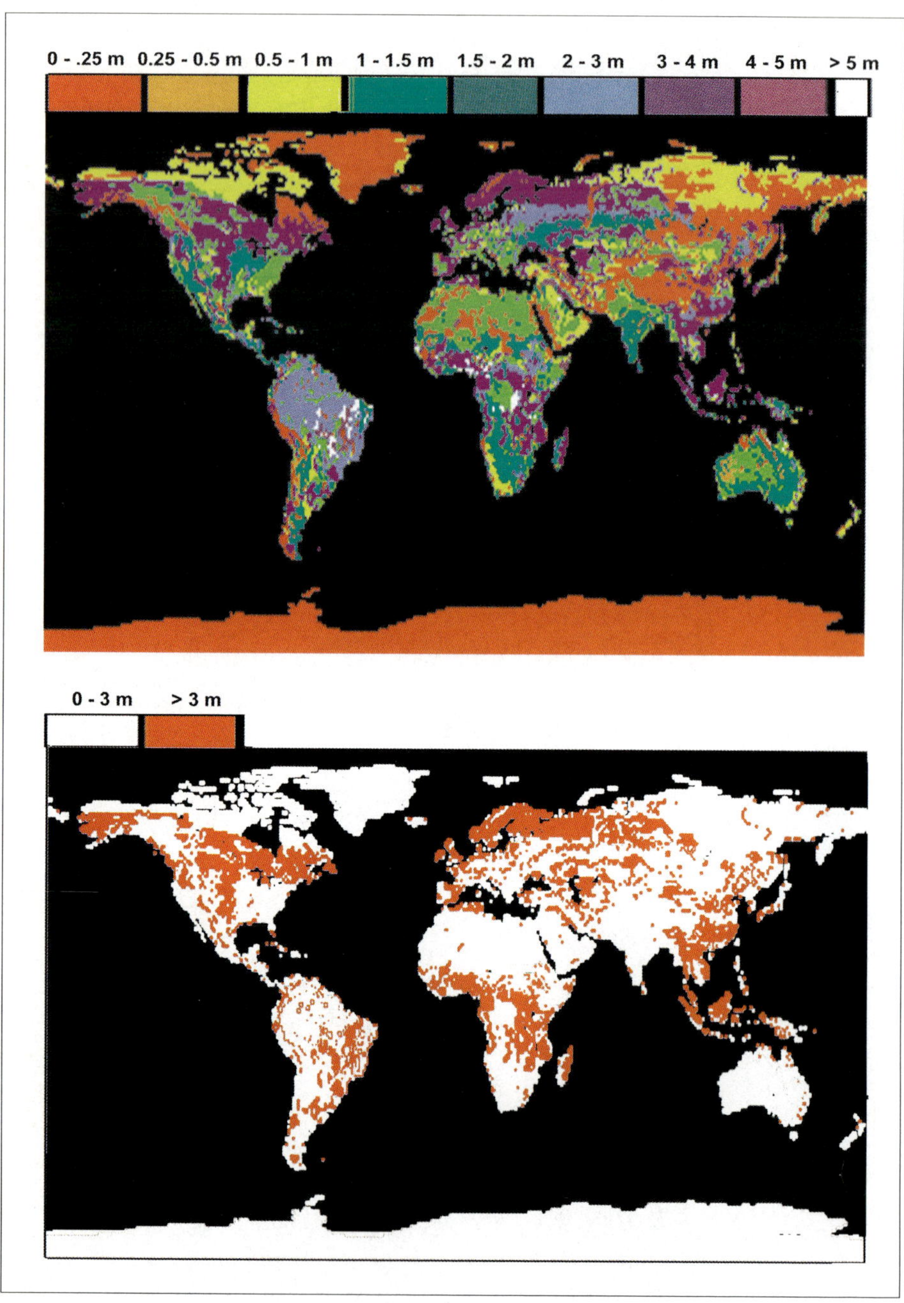

Plate 11.1

14

Do Landscape Structural and Functional Units Exist?

J.F. REYNOLDS[1] and J. WU[2]

[1]Dept. of Botany, Phytotron Building, Duke University, Durham, NC 27708–0340, U.S.A.
[2]Dept. of Life Sciences, Arizona State University West, Phoenix, AZ 85069–7100, U.S.A.

ABSTRACT

The complexity of landscapes requires that we develop simplified constructs or frameworks for understanding and predictions. Toward this goal, we pose the question, "Do landscape structural and functional units exist?" We argue in favor of the existence of landscape units, but acknowledge that there are no generic recipes for simplifying the real world for use in landscape models. While simplification is the key to achieving understanding and prediction, explicit consideration of scale multiplicity in studying complex landscapes is dictated by the hierarchical properties of landscapes. Therefore, we suggest that the hierarchical patch dynamics paradigm provides a conceptual framework to accomplish these objectives.

INTRODUCTION

At coarse scales, landscapes are usually envisioned as mosaics of various landforms, biological communities, and land uses. Such human-perceived landscapes (tens to hundreds of kilometers wide in area) tend to dominate our view of landscapes because of their relevance to pressing questions in conservation ecology, resource management, and environmental planning (Forman 1995; Pickett and Cadenasso 1995). Research at the scale of landscapes is a recent emphasis in ecology. The traditional approach has been more of a "vertical" perspective, where a system is viewed as spatially homogeneous and, hence, the internal processes and function are highlighted; in contrast, the landscape approach is more of a "horizontal" perspective since it focuses on the spatial distribution of — and interactions among — ecological entities (Rowe 1961). The vertical perspective promotes a process- or function-based approach (e.g., ecophysiology, population and ecosystem dynamics), whereas the horizontal perspective tends to encourage a structural, pattern-oriented, or geographic approach (e.g.,

Integrating Hydrology, Ecosystem Dynamics, and Biogeochemistry in Complex Landscapes
Edited by J.D. Tenhunen and P. Kabat

studies of spatial patterns of individuals, populations, and communities at relatively fine scales and biogeography at coarser scales).

The theme of this multidisciplinary Dahlem workshop was to integrate understanding of *both* the vertical and horizontal perspectives of hydrologic, ecosystem, and biogeochemical processes in landscapes — including important feedbacks, disequilibria, and inertia. This is an enormous challenge. Perhaps the greatest challenge is to design appropriate experiments and to build models that will permit us to extrapolate results from our traditional short-term and small-scale empirical studies to the larger temporal and spatial scales of landscapes. The complexity of landscapes requires that we develop simplified constructs or frameworks representing the real world.

Toward this goal, we pose the question, "Do landscape structural and functional units exist?" Presumably, an affirmative answer to this question implies that generic recipes are available for simplifying the real world for use in landscape models. While we argue that landscape structural and functional units *do* exist, such recipes *do not,* in that each case study is dependent on its specific objectives and the relevant spatial and temporal scales. There are, nevertheless, some useful tools that can greatly aid us in modeling landscape dynamics. In this chapter, we describe hierarchy theory and the hierarchical patch dynamics paradigm in this context. Our general question, "Do landscape structural and functional units exist?" leads to more detailed questions, for example: Do landscapes function as assemblages of "patches?" What criteria and methods may be used to delimit the scale at which ecosystem processes may be considered homogeneous? What determines the size or structure of landscape "patches"? Can hierarchical scaling help simplify landscape complexity? Can we determine if one landscape is "more" heterogeneous than another? Is it possible to develop landscape models based on current knowledge, which is usually not at the landscape level but at much smaller spatial and temporal scales?

LANDSCAPE STRUCTURE: HETEROGENEITY AND PATCHINESS

In a broad sense, a landscape may simply be defined as a geographic area in which ecological processes of interest are significantly affected by spatial pattern. This definition has several important implications that frame much of what we present in this chapter:

1. A landscape may be understood as an ecological criterion rather than a geographic area with a fixed spatial scale (Pickett and Cadenasso 1995).
2. The spatial scale of a landscape is dependent upon specific ecological processes or organisms under study (Kotliar and Wiens 1990).
3. Considering that spatial pattern and ecological processes operate over a range of different scales within a geographical area, the multiplicity of scale is inherent in the study of landscape ecology (Wiens 1989).

In this section, we introduce three topics crucial to the concept of landscape structural units. First, we introduce the notion of the patch as the basic element of landscapes; second, we define spatial heterogeneity in terms of what it is (and is not) and describe some of the issues and problems of quantification; and third, we describe Cantwell and Forman's (1993) scheme for reducing inherently complex landscapes to a common structure.

Patch: The Basic Element of Landscape Structure

Six computer-generated landscape maps are shown in Figure 14.1. The scale of these maps is arbitrary and for our purposes here, we assume that each represents an area of 32 km × 32 km, where each 1 km^2 cell in the grid is one of nine different *patch* types (black, white, grey, left-hatched, right-hatched, etc.). The eye immediately discerns different patterns for each map, which are not unlike some of the landscape mosaics we might see when flying across Europe in an airplane. Such patterns are the hallmark of a landscape (Urban et al. 1987). Hence, we could also define a landscape as a geographic area composed of "patches," which in the real world are such entities as forest types, farmland, residential areas, lakes, etc. When high up in the air, coarse-grained patches composed of large tracts of forests, croplands, and lakes are evident as well as some characteristics of their spatial configuration (e.g., adjacency of different patch types, shapes). When taking a bird's-eye view of these same landscapes, one may see not only the patches themselves but also their spatial pattern change as a function of the distance to the ground. Nearer the ground, as we are landing at the airport (say, *within* a km^2), the obvious patches we see are localized forest remnants, residential areas, city parks, and golf courses, and their spatial pattern may differ significantly from that we saw at coarser scales.

Patchiness exists at all spatial scales so what we define as a patch is dependent on our scale of interest and measurement system (Kotliar and Wiens 1990). For both modeling and experimental purposes, we usually consider patches to be relatively discrete and internally homogeneous units that are quantitatively and qualitatively different from their immediate surroundings. For example, imagine that map A in Figure 14.1 represents a satellite image of 1,064 hectares of a desert landscape in southern New Mexico, where the black patches represent grass communities, the cross-hatches are shrub communities, and the white patches are sand dunes. We see that some of each of the three patch types are adjacent to one another and thus form larger (1 ha) patches of irregular shapes. At a different scale, these three patch types — which show up in the satellite image as relatively homogeneous — are, in fact, not. To illustrate this, imagine that map E represents *one* of the 1 ha black (or grass) patches in map A. Map E is itself composed of three distinct patch types: black represents grass cover, cross-hatches are herbaceous plants, and white is bare soil. Each patch type in map E is $\approx$10 m^2 but again, adjacent patches of similar types form larger patch sizes of each type. Logically, we could continue to go down in spatial scale, eventually to the level of a single grass

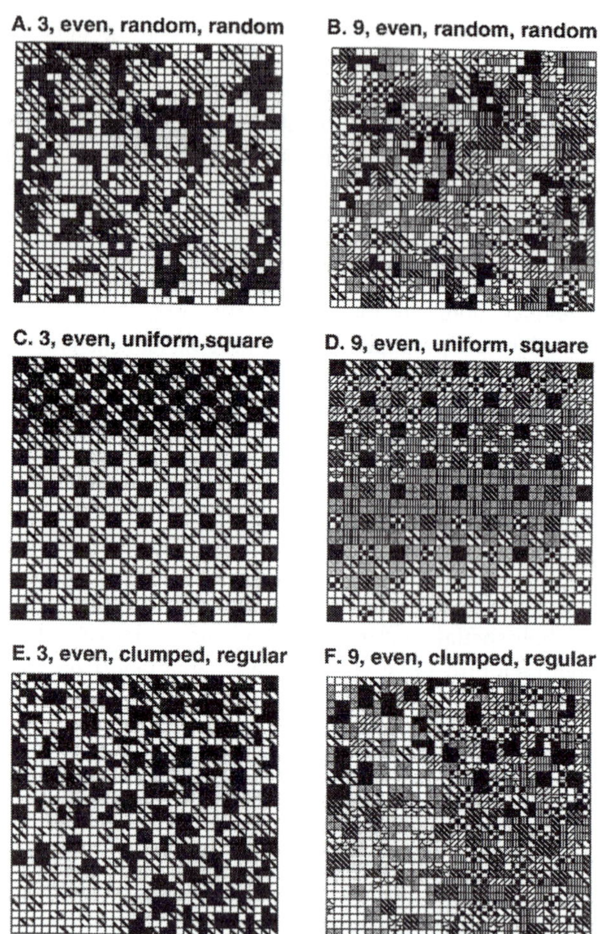

Figure 14.1 Simulated landscape maps. Levels of four map-generating components of spatial heterogeneity are indicated on the top of each map: i.e., the number of patch types (3 or 9), proportions of each patch type (even, uneven), the spatial arrangement of patches (random, uniform, clumped), and patch shape (square, regular, random). These maps show comparisons between two levels of the number of patch types, three levels of the spatial arrangement, and three levels of the patch shape. Mean patch size is four for all maps. From Li and Reynolds (1994).

clump, which itself is heterogeneous in terms of leaf nitrogen distribution, stomatal density, etc.

What causes patchiness? A wide number of both natural and anthropogenic agents play a role in creating and maintaining patchiness, including natural and human disturbances (e.g., fire, erosion, tree windfalls, clear-cutting, farming, pest outbreaks, development), land management, soil and landform factors, and biological interactions (e.g., competition, predation) (see Forman and Godron 1986; Holling 1992; Turner 1989; Wu and Loucks 1995). For convenience, we can view spatial patchiness from either abiotic or biological perspectives, both of which operate interactively across a range of spatial, temporal, and organizational scales. At continental or global scales, climatic variables — primarily temperature and precipitation — are largely responsible for the spatial pattern in vegetation. From the local ecosystem (e.g., a stand of trees) to landscape scales, patchiness is usually a result of factors such as disturbance, aspect (i.e., north- vs. south-facing slopes of a mountain), and geomorphology (e.g., floodplain of a river). Biological interactions such as competition, predation, and complex plant–soil processes can affect patchiness at finer scales.

Spatial Heterogeneity: What Is It and How to Measure It?

Are quantitative indices available to represent landscape structure? Is the patchiness evident in Figure 14.1 equivalent to landscape structure? This brings us to the concept of spatial heterogeneity.[1] While spatial heterogeneity is perhaps the single most important concept in landscape ecology, it has many different meanings and usage in the literature (Dutilleul and Legendre 1993; Kolasa and Rollo 1991; Li and Reynolds 1995; Wu and Loucks 1995). Towards resolving some of this confusion, Li and Reynolds (1994, 1995) proposed various guidelines, which we briefly summarize here. They define spatial heterogeneity as the *complexity* and/or *variability* of a system property in space and/or time. A system property is anything of interest that we wish to measure in the landscape, e.g., normalized difference vegetation index (NDVI), soil nutrients, vegetation cover, topography, and this system property can be measured either by its complexity (that is, qualitative or categorical descriptors) or its variability (quantitative or numerical descriptors) (see Figure 14.2). This is important because it says that landscapes have characteristics that can be observed and measured. Of interest to us are *structural heterogeneity*, that is, the complexity or variability of a structural property (vegetation cover, soil nutrients, elevation, etc.) and *functional heterogeneity*, that is, the complexity or variability of a functional property (gas flux, primary productivity, etc.).

As illustrated in the example of maps A and E, whatever heterogeneity we attempt to measure will be a function of the scale we choose. This can best be understood in terms of grain and extent, two of the primary scaling factors that affect complexity and

---

[1] "Spatial heterogeneity" and "patchiness" are used somewhat interchangeably in the literature. Jarvis (1995) defines patchiness as "organized heterogeneity of a number of properties," e.g., the patchiness of a landscape composed of a forest plantation, a wheat field, and a pasture results from the fact that these vegetation types are "organized" into discernible patches; if the individual plants of each vegetation type were randomly mixed and distributed over the same landscape (i.e., "unorganized"), patches would not be evident. Our consideration of spatial heterogeneity in this section gives a more explicit meaning to patchiness.

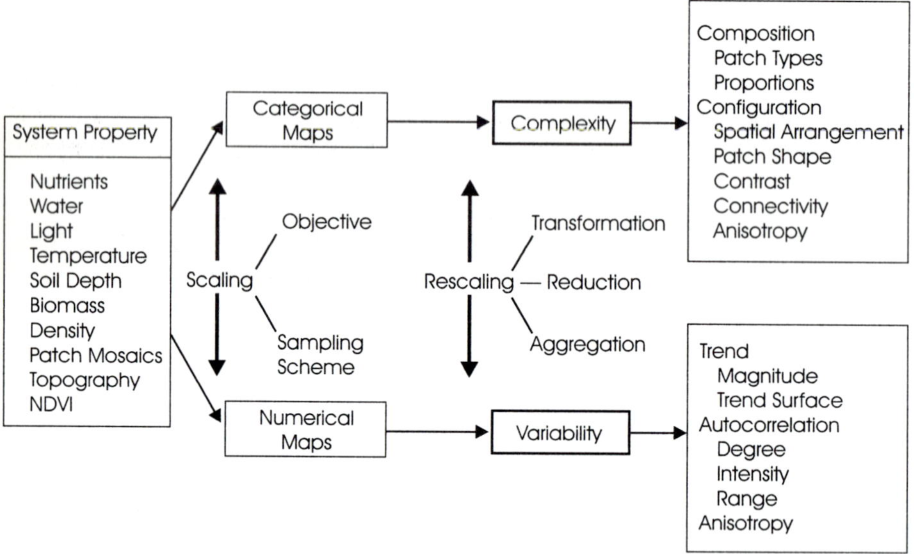

Figure 14.2 Overview of scheme for quantifying spatial heterogeneity. From Li and Reynolds (1995).

variability and, hence, heterogeneity. *Grain* is the finest resolution of data (e.g., the pixel size in Figure 14.1, minimum time step for time series data) and *extent* is the area or duration encompassed by a study. So, in any landscape study, we choose an observational scale (i.e., grain or extent) and a sampling scheme based on our research objective and the nature of the phenomenon of interest and these resultant data determine what kind of heterogeneity can be measured. It must also be emphasized that, from a data analysis viewpoint, *rescaling* of data — including transformations, data reduction and aggregation, and resampling — is also a form of scaling. Rescaling may modify grain or extent or both and, hence, plays a role in how we actually quantify heterogeneity (Figure 14.2).

Which of the six landscapes in Figure 14.1 is the "most" heterogeneous? This is a critical issue since we are interested in how landscape structure (pattern) affects (and is affected by) hydrologic, ecosystem, and biogeochemical processes. However, quantifying heterogeneity is far from a straightforward exercise. Although a large number of indices have been developed that quantify spatial heterogeneity (see Kolasa and Rollo 1991; Turner et al. 1991), what they actually are measuring is somewhat equivocal. We illustrate this with an example. As shown in Figure 14.2, landscape maps may be either categorical or numerical, depending on the data types used — and this determines what type of index or method is appropriate (Table 14.1). In the case of Figure 14.1, which contains six categorical maps, spatial heterogeneity (SH) is measured by its complexity in *composition* and *configuration* of the patches. Composition includes the number of different patch types (NPT) and the proportions of each type (PET), while

Table 14.1 Examples of data types and methods for quantifying heterogeneity. From Li and Reynolds (1995).

Data Type	Description	Indices
Point pattern	Variables or individuals of species distributed at discrete locations	• Parameter *k* of negative binomial • Nearest neighbor index • Block-size variance statistic
Geostatistical	Continuous variables sampled regularly or irregularly in space	• Variogram • Correlogram • Fractal dimension
Quantitative lattice	Numerical maps	• Variogram • Correlogram • Autocorrelation indices
Qualitative lattice	Categorical maps	• Diversity indices • Fractal dimension • Patchiness index • Contagion index • Joint-count statistic

configuration includes spatial arrangement of patches (SA), patch shape (PS), contrast between neighboring patches (NC), connectivity among patches of the same type, and anisotropy (i.e., variation in different directions). The maps in Figure 14.1 were generated by the SHAPC simulation model that controls five of these components of complexity, that is:

$$\text{SH} = f\,\{\text{NPT, PET, SA, PS, NC}, \varepsilon\} \tag{14.1}$$

where ε is the random error (for details, see Li and Reynolds 1994). To generate a particular landscape map, each of these components is set to a specific level (see legend in Figure 14.1). Using SHAPC, Li and Reynolds (1994) examined the relative contributions or influence of NPT, PET, SA, PS, and NC on different landscape indices and concluded that any definition of SH is strongly dependent on the underlying variables and the methods used, i.e., different indices depict different aspects of SH, significant interactions exist among the various components of SH, and some indices are strongly correlated.

Can we say one landscape is "more" heterogeneous than another? Two categorical maps representing mosaics of cover types are shown in Figure 14.3. These "landscapes" were created with SHAPC (PET, SA, and PS were varied in Eq. 14.1; see Table 14.2) and four spatial indices (evenness, fractal dimension, contagion, patchiness) were used to quantify their heterogeneity. Map A has higher values of evenness and fractal dimension, while map B has higher values in contagion and patchiness (Table 14.2). From this we may conclude that "landscape" A is more diverse and contains

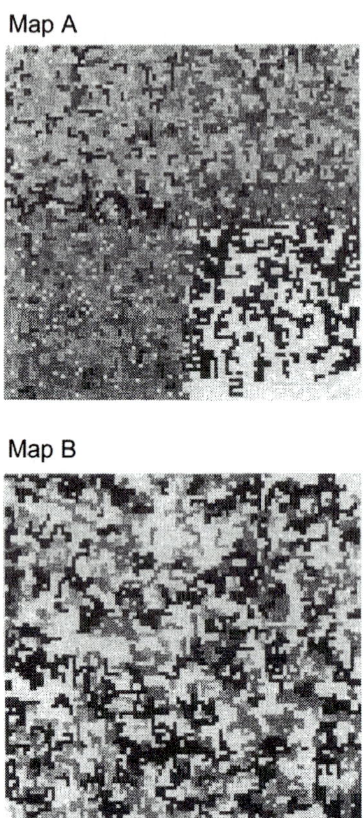

Figure 14.3 Two categorical maps generated from simulation parameters in Table 14.2. From Li and Reynolds (1995).

more irregularly shaped patches, whereas "landscape" B has larger patches and higher contrast. So, instead of asking the question of which landscape is "more" heterogeneous, it makes more sense to ask how two landscapes might differ with regard to some *specific aspect* of spatial heterogeneity of interest. For example, the various schemes used in clear-cutting forests differentially affect NPT, PET, SA, PS, and NC and, hence, will affect ecosystem dynamics in different ways. Bierregaard et al. (cited in Pickett and Cadenasso 1995) found that the number of carrion beetles and bird diversity in Amazon forests declines with the size of residual forest fragments, whereas Franklin and Forman (1987) showed that the degree of systematic "checkerboarding" remaining after clear-cutting in the U.S. Pacific Northwest was the main determinant of the susceptibility of old growth forests to catastrophic windthrow. Each component of heterogeneity (Equation 14.1) characterizes or represents a distinct aspect of spatial heterogeneity and thus it is important to define explicitly which component is of interest. More detailed treatments of this subject are given in Turner (1989), Dutilleul and

Table 14.2 Simulation parameters used in Equation 14.1 and the heterogeneity measures for the categorical maps shown in Figure 14.3. From Li and Reynolds (1995).

	Map A	**Map B**
Map Characteristics (simulation, see Eq. 14.1)		
No. of patch types (NPT)	6	6
Proportion (PET)	Even	Uneven
Spatial arrangement (SA)	Aggregated	Random
Patch shape (PS)	Random	Regular
Heterogeneity Measures		
Evenness	1.000	0.819
Contagion	0.152	0.196
Fractal dimension	1.640	1.603
Patchiness	0.296	0.550

Legendre (1993), Kolasa and Rollo (1991), Pickett and Cadenasso (1995), and Li and Reynolds (1994, 1995).

Landscape Graphs: Common Structure

Cantwell and Forman (1993) introduced the concept of landscape graphs for reducing complexity in landscapes to a common structure. Landscape graphs define common structural units that may correspond to distinctive functional characteristics. While graphs may well differ with the scale of observation, it is interesting and useful to investigate if there are common landscape graphs across scales. This method is described and illustrated in Figure 14.4. Cantwell and Forman's goals were (a) to identify common structural configurations within landscapes; (b) to examine the connectivity of patches; and (c) to identify potential links to various landscape modeling approaches. The strength of Cantwell and Forman's scheme is their emphasis on *structural links between adjacent patches*, which are where functional exchanges of mass and energy may occur and thus are important in models of landscape functional units (see below).

Using the method of Figure 14.4, Cantwell and Forman developed graphs of 25 aerial photographs that represented a range of human-dominated landscapes. The predominant patch types represented in these photographs included forestry, pastureland, cultivation, suburbia, and "natural" vegetation, e.g., a pastureland in southern Wisconsin, a tropical rainforest with shifting cultivation in the Dominican Republic, several agriculture fields in Australia, suburban Chicago, and string bogs and a spruce forest site in Canada. They found seven distinctive patterns of nodes and linkages (i.e., they were present in > 3 graphs) (Figure 14.5). Of the seven patterns, three (necklace, spider, and loop) were found in > 90% of their graphs.

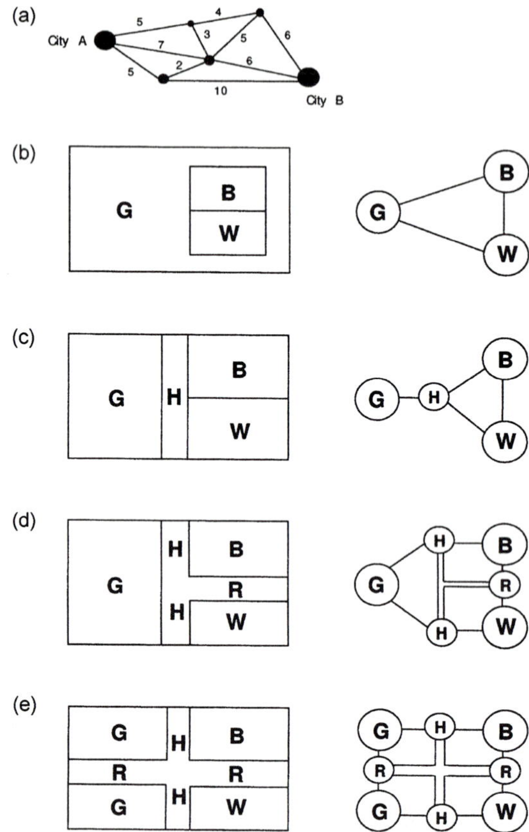

Figure 14.4 Landscape graphing methods. Landscape patch types are: G-grassland; B-bean field; W-woods; H-highway; and R-road. (a) Transportation graph with spatially-explicit linkage lengths representing distances between cities; size of the node size indicates relative city size. (b)–(c) Landscape areas and their corresponding graphs; nodes (circles) represent patch types (ecosystems, land uses) in the landscape, and linkages (lines) represent shared boundaries or points between patches. (b) Landscape graph of a matrix and two patches, recognizing the matrix as an element. (c) Landscape graph recognizing a corridor (highway) as an element. (d) and (e) Landscape graphs where a corridor network is represented as comprised of component elements, or lengths of corridor defined by intersections. (d) Landscape graph recognizing a 3-way or T-intersection, (e) Landscape graph recognizing a 4-way or X-intersection, and where one corridor axis predominates over the second. Redrawn from Cantwell and Forman (1993).

We refer the reader to Cantwell and Forman (1993) for details of the ecological interpretation and significance of these structural landscape patterns. Here we limit our discussion to a single one — the necklace — to illustrate the ecological consequences. The necklace patterns in Cantwell and Forman's human-dominated landscapes are all composed of the same landscape element (or patch) type linked together in a linear fashion, e.g., roads, hedgerows, or powerline corridors. They also found

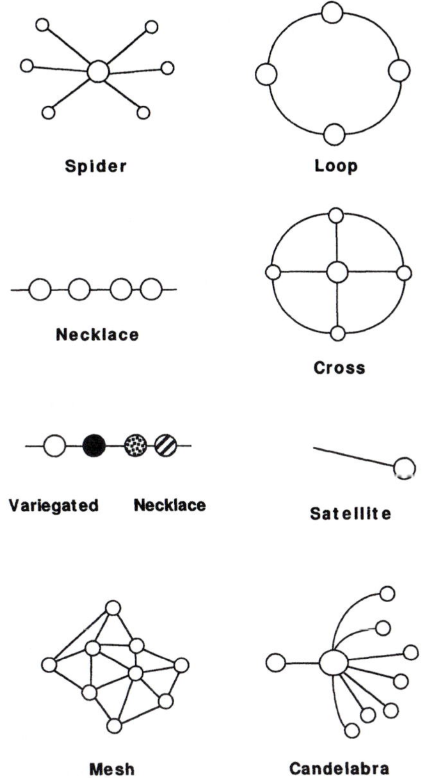

Figure 14.5 Common graph patterns identified from 25 diverse landscapes using approach described in Figure 14.4. Modified and redrawn from Cantwell and Forman (1993).

"variegated necklace" patterns where the elements are different, which is similar to what Woodmansee (1990) labeled a "flow path." In the case of coarse-scale landscapes, distinct geomorphic surfaces (e.g., alluvial fans, piedmonts) or topographic features (e.g., watersheds) often form natural boundaries for necklaces or flow paths. Several examples illustrate this principle. Kemp et al. (1997) described a toposequence established on a gently sloping, NE facing piedmont of a small mountain characteristic of the basin and range topography found in the southwestern U.S. This toposequence extends for 2.7 km from a basin floor playa (1310 m elevation, fine-textured soil), across a piedmont slope, and onto the base of a granitic mountain (1410 m elevation, coarse-textured soil). The gradients in elevation and soils across the transect, along with variable seasonal rainfall, downslope redistribution of water and organic matter, and soil texture-related variation in infiltration, water holding capacity, and moisture release characteristics, interact to generate a complex spatial and temporal gradient of patch types, available soil water, and nitrogen along the toposequence. In the mulga woodlands of eastern Australia. Ludwig et al. (1994) described a series of toposequences that are dotted with groves of mulga (*Acacia*) trees that act as "filters,"

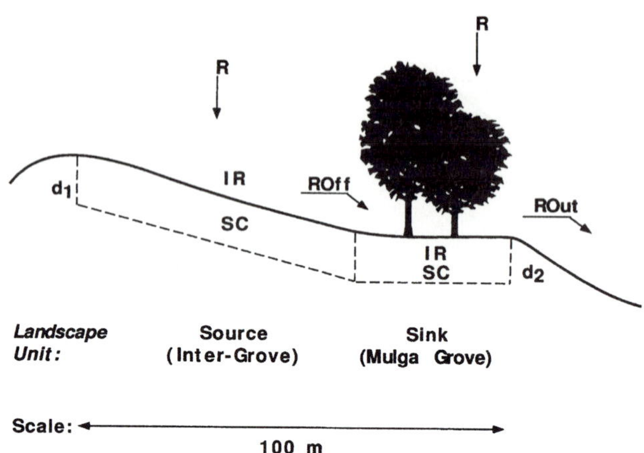

Figure 14.6 Vertical view of idealized semi-arid landscape in eastern Australia, depicting the flow of water by run-off (ROff) following rainfall events (R) when the amount and intensity of R exceeds the infiltration rate (IR) or the water storage capacity (SC) of the soil. The soils are deeper in depth within landscape sinks (d2) than under slopes (d1). Water not captured by the lowest sink runs out (ROut) of the landscape system. Redrawn from Ludwig et al. (1994).

capturing water (and nutrients, etc.) from barren inter-grove areas (see Figure 14.6). They described most of the semi-arid woodland landscapes in eastern Australia as composed of such complex toposequences, with repeating grove-intergrove units. In a small Alaskan watershed, Walker and Walker (1996) identified five patch types — crest, shoulder, upper backslope, lower backslope, and footslope — that form a necklace pattern; these patches are distinguished on the basis of differences in soils, vegetation physiognomy, geologic surface forms, and plant community types.

RELATING LANDSCAPE FUNCTION TO STRUCTURE: A HIERARCHICAL APPROACH

Given the ubiquitous existence of patchiness in ecosystems, is there a relationship between landscape structure and function? Not surprisingly, empirical evidence supports a strong, reciprocal relationship between patchiness (and the various components that comprise spatial heterogeneity) and ecological function or process. This relationship exists across a wide range of ecosystems and different types of functions, e.g., animal behavior, biogeochemical cycling, primary productivity (see Pickett and White 1985; Swanson et al. 1988; Turner 1989; Turner et al. 1995). However, while a great number of empirical and theoretical studies favor a strong link between landscape structure and function — and this is the central focus of much current research in landscape ecology (Pickett and Cadenasso 1995; Turner et al. 1995; Wu and Loucks 1995) — there are no "hard rules" available to guide our efforts in modeling complex landscapes. Instead, there are some general schemes or approaches that have emerged that, along with various modeling approaches, provide us with strong conceptual and analytical tools to address landscape functioning. In this section, we discuss how landscape structure affects function and how the hierarchical nature of this structure is a powerful tool that can be used to model function.

In the previous section we suggest that identifying *landscape structural units* involves a "synthesis" approach in which variations in patchiness are integrated over the entire landscape into a synoptic index to reflect the degree of landscape complexity or variability. In this section we argue that identifying *landscape functional units* requires more of an "analysis" approach, whereby landscape heterogeneity is partitioned hierarchically so that relatively homogeneous functional units can be distinguished. Conceivably, the same mathematical and statistical methods used to quantify landscape structural units can also be used to identify functional units when applied in a hierarchical manner.

Hierarchical Properties of Landscapes

From our discussion of patches and spatial heterogeneity, it is obvious that the structure of landscapes is hierarchical in nature (see reviews in Holling 1992; Urban et al. 1987; Wickham and Norton 1994; Woodmansee 1990). This is nicely illustrated in our example of maps E and A in Figure 14.1. Map E has unique structure at one scale (1 ha) that is not evident when viewed as a pixel in map A (which in turn has its own unique patchiness). Woodcock and Harward (1992) presented a schematic representation of forested landscapes as a nested hierarchy (Figure 14.7). Their objects or patches at the lowest scale in the hierarchy are individual trees, which may be of different sizes and species. The next scale of objects correspond to forest stands that are characterized by the collective attributes of the individual trees, e.g., the relative amount of conifers vs. hardwoods, overall density of trees, and aerial extent. A collection of stands form forest type objects, which in turn, form vegetation types. In a similar way, Reynolds et al. (1997) used a hierarchical approach to model desert landscapes (Figure 14.8). A patch at the lowest scale in their hierarchy is the average size of a single type of plant (grass clump or shrub) growing on a particular soil type (e.g., sandy-loam soil). Two general types of vegetation patches are recognized: grass and shrub islands, the latter of which is composed of a single shrub, representing a "hot spot" or "island" of biological activity within a matrix of relatively barren soil. Contiguous patches form patch mosaics where the strength of individual interactions is in part determined by spatial proximity (see Aber et al., this volume). In this system, they view landscapes as consisting of several patch mosaics (Figure 14.8).

Previously, we defined functional heterogeneity as the complexity or variability of a functional property, e.g., gas flux or primary productivity (see Figure 14.2). Thus, nested structural hierarchies are useful ways to "organize" and model landscape function. This approach builds on both the "vertical" and "horizontal" perspectives of viewing ecosystems and landscapes (see **INTRODUCTION**) where, at each level in the hierarchy, emphasis is placed on internal structure and function (vertical) whereas the

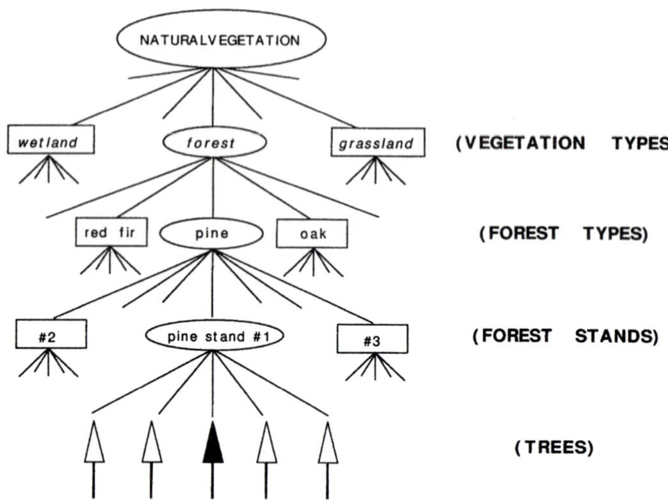

Figure 14.7 Schematic representation of forested landscapes as a nested hierarchy. Redrawn from Woodcock and Harward (1992).

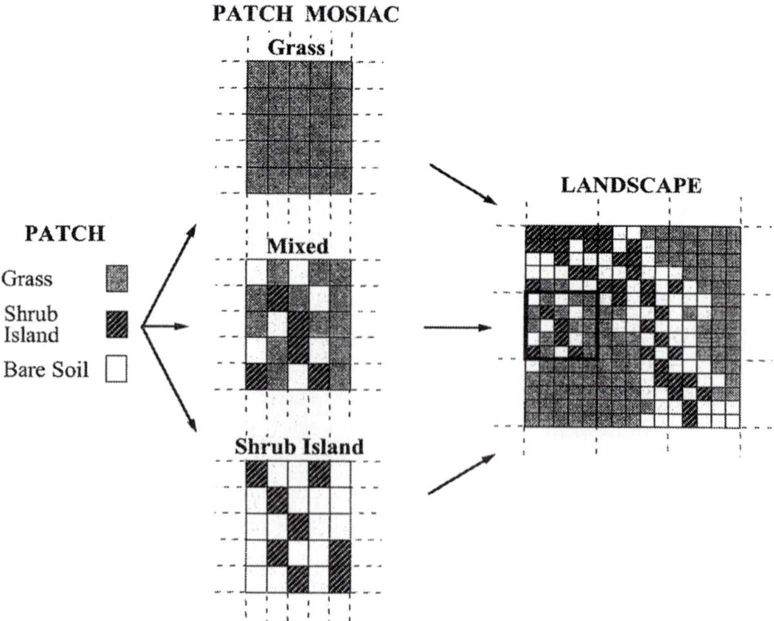

Figure 14.8 Landscape hierarchy for an arid ecosystem in southern New Mexico. Patches (ca. 1–10 m^2) form patch mosaics (ca. 1 ha – 1 km^2), the behavior of which is a function of its composition (number and proportion of different types of patches) and configuration (e.g., spatial distribution, shapes, contrast, etc.). Landscapes are composed of patch mosaics. From Reynolds et al. (1997).

horizontal perspective focuses attention to the spatial distribution of, and interactions among, system components (Rowe 1961). In Woodcock and Harward's scheme for forest landscapes, each object in the hierarchy has relevant factors that are important in functioning at each particular scale, e.g., land use and fire history are important at the forest stand level, light availability is critical at the individual tree scale for recruitment, and carbon dioxide concentration affects photosynthesis at the leaf level. In Reynolds et al.'s hierarchy, the behavior of an individual patch is determined by the functional and structural properties of the plants (vertical perspective), and interactions of the plants with their immediate abiotic and biotic environment. The behavior of a patch mosaic is a function of its composition, configuration, and dominant flow paths (horizontal perspective). These structural characteristics are key in determining the functioning of a particular mosaic in aridlands, e.g., the effect of vegetation on hydrologic flow, seedling establishment, and exchanges of water, organic matter, propagates, nutrients, sediments, etc. (see review in Reynolds et al. 1997).

In Table 14.3 we show a nested hierarchical scheme developed by Reynolds et al. (1996) for integrating horizontal and vertical perspectives in arctic tundra landscapes in Alaska. The smallest landscape spatial unit is at the scale of an individual plant. The functioning of plants is determined by the strength of couplings to the environment via numerous plant–soil–atmosphere–animal vectors. At the next level, plants plus soils, microbes, animals, etc. comprise "patch" ecosystems, which are relatively discrete and internally homogeneous. At the next level, a series of connected and inter-related patches may take a number of different spatial configurations, e.g., necklaces, spiders, meshes (see Figure 14.5). In tussock tundra landscapes on the North Slope of Alaska, variegated necklaces tend to predominate (Reynolds et al. 1996; Shaver et al. 1991; Walker and Walker 1996). Field studies at the level of flow paths correspond to examining the degree of "connectedness" between patches, which is critical since ecosystem processes are directly affected by transfers of materials across system boundaries. Typically, water, atmospheric turbulence, and animals are the main vectors of transport of material and energy between patches (Table 14.3). The next level in Reynolds et al.'s hierarchy is the landscape, which is viewed in hydrological terms as an "integrated flow system." An integrated flow system is the assemblage of individual plants, patches, and flow paths that deliver mass and energy (mainly by gravitational processes) to a "distribution flow line" or stream and respond similarly when subjected to a precipitation event. Movement of mass along flow lines is distinguished from flow paths in that the former is not between adjacent patches. The largest landscape spatial unit is a region, which is a mixture of integrated flow systems that make up the scale of interest. The landscape spatial units in Table 14.3 illustrate both vertical (functional) and horizontal (structural) components of landscapes and, via the various coupling vectors, these exist as interactive hierarchies. While one may argue about whether these units are inherent entities or results of interactions between human perception and the objective world, in either case explicitly identifying these hierarchical levels is essential in order for us to simplify and understand functioning in complex landscapes.

Table 14.3 Landscape spatial units (LSU) and associated water, energy, and nutrient functions used to model arctic landscapes. Modified from Reynolds et al. (1996).

	Structure		Function and coupling vectors	
LSU	Spatial scale[a] (m^2)	Typical Components	Water	Energy/Nutrients
Plant	10^{-4}–10^{-1}	• Leaves • Stems • Roots • Soil volume	• Soil water • Transpiration • Water uptake	• Photosynthesis • Respiration • Nutrient uptake • Herbivory
Patch	10^{0}–10^{4}	• Plants • Soil • Microbes • Animals	• Precipitation • Infiltration • Run-off • Evapo-transpiration • Soil water	• Net carbon balance •Nutrient cycling •Decomposition •Trace gas flux
Flow path	10^{2}–10^{5}	• Patch ecosystems • Connectivity patterns (see Figure 14.5)	• Soil water flux and discharge [Hillslope Darcian flow dominant]	• Mass flux of sediment • Mass flux of dissolved nutrients • Aeolian transport
Landscape (= Integrative flow system)	10^{6}– $4{\times}10^{6}$	• Flow path ecosystems • Groundwater; channel storage		
Region[b]	10^{7}–10^{10}	• Landscapes (= Integrative flow systems); lakes; rivers	• Channel flow [Turbulent flow dominant]	• Hydrologic transport of sediments • Hydrologic transport of dissolved nutrients • Aeolian transport • Migration

[a] Examples of "typical" values based on Osmond et al. (1980), Woodmansee (1990), and Walker and Walker (1991).
[b] "Mesoscale" in Walker and Walker (1991), which includes second-order watersheds.

Hierarchical Patch Dynamics

Since the late 1970s, patch dynamics has gradually been recognized as a general conceptual framework for ecological studies at different levels of organization, facilitating the integration of the horizontal with vertical perspectives. Recent developments in hierarchy theory, on the other hand, have elevated the vertical perspective to a new level of sophistication that recognizes the multiplicity of scale. One of the most significant contributions of hierarchy theory is its role in making researchers acutely aware of the importance of scale (O'Neill 1996) (see discussion below). The structure, function, and dynamics of complex landscapes are determined by individual patches and their interactions at — and usually across — different hierarchical levels. Given the nature of patches — and the causes and mechanisms of patch formation described above — we can view ecological systems as hierarchical, dynamic patch mosaics. The hierarchical patch dynamics paradigm (HPDP, Wu and Levin 1994; Wu and Levin 1997; Wu and Loucks 1995), which explicitly integrates hierarchy theory (Pattee 1973) and the patch dynamics perspective, enhances understanding pattern–process–scale relationships in landscapes by providing both an organizational and operational framework. The main elements of HPDP include:

1. *Ecological systems may be viewed as nested hierarchies of patch mosaics with discrete levels.* In contrast with the traditional individual–population–community/ecosystem hierarchy, HPDP takes a naturally defined spatial unit — the patch — as a structural and functional unit that is scale and process dependent.
2. *Dynamics of ecological systems may be viewed as the composite dynamics of interactive patches at different (usually adjacent) hierarchical levels or scales.* Phase changes of individual patches at local scales and pattern changes of patch mosaics at broader scales, as constrained by higher levels, together give rise to system dynamics. For example, the dynamics of a regional landscape may be understood as the result of the dynamics of its nested components, various ecosystems, and the exchanges of energy and materials among them via topographical, hydrologic, and other physical and biological processes. In general, higher levels impose top-down constraints to lower levels by having slower or low frequency processes, while lower levels provide mechanistic explanations for and give apparent integrity to higher levels through active interactions among components (holons). Thus, one can only see the meanings of lower-level processes at higher levels, and understand mechanistically higher-level phenomena by examining lower levels.
3. *Pattern and processes are reciprocally related, and both of them and their relation are also scale dependent.* Pattern — be it spatial or temporal — is inseparable from scale. Different patterns and processes are usually distinctive in their characteristic scales at which they operate. The challenge is to identify these scales, and properly link processes and patterns.
4. *Nonequilibrium and stochastic processes are predominantly common in ecological systems over different scales.* To understand and predict ecological dynamics, transient phenomena (frequently at finer scales) must be considered. For example, variability often is more insightful than means. Also, nonequilibrium and stochastic processes do not necessarily negate stability (see below).
5. *While homeostatic stability essentially does not exist in ecological systems (except individual organisms), persistent ecological systems usually exhibit metastability (homeorhetic, quasi-equilibrium states).* An important mechanism of achieving this metastability in hierarchical systems is incorporation, whereby nonequilibrium patch processes at one level translate to a quasi-equilibrium state at a higher level. For example, the steady state dynamics at the landscape

> level may be composed of transient dynamics at the component ecosystem level such as in the case where long-term fire patterns may consist of many short-term destabilizing forest fires. In contrast to the stability that derives from an assumed self-regulation in a closed system, the concepts of incorporation and metastability emphasize multiple-scale processes and the consequences of heterogeneity.

The HPDP has several implications for modeling complex landscape models (see Wu 1999). First, a landscape can be modeled as a spatially nested hierarchy composed of patch ecosystems as in the examples shown in Figures 14.6–14.8. The spatial patches are defined based on natural boundaries (e.g., soil types, topography, hydrologic units, or disturbance regimes) at a particular scale (or range of scale) where most, if not all, processes of interest respond. Different processes will operate at different characteristic scales, which in turn dictate the average size of patches relevant to them and thus give meanings to patchiness at respective scales. Hierarchy theory also suggests that ecological systems are, but only, near-decomposible (Pattee 1973; Wu 1999). Conceivably, different ecological systems may have different degrees of decomposability. In general disaggregation errors should be expected when systems are modeled hierarchically.

Second, hierarchy theory suggests that only three levels (or scales) usually are necessary to be considered in a model in lieu of completeness and parsimony (e.g., O'Neill 1988). However, there are exceptions to this general rule where effects at one level can penetrate through several levels above or below (see O'Neill 1988). It is conceivable that one may be allured to include more levels in one model due to, for example, ample data existing at finer scales. However, increasing the number of levels in model usually leads to greater model complexity, which reduces its comprehensibility and explanatory power and introduces a number of other potential sources of errors (parameter estimation, etc.) (Reynolds et al. 1993).

Third, it is not necessary (and not ecologically defensible in most cases) to assume the existence of a stable equilibrium when one builds a model to simulate a landscape system that appears to be stable. Indeed, assumptions of equilibrium and homogeneity may have hindered our understanding of spatially complex ecological systems (see discussion in Wu and Loucks 1995).

Hierarchical Modeling Approaches to Landscape Function

At all scales, the structural and physiological properties of patches will be different. Hence, we expect that functional attributes — such as carbon dioxide and water vapor fluxes — will differ from one patch to another (Jarvis 1995). If we were able to measure functional attributes for each patch simultaneously, simply adding them up would provide total landscape estimates. This is probably impractical at present for very large, complex landscapes. Using some of the experimental methods described in Valentini et al. (this volume), estimates of net ecosystem exchange (NEE) may be taken at

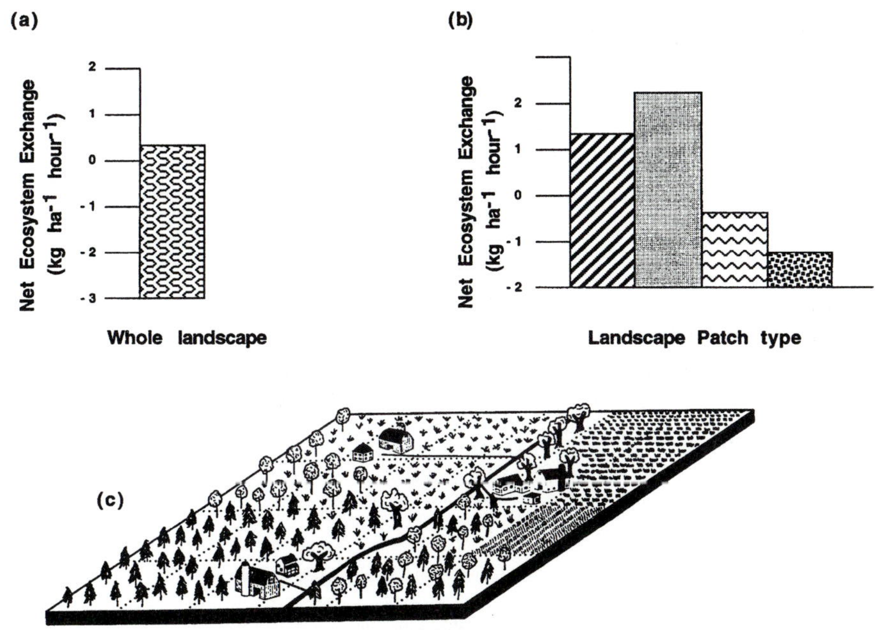

Figure 14.9 Three types of models of landscape change, distinguished by their level of aggregation: (a) Non-spatial, patch implicit, (b) Quasi-spatial, patch explicit, and (c) Spatially explicit. Modified and redrawn from Baker (1989).

different scales (patch to landscape) using towers or aircraft-borne instrumentation. Jarvis (1995) provides a thorough overview of these and other experimental methods as well as some of the compromises involved in trying to scale experimental results up to a landscape by summation, averaging, and aggregation.

How we model landscape function is dictated by how much structural detail we want to consider, that is, by the level of "lumping" or aggregation attempted. We distinguish three general types of models (modified from Baker 1989). *Non-spatial, patch implicit* models are those that ignore spatial structure and attempt to model some landscape variable in aggregate, e.g., NEE (Figure 14.9a). These are traditional point models used in population, community, and ecosystem ecology but applied at a larger scale. *Quasi-spatial, patch explicit* models (e.g., partial differential equation models with inner-patch dynamics) are models that include more details of the landscape, such as the relative contributions of each patch type to the modeled variable of interest (Figure 14.9b). *Spatially explicit* (including raster and vector based) landscape models are the most detailed: they consider explicit spatial locations and configurations of the patches (Figure 14.9c). Both nonspatial and quasi-spatial models may be submodels of the spatial landscape model, depending on the degree of detail incorporated. In this example, as more and more details are added, estimates of NEE as a function of patch configuration and environmental driving variables may be improved, but we must determine how to scale these results to larger landscapes.

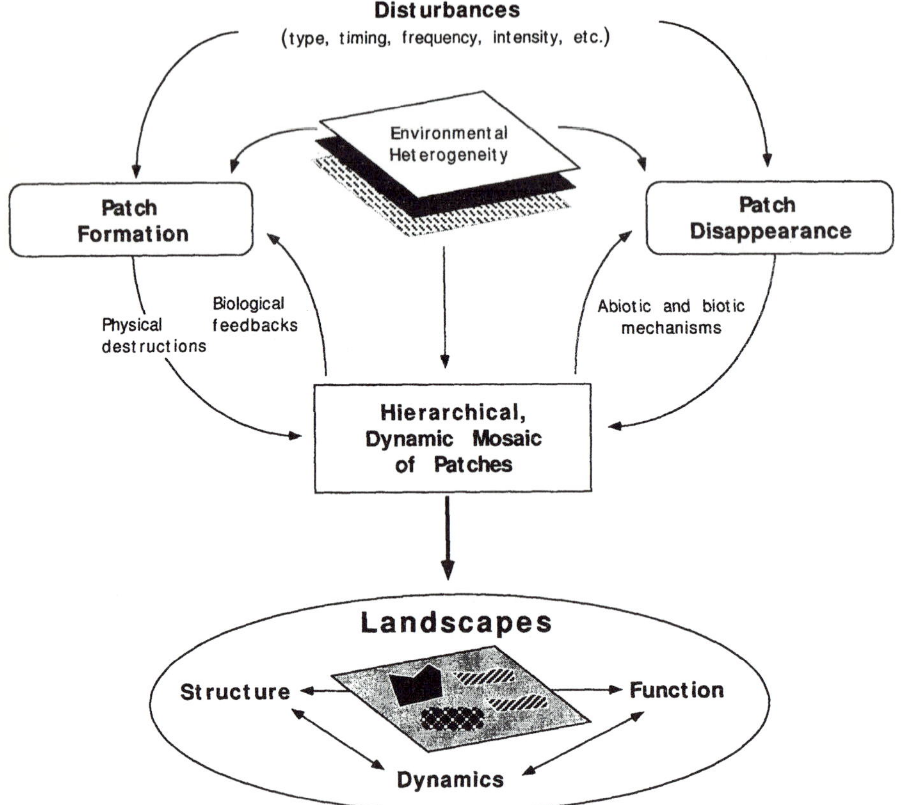

Figure 14.10 A hierarchical patch dynamics modeling framework, illustrating how patch-level processes scale up to the landscape level. Redrawn from Wu and Levin (1997).

A general bottom-up hierarchical approach to modeling patchy ecological systems involves accounting for the dynamics of pattern and processes on different levels (e.g., local patches, patch aggregates, and the landscape). By incorporating pattern and processes that operate on larger scales, these local patch models can be scaled up to a landscape-level patch dynamics model (Figure 14.10). Although this conceptual framework holds for many systems and is useful to model-building, the biological connotation of patch dynamics and mathematical details may vary greatly depending on questions to be addressed. This conceptual framework is evident in many spatially explicit landscape models (Baker 1989; Wu and Levin 1994, 1997).

Finally, we present an example of HPDP from Reynolds et al. (1993) of a "three-level-chain" hierarchical modeling scheme to illustrate how mechanistic information at lower hierarchical levels may be used at higher levels via a "filtering" or simplification process. The goal was to examine *stand* level responses of chaparral ecosystems to altered resource availabilities, e.g., nitrogen (N), solar radiation, carbon dioxide, etc. Reynolds et al. used a mechanistic plant growth model (GePSi), parameterized for various chaparral species, to predict maximum relative growth rate (Rmax) of these species as a function of light, CO_2, and N. Simulations of GePSi (daily time step) were conducted for a wide range of resource availabilities to calculate a response surface for Rmax, from which a simple hyperbolic model was then fit. Next, this empirical function for Rmax was substituted into PHENPLT, a simple phenomenological model of plant growth contained in STAND, a community model of long-term (annual) stand dynamics of chaparral ecosystems. In the original version of STAND, Rmax was fixed and not responsive to changes in resource availabilities; although the new Rmax is an empirical function of CO_2, solar radiation, carbon dioxide, and N it was developed from data generated by GePSi, which is a detailed model with high extrapolation potential. Stand level effects of altered resource availability could not have been investigated in the original version of STAND because plant growth responses to resources were not represented. On the other hand, population and community effects of resource availability could not be addressed by GePSi since community level processes like mortality and competition among plants are not included. Each model (GePSi and STAND) is best at answering questions at its own level of focus. The HPDP approach described here permits us to include information from a mechanistic model at a lower level in the hierarchy into a model at a higher level without directly incorporating the complete model structure. Following a similar hierarchical approach, Williams et al. (1997) developed a model to predict terrestrial gross primary productivity in diverse environments and ecosystems based on a more detailed model.

SUMMARY

In the **INTRODUCTION**, we noted the many gaps in our current understanding of hydrologic, ecosystem, and biogeochemical processes in complex terrestrial landscapes. This lack of understanding presents an interesting paradox. On one hand, while empirical studies cannot possibly provide the necessary understanding — which suggests a central role for modeling — the paucity of data also suggests that it is premature to develop integrated models of complex landscapes. There is a danger that in the absence of understanding, untested hypotheses can easily become incorporated into models (and then forgotten) and important processes that operate at various levels of biological organization are ignored with unknown consequences.

On the other hand, we have no choice since the questions being posed by resource managers and government policy-makers at local, regional, and global scales require the development of such integrated models. Consequently, models must represent an uneven blend of "state-of-the-art" knowledge and numerous assumptions, which are based on both practical and theoretical considerations and guesses. By linking these models with emerging technologies like remote sensing and geographical information systems, it will be possible to simulate complex scenarios of environmental–biotic interactions under differing sets of assumptions. As our knowledge improves, these models will improve, although numerous pitfalls exist since model testing and validation will often be inadequate or impossible at broader spatial and temporal scales.

In this chapter, we have argued that complex landscapes have structural and functional units at different scales. This argument is based on both theoretical and empirical evidence. According to the hierarchical patch dynamics paradigm, landscapes can be perceived as near-decomposable, nested hierarchies, in which hierarchical levels correspond to structural and functional units at distinct spatial and temporal scales. We have presented examples from a range of systems that seem to support the existence of these units. The process of identifying structural and functional units involves finding the characteristic scales of ecological processes of interest and decomposing landscape systems. The objectives of doing so are twofold: (a) to simplify the complexity of landscapes by providing a hierarchical structure to them, and (b) to promote multiple-scale approaches by emphasizing the equal importance of top-down constraints and bottom-up mechanisms. While simplification is the key to achieving understanding, the consideration of scale multiplicity in studying complex landscapes is dictated by the hierarchical properties of landscapes. We believe that the hierarchical patch dynamics paradigm provides a conceptual framework to accomplish these two objectives. We also have discussed how to use this paradigm to guide model building of landscapes by presenting a general hierarchical patch dynamic modeling approach. While this approach has been used in landscape modeling in recent years, its potentials and pitfalls, especially when used for developing integrated landscape models, are yet to be explored.

ACKNOWLEDGMENTS

We wish to acknowledge the contribution of Harbin Li to the ideas presented here. We appreciate the critical comments from our Dahlem colleagues, which led to significant improvements in this paper. This research was supported by the Jornada Long Term Ecological Research (LTER) project under NSF grant DEB 94–11971. During the preparation of this paper, J.W. was partially supported by USDA grant 95–37101–2028, the Central Arizona-Phoenix Long-term Ecological Research (LTER) project under the NSF grant DEB 97–14833, and research grants from Arizona State University.

REFERENCES

Baker, W.L. 1989. A review of models of landscape change. *Landsc. Ecol.* **2**:111–133.

Cantwell, M.D., and R.T.T. Forman. 1993. Landscape graphs: Ecological modeling with graph theory to detect configurations common to diverse landscapes. *Landsc. Ecol.* **8**:239–255.

Dutilleul, P., and P. Legendre. 1993. Spatial heterogeneity against heteroscedasticity: An ecological paradigm versus a statistical concept. *Oikos* **66**:152–167.

Forman, R.T.T. 1995. Landscape Mosaics: The Ecology of Landscapes and Regions. Cambridge: Cambridge Univ. Press.

Forman, R.T.T., and M. Godron. 1986. Landscape Ecology. New York: Wiley.

Franklin, J.F., and R.T.T. Forman. 1987. Creating landscape patterns for forest cutting: Ecological consequences and principles. *Landsc. Ecol.* **1**: 5–18.

Holling, C.S. 1992. Cross-scale morphology, geometry, and dynamics of ecosystems. *Ecol. Monogr.* **62**:447–502.

Jarvis, P.G. 1995. Scaling processes and problems. *Plant Cell Env.* **18**:1079–1089.

Kemp, P.R., J.F. Reynolds, Y. Pachepsky, and J.L. Chen. 1997. A comparative modeling study of soil water dynamics in a desert ecosystem. *Water Res.* **33**:73–90.

Kolasa, J., and C.D. Rollo. 1991. The heterogeneity of heterogeneity: A glossary. In: Ecological Heterogeneity, ed. J. Kolasa and S.T.A. Pickett, pp. 1–23. Berlin: Springer.

Kotliar, N.B., and J.A. Wiens. 1990. Multiple scales of patchiness and patch structure: A hierarchical framework for the study of heterogeneity. *Oikos* **59**:253–260.

Li, H., and J.F. Reynolds. 1994. A simulation experiment to quantify spatial heterogeneity in categorical maps. *Ecology* **75**: 2446–2455.

Li, H., and J.F. Reynolds. 1995. On definition and quantification of heterogeneity. *Oikos* **73**: 280–284.

Ludwig, J.A., D.J. Tongway, and S. Marsden. 1994. A flow-filter model for simulating the conservation of limited resources in spatially heterogeneous semi-arid landscapes. *Pacific Conserv. Biol.* **1**: 209–213.

O'Neill, R.V. 1988. Hierarchy theory and global change. In: Scales and Global Change, ed. T. Rosswall, R.G. Woodmansee, and P.G. Risser, pp. 29–45. New York: Wiley.

O'Neill, R.V. 1996. Recent developments in ecological theory: Hierarchy and scale. In: GAP Analysis: A Landscape Approach to Biodiversity Planning, ed. J.M. Scott, T.H. Tear, and F.W. Davis, pp. 5–13. Bethesda, MD: American Society of Photographic Remote Sensing.

Osmond, C.B., O. Bjorkman, and D.J. Anderson. 1980. Physiological Processes in Plant Ecology. Berlin: Springer.

Pattee, H.H., ed. 1973. Hierarchy Theory — The Challenge of Complex Systems, p. 156. New York: George Braziller.

Pickett, S.T.A., and M.L. Cadenasso. 1995. Landscape ecology: Spatial heterogeneity in ecological systems. *Science* **269**:331–334.

Pickett, S.T.A., and P.S. White, eds. 1985. The Ecology of Natural Disturbance and Patch Dynamics. San Diego: Academic.

Reynolds, J.F., D.W. Hilbert, and P.R. Kemp. 1993. Scaling ecophysiology from the plant to the ecosystem: A conceptual framework. In: Scaling Processes between Leaf and the Globe, ed. J. Ehleringer and C. Field, pp. 127–140. New York: Academic.

Reynolds, J.F., J.D. Tenhunen, P.W. Leadley, H. Li, D.L. Moorhead, B. Ostendorf, and F.S. Chapin. 1996. Patch and landscape models of arctic tundra: Potentials and limitations. In: Landscape Function and Disturbance in Arctic Tundra, ed. J.F. Reynolds and J.D. Tenhunen, pp. 293–324. Ecological Studies Series 120. Berlin: Springer.

Reynolds, J.F., R.A. Virginia, and W.H. Schlesinger. 1997. Defining functional types for models of desertification. In: Plant Functional Types: Their Relevance to Ecosystem Properties and Global Change, ed. T.M. Smith, H.H. Shugart, and F.I. Woodward, pp. 194–214. Cambridge: Cambridge Univ. Press.

Rowe, J.S. 1961. The level-of-integration concept and ecology. *Ecology* **42**:420–427.

Shaver, G., K. Nadelhoffer, and A. Giblin. 1991. Biogeochemical diversity and element transport in a heterogeneous landscape, the North Slope of Alaska. In: Quantitative Methods in Landscape Ecology, ed. M.G. Turner and R.H. Gardner, pp. 105–125. Ecological Studies Series 82. Berlin: Springer.

Swanson, F.J., T.K. Kratz, N. Caine, and R.C. Woodmansee. 1988. Landform effects on ecosystem patterns and processes: Geomorphic features of the Earth's surface regulate the distribution of organisms and processes. *BioScience* **38**:92–98.

Turner, M.G. 1989. Landscape ecology: The effect of pattern on process. *Ann. Rev. Ecol. Syst.* **20**:171–198.

Turner, M.G., R.H. Gardner, and R.V. O'Neill. 1995. Ecological dynamics at broad scales. *BioScience* **45**:S29–S35.

Turner, S.J., R.V. O.Neill, W. Conley, M.R. Conley, and H.C. Humphries. 1991. Pattern and scale: Statistics for landscape ecology. In: Quantitative Methods in Landscape Ecology, ed. M.G. Turner and R.H. Gardner, pp. 17–49. Ecological Studies Series 82. Berlin: Springer.

Urban, D.L., R.V. O'Neill, and H.H. Shugart. 1987. Landscape ecology. *BioScience* **37**:119–127.

Walker, D.A., and M.D. Walker. 1991. History and pattern of disturbance in Alaskan arctic terrestrial ecosystems: A hierarchical approach to analysing landscape change. *J. Appl. Ecol.* **28**:244–276.

Walker, D.A., and M.D. Walker. 1996. Terrain and vegetation of the Imnavait creek watershed. In: Landscape Function and Disturbance in Arctic Tundra, ed. J.F. Reynolds and J.D. Tenhunen, pp. 73–108. Ecological Studies Series 120. Berlin: Springer.

Wickham, J.D., and D.J. Norton. 1994. Mapping and analyzing landscape patterns. *Landsc. Ecol.* **9**:7–23.

Wiens, J.A. 1989. Spatial scaling in ecology. *Fun. Ecol.* **3**:385–397.

Williams, M., E.B. Rastetter, D.N. Fernandes, M.L. Goulden, G.R. Shaver, and L.C. Johnson. 1997. Predicting gross primary productivity in terrestrial ecosystems. *Ecol. Appl.* **7**:882–894.

Woodcock, C., and V.J. Harward. 1992. Nested-hierarchical scene models and image segmentation. *Intl. J. Remote Sens.* **13**:3167–3187.

Woodmansee, R.G. 1990. Biogeochemical cycles and ecological hierarchies. In: Changing Landscapes: An Ecological Perspective, ed. I.S. Zonneveld and R.T. Forman, pp. 57–71. Berlin: Springer.

Wu, J. 1999. Hierarchy and scaling: Extrapolating information along a scaling ladder. *Can. J. Remote Sens.*, in press.

Wu, J.G., and S.A. Levin. 1994. A spatial patch dynamic modeling approach to pattern and process in an annual grassland. *Ecol. Monogr.* **64**:447–464.

Wu, J.G., and S.A. Levin. 1997. A patch-based spatial modeling approach: Conceptual framework and simulation scheme. *Ecol. Mod.* **101**:325–346.

Wu, J.G., and O.L. Loucks. 1995. From balance of nature to hierarchical patch dynamics: A paradigm shift in ecology. *Qtly. Rev. Biol.* **70**:439–466.

15

Effect of Landscape Fragmentation, Disturbance, and Succession on Ecosystem Function

I.R. NOBLE

Ecosystem Dynamics, Research School of Biological Sciences, Institute of Advanced Studies, Australian National University, Canberra 0200, Australia

ABSTRACT

Landscapes are mosaics of patches each with potentially different effects on material flows and balances. Thus, net material fluxes are dependent on the structure and pattern of landscape units. Landscape pattern derives from both exogenous factors, such as topography and soil types, and endogenous factors, such as spatially correlated disturbances, propagule dispersal, and the material flows themselves. If we aim to describe and predict the role of landscapes in moderating material flow under changing conditions, it appears that models of landscape processes are likely to be a greater limitation than data availability per se. Even simple models show that self-generated patterning within a landscape can have major effects on the predicted dynamics of the communities within it and the associated fluxes. There is some progress towards developing rules for particular phenomena associated with landscape processes but these rules are far from forming an integrated set. The challenge in developing models of the complexity of landscape processes is to separate those components that affect the phenomena of interest from those that are noise. Hierarchy theory has provided some helpful insights into how to do this but there is still no clear hierarchy of spatial and temporal scales that link the patch to the global scale via landscapes.

INTRODUCTION

We live in landscapes; we manage landscapes. We often describe the environment around us in terms of landscapes. Yet landscapes have long been a scientific blind spot (Figure 15.1). The scientific description and classification of landscapes is weak and our understanding of their role in ecosystem functioning is poor. In this chapter I ask,

Integrating Hydrology, Ecosystem Dynamics, and Biogeochemistry in Complex Landscapes
Edited by J.D. Tenhunen and P. Kabat

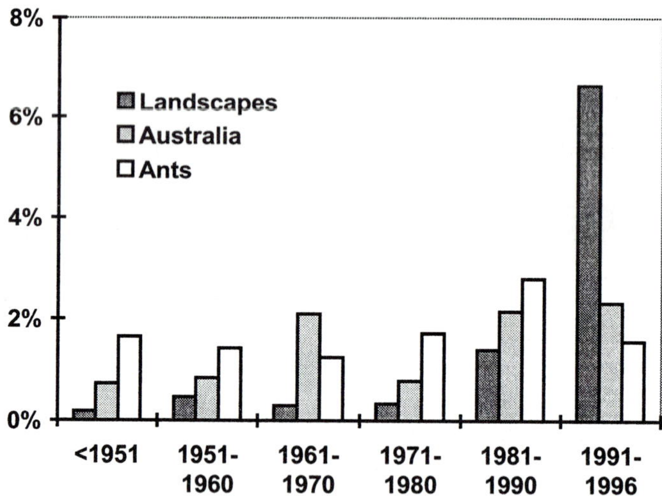

Figure 15.1 The percentage of papers with the term landscape in their title, abstract, or keywords in the journals of the Ecological Society of America compared with two other topics. Papers on landscape issues have risen dramatically this decade. During this same period several journals specifically related to landscapes have also started.

how does the ecological complexity of landscapes, i.e., their fragmentation, diversity, and patterns, affect important ecosystem functions, such as water and energy exchange, lateral water flow, and nutrient retention and loss?

First, I discuss the role of landscape complexity in horizontal and vertical transport of water and nutrients. I seek generalizations rather than enumerating the multitude of differences. Communities of different composition and structure affect the redistribution of water and nutrients in different ways. Different successional stages have different but relatively predictable effects on redistribution. Thus, landscapes are made up of a mosaic of cells with different permeabilities to water and nutrients. However, landscape complexity causes problems in interpreting measurements of fluxes and in scaling up from models of point and patch scales to larger areas.

Second, I discuss some of the factors controlling the complexity of landscapes and in particular the role of self-generating patterns arising from disturbance processes. I ask whether we can develop useful models of landscape processes and find the necessary data sets to initialize and drive them. Finally, I discuss whether we have made any advances in the application of hierarchy theory to scaling issues associated with global change. Although this chapter points out many weaknesses in our understanding of landscape processes, its goal is to focus attention on these areas and to point to opportunities for research.

HOW DOES LANDSCAPE PATTERN AND SUCCESSIONAL STAGE AFFECT NUTRIENT AND WATER REDISTRIBUTION?

Landscapes as Filters and Concentrators

There are many examples of components of the landscape acting as filters and concentrators of fluxes of water and nutrients. Both naturally occurring and human-maintained elements of a landscape may be involved.

Arid ecosystems are a particularly good example of where the redistribution and concentration of water, nutrients, and sediments are important at several scales in achieving adequate growing conditions for plants. At the smallest scale, each individual shrub in an arid shrubland can act as a "fertile island" where nutrients and water are concentrated (Noy Meir 1985). Fallen logs and minor changes in topography act as concentrators at slightly larger scales (Ludwig and Tongway 1995).

At a larger scale, the "stripe process" occurs in many flat (0.6° slope) arid regions of the world (Thiery et al. 1995). In these systems vegetation is concentrated in parallel, narrow bands and the soil between the bands of vegetation is usually almost bare. In heavy rainfalls, water quickly runs off bare areas but is trapped in vegetated areas effectively doubling the amount of water received in the stripes (Montaña 1992). The pattern of stripes can develop initially either from the loss of surrounding vegetation in previously vegetated areas or from the spread of vegetation around an initial colonizer in previously bare areas (White 1971).

At an even larger scale, arid lands are often a mosaic of "erosion cells," each consisting of erosion, transfer, and sink areas (Pickup 1985). Changes in grazing patterns brought about by human actions can affect each of these concentrating mechanisms by removing plant cover or breaking up soil crusts, thus initiating new erosional cells and altering landscape patterns.

Many other examples of patterned landscapes can be found. In temperate areas, tree strips, especially those in shelterbelts or naturally occurring riparian woodlands, can greatly reduce water flows and nutrient transfers (Ryszkowski and Kedziora 1993; Correll et al. 1992). "Fir waves," somewhat analogous to the stripe phenomenon, are found in alpine and boreal ecosystems (Sprugel 1976).

Succession and Resource Redistribution

In extensive areas of relative climatic and edaphic uniformity, the successional stages of communities affect nutrient accumulation and redistribution. Using a classical secondary successional model for simplicity of description, certain logical consequences can be deduced.

Virtually all disturbances capable of resetting a successional sequence will result in a pulse of nutrient loss. Fires release nutrients to the atmosphere and expose topsoil to erosion (Uhl and Jordan 1984) while blowdowns, landslides, and flooding also expose mineral soil and fine debris to a high risk of erosion (Matson et al. 1987). The immediate end point for these nutrients depends on the location (i.e., the surrounding landscape) and size of the disturbance. In some cases, most nutrients are trapped in surrounding, undisturbed vegetation while in other cases streams and wind redistribute nutrients more widely.

Nutrient entrapment and accumulation begins quickly in most pioneer communities although the rate and trajectory of buildup varies between communities. The scarce elements, N and P, tend to be retained in recently disturbed ecosystems more than some of the more abundant elements (Schlesinger 1991). Spatial variation in nutrient concentration appears to vary most in mid-successional communities (Gross et al. 1995). As biomass rebuilds, retention continues until a late successional phase of quasi-equilibrium biomass and species composition is reached. At this point, nutrient inputs approximate outputs and net transport of nutrients from an ecosystem will apparently increase. For example, streams draining old-aged, forested ecosystems have higher concentrations of NO^{3-}, K^+, and other plant nutrients than do streams draining intermediate-aged successional ecosystems (Vitousek 1977). However, outputs will reflect both inputs (e.g., atmospheric deposition; Hedin et al. 1995) and patterns of precipitation (Martin 1979).

A similar story applies to the flux of water through landscape elements of different successional stages. Runoff tends to be high in recently disturbed communities. Some studies indicate that the increase in water outflow is proportional to the leaf area removed (Knight et al. 1985). In landscapes of mixed structure (e.g., savannas or woodlands), the tree component traps and uses a far greater proportion of the available water than do neighboring grassland communities. Joffre and Rambal (1993) found that in sites in Spain, water yield from under-tree canopies was negligible and all precipitation was lost by evapotranspiration when annual precipitation was < 570 mm. Outside the tree canopy, water yield occurred as soon as annual precipitation exceeded 250 mm.

Thus, a landscape is a mosaic of different vegetation units, each with potentially different transfer and retention properties for water and nutrients. In landscapes with patches of vegetation undergoing successional change, the properties of the patches will change with successional stage. The very structure of the landscape itself is in turn affected by these retention and transfer properties. However, these are only a few of the processes that determine landscape patterns and structure.

ROLE OF EXOGENOUS AND ENDOGENOUS PROCESSES IN LANDSCAPE PATTERNING AND COMPLEXITY

A major question in landscape studies is how much are landscape patterns a product of the underlying topography and soils (here called exogenous because significant changes occur only over periods of decades to millennia) and how much are they a result of faster, endogenous processes, such as dispersal and disturbances, that lead to self-generated patterns. Ecological research has focused on seeking the relationships between landscape and exogenous factors, such as topography and soil. It has been demonstrated repeatedly that exogenous factors play a significant role. Reiners and Lang (1979) described three levels of influences on landscape pattern in alpine fir communities. Elevation, topography, and soils control the range of feasible community types within a particular climatic zone. Overlaid on this first-order constraint are the effects of disturbances, such as avalanches and blowdowns. Within the patterns generated by the above processes are more subtle textures arising from endogenous processes within the landscape. These include tree-fall gaps, "fir waves," and vegetation stripes. From palaeoecological studies of the past 400 yr, Swain (1980) showed that the mosaic of vegetation patterns in the Boundary Waters Canoe Area in northeastern Minnesota, U.S.A., appears to be related to topography and presence of large lakes (Reiner and Lang's first level). The patterns, however, were also affected by the ways in which these features influence periodic forest fires (Reiner and Lang's second and third levels).

A number of other authors have suggested that patterns developed by endogenous processes (self-generated patterns) can be slow to change, based on some field studies (e.g., Paine and Levin 1981; Wallin et al. 1994) and many modeling studies (e.g., Noble and Gitay 1996), which show that once self-reinforcing patterns are generated they change only slowly. However, there is still much to be explored in this area. The modeling studies are far from complete in that only a small part of their rich behavior has been investigated. In particular, the sensitivity of endogenously generated patterns to variation in the underlying physical substrate (topography and soils) and to the pattern of initiation and spread of disturbances has not yet been adequately explored.

Noble and Gitay (1996) showed that in systems with high fire frequencies, such as the Australian eucalypt forests, endogenously developed landscape patterns can be persistent; however, they are sensitive to topographic features that affect fuel loads and fuel dryness. Figure 15.2 shows a comparison between several levels of model complexity. The model is the VASL model, as described in Noble and Gitay (1996), and is based on the vital attributes model of Noble and Slatyer (1980). The system simulated is a cool temperate rainforest and eucalypt system in Tasmania. In this system, the basic successional pathway is from a grass community, through a shrubby stage and eventually to a eucalypt forest with rainforest as an understorey. The eucalypts cannot regenerate without fire; thus, a long fire-free period leads to a community of pure rainforest. The model is a multi-pathway succession model with particular sequences of fires causing local extinctions of some species. The model incorporates dispersal from neighboring sites so that species locally extinct within a patch can reinvade. Communities vary in the likelihood that they will carry fire, with grass communities being the most fire prone and rainforests the least. The model has been parameterized based on expert knowledge and field studies in the area. The model allows comparison between a point version (i.e., many independent replicates of a single stand) and a landscape version based on a 50×50 grid. Fires propagate in a realistic manner across the landscape depending on the type of community in each cell and on a background fire proneness dependent upon the topography of the landscape. In each run, the initial

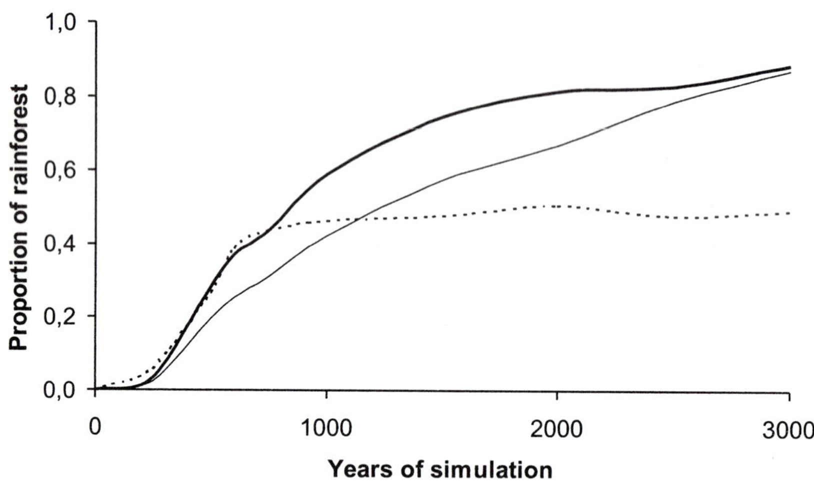

Figure 15.2 A comparison of three levels of landscape complexity underlying the same succession model. The model is of a temperate eucalypt-rainforest system (Noble and Gitay 1996). The point model (dashed line) shows the proportion of rainforest based on multiple runs for a single stand. At the next level of complexity (thin line), a 50 × 50 grid landscape is used in which fires propagate contagiously with the probability of spreading depending on the vegetation types encountered and seed dispersal. This landscape has no topographic features, i.e., it is a homogeneous area. The final level (thick line) introduces a ridge-gully system to the landscape with variation in probability of burning ranging between 80% and 120%. This variation in the probability of burning is relatively small compared to the differences between communities.

vegetation is a uniform cover of a mid-successional community with all major species present.

A comparison of the point and landscape-without-topography results shows the large effects that arise from taking landscape processes into account. The point model predicts the eventual attainment of equilibrium with about half the area covered by rainforest. When converted to a landscape model, rainforest builds up more slowly due to the effect of patches of fire-prone communities developing, which hinder the buildup of rainforest. It is common to find that after 1000 years some cells have been burnt more than 50 times while other cells nearby have never been burnt. However, as more and more of the landscape becomes dominated by rainforest, fire spread is inhibited and eventually rainforest covers almost the entire landscape — a very different outcome from the point model. When a simple topography consisting of a series of three ridges and gullies running across the landscape perpendicular to the most common direction of fire spread is introduced, the community dynamics change again. Even though the topography changes the probability that a fire will be carried through a cell by only ± 20% (in contrast, a grass community is 13 times more likely to carry a fire than a rainforest and a eucalypt-dominated community four times more likely), the net effect is to limit the creation of large patches of flammable communities. This allows rainforest to build up as quickly as in the point model.

These specific results are peculiar to this particular representation of the biological succession on the site, to the fire spread model, and to the assumptions about the landscape. However, experience with other communities and other forms of landscape structure show that the wide differences between point representations, homogeneous landscapes, and structured landscapes are common. This must be taken as a warning to be cautious in developing overly simplified representations of landscape phenomena. In many studies of global change, point models have been taken as representative of large regions, and in some catchment studies, the landscape structure within the catchment is overly simplified.

CAN WE DEVELOP USEFUL MODELS OF COMPLEX LANDSCAPES?

Models of Self-generating Patterns

There are many models of the interactions between components of landscapes. The simplest models use cellular automata with simple rules of interaction; the most complex use patch models to represent each landscape element with detailed representation of the exchange of propagules and resources between elements (see Gardner et al., unpublished, for a review of the classes of landscape models associated with the effects of fire on landscapes). The models show a huge variety of behaviors, but almost all point to the conclusion that a vast range of self-generating patterns can be developed from quite simple interaction rules between the elements of the landscape. The patterns developed are very sensitive to the underlying assumptions. Most studies have stopped at the point of demonstrating the self-generating pattern process and describing coarse outcomes (as in the example given above). There is probably little chance of coming to grips with the phenomena associated with self-generating patterns unless rules describing their behavior can be developed (cf. Pacala and Duestchman 1995), and these rules must be interpretable in ecological terms.

Can Rules Be Derived for Phenomena at the Landscape Scale?

There are some promising developments in this area, but progress is piecemeal and it will be some time before a comprehensive understanding of the roles of exogenous and endogenous process in landscape patterning is achieved. However, certain principles are beginning to appear. I present a few examples here.

There are many studies based on both analytical and simulation models that conclude that spatial heterogeneity enhances ecosystem stability and species persistence (Kareiva 1994), yet fragmentation of suitable habitats is also a threat to biodiversity. From a review of studies on birds and mammals, Andren (1994) concluded that the effect of habitat fragmentation on landscapes with more than 30% of suitable habitat was primarily that of habitat loss. However, in landscapes with even less suitable habitat, patch size and isolation can increase the effect of habitat loss, and the loss of species, or decline in population size, may be greater than expected from habitat loss alone.

Similar broad principles have been determined for understanding the role of dispersal in the migration of species across heterogenous landscapes (Pease et al. 1989; With and Crist 1995; Malanson and Armstrong 1996; Pitelka et al. 1997). Gardner et al. (1989) developed some guidelines to handle the effect of scale changes on percolation modeling. Unfortunately, many real-life examples fall into their category where they conclude extrapolation is risky or impossible. Figure 15.3 shows such an example. The fire spread model (Noble, unpublished) is a simple percolation model, and the outcome varies greatly with the pixel size selected.

A useful insight into the effect of scale was obtained by Costanza and Maxwell (1994), who analyzed land-use data by resampling maps at several different spatial

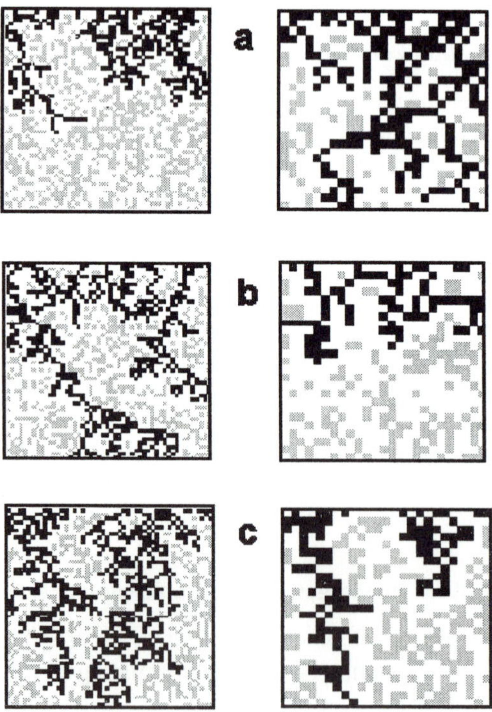

Figure 15.3 Effect of changing pixel size on a fire spread model (Noble, unpublished). Each left-right pair represents the same landscape but with the right image sampled in blocks of 2 ×2 of the pixels in the left image. The fire is ignited along the top edge in each case and same simple spread model is used in both images. The model allows fire to spread from a burning cell to adjacent cells containing fuel that are ahead of the currently burning cell (down the page) or along side it (i.e., to the left or right). Black pixels are burnt while grey pixels were flammable but unburnt. The outcomes of the fire spread vary enormously and in no consistent manner.

resolutions and measuring predictability at each. They defined spatial auto-predictability [P(a)] as the reduction in uncertainty about the state of a pixel in a scene given knowledge of the state of adjacent pixels in that scene (i.e., a measure of the strength of the pattern in the data), and spatial cross-predictability [P(c)] as the reduction in uncertainty about the state of a pixel in a scene given knowledge of the state of corresponding pixels in other data sets of the same scene (i.e., a measure of the ability of other data to model the target pattern). In general they found that P(a) increased with increasing log of the resolution (as more information is included), while P(c) fell or remained stable as resolution increased (because it is easier to model aggregate information than fine grain detail). Thus, in many practical situations, choosing an "optimal" resolution for a particular modeling problem will involve balancing the benefit in terms of increasing data predictability [P(a)] as one increases resolution, with the cost of decreasing model predictability [P(c)].

Thus, although there are only a few examples of simplifying principles or rules for landscape phenomena, progress has been promising.

WILL WE EVER HAVE SUFFICIENT DATA TO DESCRIBE REAL LANDSCAPES?

Given the role of exogenous factors in determining landscape patterns, is it feasible to collect data of sufficient scope (i.e., inclusiveness of parameters) and with sufficient resolution to drive models of landscape change?

How Much Data Do We Need for Local Predictions?

There are numerous examples of correlations between vegetation and climate, soils and topography, and there is an extensive literature on the best models and methodology for deriving such relationships (e.g., Austin et al. 1994). These relationships have been based on a wide range of spatial scales, and again there are a number of studies of the effect of sampling at different scales (e.g., Obeysekera and Rutchey 1997). However, as topographic and soils data of ever finer resolution have become available, disturbances such as fire, blowdown, and land and snow slides have been increasingly recognized as important in describing differences between observed landscapes and simple correlations between vegetation and physical parameters (e.g., Walsh et al. 1994).

Modeling the relationship between fire patterns and landscape is proving to be challenging. Fire modelers have had only limited success in attempting to forecast real-time fire spread patterns mainly due to variation in weather patterns (including self-generated wind patterns at the flame front) and fuel loads. It is unlikely that with even finer resolution data sets much improvement will be achieved. Attempts to predict longer-term fire frequencies over entire landscape may be more successful; however, relatively few comprehensive studies have been done (Keane et al. 1996; Cary, unpublished). In a review of this area, Gardner et al. (unpublished) concluded that our present ability to "anticipate the effects of shifting fire regimes remains surprisingly limited." Among the limiting factors they list the sensitivity of models to fuel connectivity patterns and to the development of fire dependent vegetation communities. Both can lead to self-generated patterns of fuel distribution and fire occurrence.

Thus, it appears that at fine scales data availability will remain a limitation. Even if more processes are incorporated in the models, it will be difficult to provide sufficiently precise data on such initial conditions as vegetation and fire-fuel distributions to avoid chaotic processes dominating the results.

How Much Data Do We Need for Regional/Global Predictions?

Another question is how detailed land-surface schemes need to be to account for spatial variability in regional and global scale models?

White and Running (1994) investigated whether areal average flux of carbon and water can be scaled linearly over a complex landscape by applying models to data derived at 30 m and 1 km scales for a complex landscape in Glacier National Park. They concluded that estimated daily photosynthesis and ANNP were highly correlated between modeling scales with maximum differences in ANPP predictions of ca. 0.5 t C/ha/yr. Evapotranspiration and hydrologic outflow were not as highly correlated between different modeling scales probably due to the structural differences in the hydrologic models used at different scales. Noilhan and Lacarre (1995) came to a similar conclusion based on the HAPEX-MOBILHY field campaign in early summer of 1986. They found that despite the nonlinear dependence of surface fluxes on both vegetation and soil water content, the effective surface fluxes computed with a 1-D column model matched the areal-averaged fluxes estimated from 3-D mesoscale model results with a relative error less than 10%. Again the match was poorer for the soilwater transfers than for the vegetation behavior.

Li and Avissar (1994) investigated the impact of subgrid-scale variability of land characteristics on land-surface energy fluxes simulated in atmospheric models (e.g., GCMs) with a land-surface model that represents the land surface as a mosaic of patches. They found that, for characteristics that are nonlinearly related to the energy fluxes, the error in calculating the energy fluxes based on the mean, rather than the distribution, increased as the skewness of the subgrid-scale variability increased. Spatial variability of surface roughness, leaf area index, stomatal conductance, and soil-surface wetness were found to be important while the relationship between albedo and surface energy fluxes was almost linear, and therefore, using a mean value of this characteristic was appropriate.

Gutowski et al. (1991) examined the scale dependence of the surface energy balance by averaging over spatial domains ranging from the entire globe to regions encompassing just a few model grid points. They concluded that in the (then) current climate simulations, the basic large-scale mechanisms of greenhouse warming were not very sensitive to the precise surface balance of heat. Increasing horizontal resolution would not improve the consistency of regional-scale climate simulations in these models unless discrepancies in global-average longwave radiation were resolved. Differences between models in simulating effects of moisture in the atmosphere and in the ground appear to be an important cause of differences in surface energy budgets on all scales.

Again it appears that data availability *per se* is not a limiting factor. Relatively coarse data sets can provide reasonable approximations of the results from finer resolution data. The models themselves are likely to be the greater limitation.

DISCUSSION

Most ecologists realize that the interplay of dispersal, disturbance, and spatial mosaics can profoundly alter the outcome of community dynamics. Indeed, an appreciation of the importance of spatial scale and spatial heterogeneity is well established in natural history. Nevertheless, one must choose a scale at which to observe the system. There is no single correct scale for any system, but not all scales are equally effective (Levin 1992). Similarly, there is no single level of complexity for modeling. Models that are too simplified are almost certain to omit critical detail that is responsible for phenomena of importance in the application of the model (e.g., the exclusion of landscape processes in current dynamic global vegetation models, DGVMs). However, models that are too detailed are likely to require far more parameters and functional forms than available data justify (Levin 1992). The key is to separate the components of variability and complexity into those that affect the processes of interest and those that are noise.

Application of Hierarchy Theory

At a meeting held almost a decade ago and influential in the structuring of the International Geosphere-Biosphere Programme (IGBP), O'Neill (1988) presented a set of nine criteria relevant to the application of hierarchy theory to the scaling problem in global change studies (Table 15.1). It is interesting to revisit these principles after a decade of additional research to see whether they have proven useful and what progress has been made.

Find the Fundamental Hierarchy

The hierarchy that currently dominates much of biological science may not be the most appropriate to global change research. A perusal of the structure of biology institutions and journal titles indicates that modern biology is dominated by the hierarchy: molecule, cell, community/ecosystem, and (maybe) global. This is notable for the absence of certain levels, for example, the organism and the landscape. It may not be the correct hierarchy for global change, especially when interactions with other disciplines are taken into account. For example, in dealing with the flux of water and nutrients, a better hierarchy may be cell, individual, landscape unit, catchments, and globe.

Table 15.1 O'Neill's (1988) nine criteria relevant to the application of hierarchy theory to the scaling problem in global change studies.

1. Find the fundamental hierarchy
2. Find the fundamental level
3. Find coherent levels (i.e., ones that work as an object of study)
4. Translate principles between levels
5. Find state variables that translate across levels
6. Accept innovative approaches
7. Learn how higher levels constrain lower levels
8. Learn when the rapid dynamics of lower levels cannot be ignored
9. Find critical points in parameter space (thresholds)

In studying the impact of global change on human society, the hierarchy may start with cells and individuals, and, in preindustrial societies, continue with family/clan-groups (operating at the level of landscape units?) to tribes (catchments?), and wider institutions (global?). However, any link between biophysical hierarchies and human hierarchies may have been lost with the dramatic changes in communication systems over the past century or so. There will be no single, fundamental hierarchy for all of global change science, but relatively little effort has been made to question whether better structures can be found. Most scientists have stayed within their discipline hierarchy, with a few attempts to patch it, for example, by inserting the landscape and regional levels between the community/ecosystem and global levels in GCTE (Global Change in Terrestrial Ecosystems Core Program of IGBP).

Holling (1992) has suggested that a small set of processes may structure ecosystems in time and space. These processes establish a small number of dominant temporal frequencies that entrain other processes and generate characteristic spatial scales. They also lead to discontinuities in body mass of both birds and mammals. He suggests that the temporal and spatial scales of ecosystems are dominated by three major classes of processes. Vegetative processes (i.e., plant growth, seed production, nutrient mobilization and immobilization, foraging) determine the texture at microscales of centimeters to tens of meters and days to decades. Geomorphological processes dominate the macroscales of 100s to 1000s km and centuries to millennia. The mesoscale remains problematic. It fits between the other two, covering the patch to landscape scales, and is dominated by contagious processes such as fire, pest outbreaks, and water flow and is most subject to disturbance by human activity. Although Holling's approach is novel, the conclusions are familiar from less well-developed arguments. However, his arguments do suggest opportunities for experimental testing of the link between body size and landscape grain.

Find the Fundamental and Coherent Levels (i.e., ones that work as an object of study)

With little progress in seeking fundamental hierarchies, it might be thought that little progress could be made in locating fundamental levels, but this is not necessarily the case. For example, physics made great progress by focusing on the atom as a fundamental level for many of its subdisciplines. In global change science, some levels are gaining more attention. Clearly, the global level is a common priority for most disciplines, but in ecology the patch level has assumed increasing importance. I suspect that it is a coherent level in O'Neill's terminology[1], in that it integrates the responses of individuals in a realistic context.

Find Principles and State Variables That Translate between Levels

Here the goal is to find principles and state variables that are meaningful on several lcvcls (at lcast thrcc, a basc lcvcl and at lcast onc abovc and onc bclow). An cxamplc of a state variable that translates between levels is the moisture content of the air. At some levels, e.g., the individual plant or leaf, it directly affects performance; at higher levels, aggregations of individuals (e.g., forests) can affect the moisture content of air masses. The moisture content of the air, when linked with other physical information can be transformed into other useful drivers, such as precipitation. However, simply identifying such state variables is not sufficient. Global change science is rife with mismatches between the measurement or estimation of common variables — air moisture and precipitation being one.

Learn How Higher Levels Constrain Lower Levels

A common insight in hierarchy theory derived from engineering principles is that higher levels operate at much larger spatial scales and much slower temporal scales than lower levels. Thus, their role is as a relatively constant constraint on the operation of processes at lower levels.

Learn When the Rapid Dynamics of Lower Levels Cannot Be Ignored

The problem with landscapes is that their temporal scale is both fast and slow relative to the global scale (slow in the sense that many landscape changes take decades to centuries to unfold — much slower than many of the interesting global phenomena — but fast enough to be relevant to longer-term planning). These relatively slow processes are affected by, and often initiated by, very fast processes such as disturbances (Holling's 1992 brittleness concept). In these situations there is no clear series of levels

---

[1] Coherence may best be described by example. It is very difficult, and of little value, to predict the movement behavior of an individual animal. However, the problem is more soluble and usually more valuable if the movement behavior of a herd of animals is the level of attention.

each with increasing spatial and temporal scales. Is this a flaw in hierarchy theory, or is it simply a failure on our part to find better descriptions of hierarchies, that is, in O'Neill's terms a failure to accept (find) innovative approaches (Table 15.1, Principle 6)?

Find Critical Points in Parameter Space (Thresholds)

Often the task in global change studies is to predict marginal and threshold behavior of complex systems. Many models that prove satisfactory in normal operational ranges are inadequate when applied to larger scales. For example, it is no use assuming that an exponential dispersal function is adequate because it covers 99.99% of dispersal events, when it is the remaining 0.01% of events that determine rates of migration over larger distances.

Just as there are yet no simple rules to translate spatial data and process models into emergent landscape phenomena, there is not a simple set of rules to translate hierarchy theory into operational principles. Despite the rapid increase in publications on landscape matters, "spatial complications" are often used as a catch-all by ecologists for explaining surprising results, or as a condemnation of ecological theory by accusing it of oversimplification through its neglect of spatial variation. Much more could be gained from appropriate experiments, both in the field and with models, that explicitly test major hypotheses emerging from recent theoretical explorations of spatial processes and scaling rules.

REFERENCES

Andren, H. 1994. Effects of habitat fragmentation on birds and mammals in landscapes with different proportions of suitable habitat: A review. *Oikos* **71**:355–366.

Austin, M.P., A.O. Nicholls, M.D. Doherty, and J.A. Meyers. 1994. Determining species response functions to an environmental gradient by means of a beta-function. *J. Veg. Sci.* **5**:215–228.

Correll, D.L., T.E. Jordan, and D.E. Weller. 1992. Nutrient flux in a landscape: Effects of coastal land use and terrestrial community mosaic on nutrient transport to coastal waters. *Estuaries* **15**:431–442.

Costanza, R., and T. Maxwell. 1994. Resolution and predictability: An approach to the scaling problem. *Landsc. Ecol.* **9**:47–57.

Gardner, R.H., R.V. O'Neill, M.G. Turner, and V.H. Dale. 1989. Quantifying scale-dependent effects of animal movement with simple percolation models. *Landsc. Ecol.* **3**:217–227.

Gross, K.L., K.S. Pregitzer, and A.J. Burton. 1995. Spatial variation in nitrogen availability in three successional plant communities. *J. Ecol.* **83**:357–367.

Gutowski, W.J., D.S. Gutzler, and W.C. Wang. 1991. Surface-energy balances of three general-circulation models: Implications for simulating regional climate change. *J. Clim.* **4**:121–134.

Hedin, L.O., J.J. Armesto, and A.H. Johnson. 1995. Patterns of nutrient loss from unpolluted, old-growth temperate forests: Evaluation of biogeochemical theory. *Ecology* **76**:493–509.

Holling, C.S. 1992. Cross-scale morphology, geometry, and dynamics of ecosystems. *Ecol. Monogr.* **62**:447–502.

Joffre, R., and S. Rambal. 1993. How tree cover influences the water balance of mediterranean rangelands. *Ecology* **74**:570–582.

Kareiva, P. 1994. Space: The final frontier for ecological theory. *Ecology* **75**:1.

Keane, R.E., K.C. Ryan, and S.W. Running. 1996. Simulating the effects of fire on northern Rocky Mountain landscapes with the ecological process model FIRE-BCG. *Tree Physiol.* **16**:319–331.

Knight, D.H., T.J. Fahey, and S.W. Running. 1985. Water and nutrient outflow from contrasting lodgepole pine forests in Wyoming. *Ecol. Monogr.* **55**:29–48.

Levin, S.A. 1992. The problem of pattern and scale in ecology. *Ecology* **73**:1943–1967.

Li, B., and R. Avissar. 1994. The impact of spatial variability of land surface characteristics on land surface heat fluxes. *J Clim.* **7**:527–537.

Ludwig, J.A., and D.J. Tongway. 1995. Spatial organisation of landscapes and its function in semi-arid woodlands, Australia. *Landsc. Ecol.* **10**:51–63.

Malanson, G.P., and M.P. Armstrong. 1996. Dispersal probability and forest diversity in a fragmented landscape. *Ecol. Mod.* **87**:91–102.

Martin, C.W. 1979. Precipitation and streamwater chemistry in an undisturbed forested watershed in New Hampshire. *Ecology* **60**:36–42.

Matson, P.A., P.M. Vitousek, J.J. Ewel, M.J. Mazzarino, and G.P. Robertson. 1987. Nitrogen transformations following tropical forest felling and burning on a volcanic soil. *Ecology* **68**:491–502.

Montaña, C. 1992. The colonisation of bare areas in two-phase mosaics of an arid ecosystem. *J. Ecol.* **80**:315–327.

Noble, I.R., and H. Gitay. 1996. A functional classification for predicting the dynamics of landscapes. *J. Veg. Sci.* **7**:329–336.

Noble, I.R., and R.O. Slatyer. 1980. The use of vital attributes to predict successional changes in plant communities subject to recurrent disturbance. *Vegetatio* **43**:5–21.

Noilhan, J., and P. Lacarrere. 1995. GCM grid-scale evaporation from mesoscale modeling. *J. Clim.* **8**:206–223.

Noy Meir, I. 1985. Desert ecosystem structure and function. In: Ecosystems of the World 12A: Hot Deserts and Arid Shrublands, ed. M. Evenari, I. Noy Meir, and D.W. Goodall, pp. 93–103. Amsterdam: Elsevier.

Obeysekera, J., and K.R. Rutchey. 1997. Selection of scale for Everglades landscape models. *Landsc. Ecol.* **12**:7–18.

O'Neill, R.V. 1988. Hierarchy theory and global change. In: Scales and Global Change, ed. T. Roswall, R.G. Woodmansee, and P.G. Risser, pp. 29–45. New York: Wiley.

Pacala, S.W., and D.H. Deutschman. 1995. Details that matter: The spatial distribution of individual trees maintains forest ecosystem function. *Oikos* **74**:357–365.

Paine, R.T., and S.A. Levin. 1981. Intertidal landscapes: Disturbance and the dynamics of pattern. *Ecol. Monogr.* **51**:145–178.

Pease, C.M., R. Lande, and J.J. Bull. 1989. A model of population growth, dispersal, and evolution in a changing environment. *Ecology* **70**:1657–1664.

Pickup, G. 1985. The erosion cell: A geomorphic approach to landscape classification in range assessment. *Austral. Rangel. J.* **7**:114–121.

Pitelka, L.F., R.H. Gardner, J. Ash, S. Berry, H. Gitay, I.R. Noble, A. Saunders, R.H.W. Bradshaw, L. Brubaker, J.S. Clark, M.B. Davis, S. Sugita, J.M. Dyer, R. Hengeveld, G.

Hope, B. Huntley, G.A. King, S. Lavorel, R.N. Mack, G.P. Malanson, M. McGlone, I.C. Prentice, and M. Rejmanek. 1997. Plant migration and climate change. *Am. Sci.* **85**:464–473.

Reiners, W.A., and G.E. Lang. 1979. Vegetational patterns and processes in the balsam fir zone, White Mountains, New Hampshire. *Ecology* **60**:403–417.

Ryszkowski, L., and A. Kedziora. 1993. Energy control of matter fluxes through land water ecotones in an agricultural landscape. *Hydrobiologia* **251**:239–248.

Schlesinger, W.H. 1991. Biogeochemistry: An Analysis of Global Change. San Diego: Academic.

Sprugel, D.G. 1976. Dynamic structure of wave-regenerated *Abies balsamea* forests in the north-east United States. *J. Ecol.* **64**:889–910.

Swain, A.M. 1980. Landscape patterns and forest history in the Boundary Waters Canoe Area, Minnesota: A pollen study from Hug Lake. *Ecology* **61**:747–754.

Thiery, J.M., J.M. Dherbes, and C. Valentin. 1995. A model simulating the genesis of banded vegetation patterns in Niger. *J. Ecol.* **83**:497–507.

Uhl, C., and C.F. Jordan. 1984. Succession and nutrient dynamics following forest cutting and burning in Amazonia. *Ecology* **65**:1476–1490.

Vitousek, P.M. 1977. The regulation of element concentrations in mountain streams in the northeastern United States. *Ecol. Monogr.* **47**:65–87.

Wallin, D.O., F.J. Swanson, and B. Marks. 1994. Landscape pattern response to changes in pattern generation rules: Land-use legacies in forestry. *Ecol. Appl.* **4**:569–580.

Walsh, S.J., D.R. Butler, T.R. Allen, and G.P. Malanson. 1994. Influence of snow patterns and snow avalanches on the alpine treeline ecotone. *J. Veg. Sci.* **5**:657–672.

White, J.D., and S.W. Running. 1994. Testing scale dependent assumptions in regional ecosystem simulations. *J. Veg. Sci.* **5**:687–702.

White, L.P. 1971. Vegetation stripes on sheet wash surfaces. *J. Ecol.* **59**:615–622.

With, K.A., and T.O. Crist. 1995. Critical thresholds in species' responses to landscape structure. *Ecology* **76**:2446–2459.

16

Can We Close the Water/Carbon/Nitrogen Budget for Complex Landscapes?

J.D. ABER

Complex Systems Research Center, Institute for the Study of Earth, Oceans, and Space, University of New Hampshire, Durham, NH 03824, U.S.A.

ABSTRACT

How well we can close the cycles of water, carbon (C), and nitrogen (N) might be answered very differently depending upon your disciplinary perspective. Trace gas fluxes, which are critical in determining biosphere–atmosphere interactions, represent very minor components of most ecosystem cycles. Cycles for which input and outputs are well-defined relative to the large internal recycling fluxes within ecosystems might seem very "open" or poorly defined relative to atmosphere–biosphere exchanges.

For carbon, eddy correlation, when carefully applied and compared with measurements of internal processes, provides as accurate an estimate of total ecosystem C balance as we are likely to obtain. Different components of the gaseous C balance are measurable with different degrees of precision. Budgets for CO_2 can be fairly well closed, while those for CH_4 and isoprenes, for example, are less well specified.

For water, precipitation inputs have been measured and mapped for most parts of the world. Evapotranspiration (ET) has also been widely estimated, but direct measurements are few outside of gauged watershed systems. The physical and biological simplicity of the water cycle may underlie the frequent use of model results in place of measurements in many studies. If we believe the models, we can close the budget in most places.

The nitrogen budget is the most difficult of the three cycles to close in that N inputs and outputs occur in many forms with relatively equal importance. Unlike water and carbon, there are no large gross fluxes into and out of the ecosystem. Rather there is a large and slowly recycling internal pool maintained by a variety of relatively small fluxes which are sensitive to a different array of controlling factors.

Accurate measurement of a process should not be confused with understanding the mechanisms which result in the measured rate, nor with an ability to extrapolate those measurements over space or time. At this point in the study of ecosystem biogeochemistry, we tend to

Integrating Hydrology, Ecosystem Dynamics, and Biogeochemistry in Complex Landscapes
Edited by J.D. Tenhunen and P. Kabat

overvalue measurements and undervalue synthesis, understanding, and prediction. While we can carry out measurements of most ecosystem processes, our understanding of the controls over those processes, and our ability to predict them, is much more variable.

I propose that adjacency problems are generally small, and that they may affect measurements more than the processes measured.

INTRODUCTION

The question considered in this chapter may be answered very differently depending upon your disciplinary perspective. From an ecosystem point of view, most of the quantitatively important fluxes for water, carbon (C) and nitrogen (N) can be measured with some precision. Residual terms (e.g., net fluxes of trace gases such as CH_4) are small compared with either total ecosystem gross fluxes (e.g., gross photosynthesis) or storage terms (e.g., total soil organic C). From an atmospheric point of view, however, where net fluxes are critical, where both gross internal fluxes and large total storage terms are invisible, and where small fluxes are multiplied over very large areas, these small residual terms constitute strong forcing functions. Budgets that are reasonably closed for ecosystems may be very poorly defined relative to the atmosphere.

These different points of view may lead to very different analyses of the state of the science, and very different recommendations for approaches to improving our understanding. Do we want to close current budgets in specific locations using the most precise measurement systems available? Many global studies approach the ecosystem in this way, using fairly simple and unsophisticated biome lumping or spatial interpolation routines to obtain global estimates of current cycling rates from a set of point samples. In this approach, the ecosystems are black boxes to be measured, not understood or modeled. Or, do we want to develop an understanding of the mechanisms that underlie the measurements of exchange rates in order to predict responses to changes in climate, pollution loading and land use, as well as naturally occurring gradients? This implies the use of physiologically based models in combination with geographic information systems (GIS) to extrapolate site-based findings to larger spatial scales and over time. Direct measurement emphasizes current flux rates for resources individually, while understanding emphasizes interactions between resources, such as water use and nutrient use efficiencies, and relationships along major environmental gradients.

The trade-off between precise measurement and general understanding is a classic one in science. I will argue that, while current global rates of cycling might best be quantified by a large number of precise measurements, extrapolating those measurements into a changing future is not possible. To predict, we need insights into the chemical and physiological interactions among water, C, and N, which in fact control the cycling of all three. However, there are limits on the level of complexity that this understanding can reach, since the resulting models must be simple enough to be driven by the types of spatial data that are available regionally or globally. Producing such simple yet generalized models requires that much more creativity be invested in ecosystem analysis. It can (and will) be argued that models also provide better interpolations between data points than statistical methods, such that mechanistic models are required to complete the time series of site-specific balance measurements.

Therefore, closing the water, C, and N budgets over landscapes will require enough process-level understanding to develop simple models that capture most of the dynamics of these three resources across the major controlling environmental gradients, and under predicted future environmental conditions. From this perspective, whole-ecosystem measurements serve not only as site-specific values for given fluxes, but also as validation data for process models which are then used to both interpolate the system-level measurements over time and space and to predict changes into the future. The combination of the two approaches is more powerful than either alone, or the sum of each separately.

Simultaneously closing three budgets for any landscape adds complexity in terms of the number of fluxes and processes that must be measured and understood. However, it also offers some advantages. Not only are certain fluxes of one resource required to measure the other (e.g., stream flow for losses of dissolved N), but data on all three allow cross-checks of processes and measurements (e.g., does gross photosynthesis provide enough C to drive measured rates of gross N immobilization), and provides a far more rigorous set of validation data for models. Models that can predict measured budgets of water, C, and N should be more robust than those validated against only one of the three.

With these hopefully controversial assertions as a background, this chapter reviews briefly the state-of-the-art in the measurement and prediction of water, C, and N balances at the landscape scale. I begin with a comparative overview of each cycle and the methods used to measure input/output balances and internal cycling at the stand or small watershed scale, along with an assessment of the accuracy and precision of each, essentially answering the closure question for each resource individually at this scale. This will be followed by a discussion of the use of models to integrate and extrapolate site-level research, and a discussion of the degree of "inter-pixel" interaction.

A GENERAL COMPARISON OF BUDGETS

The cycles of water, C, and N differ significantly in the quantity and form of atmospheric and hydrologic exchanges. Inputs and outputs of C are almost exclusively in gaseous form. Inputs of water are almost entirely through wet deposition, while losses are split between gaseous and liquid forms. Unlike N, internal recycling of water or C is not important in determining total availability to plants. For N, availability is determined mainly through internal mineralization and recycling (with exceptions being areas of primary succession or those dominated by plants hosting symbiotic N-fixing species). Over the time scales of interest here, none of these three has an important geologic component.

THE CARBON CYCLE

Overview

For C (Figure 16.1), inputs are dominated by a single process, gross photosynthesis (also called gross primary production, GPP). Inputs of dissolved forms are trivial by comparison. GPP fuels the two major output mechanisms: respiration by autotrophs (plants, R_A) and by heterotrophs (almost entirely microbial, R_H). Internal storage and cycling mechanisms include the production of biomass (net primary production, NPP [= GPP-R_A]), which drives the production of necromass or litter. The decomposition of litter is a major component of R_H, and also leads to the production of humus, a long-term storage product that is only slowly returned to the atmosphere by decomposition (also part of R_H). Net fluxes of non-CO_2 trace gases represent a small fraction of the total ecosystem flux of C and yet are critical to atmospheric chemistry and climate change forcing. Methane can be produced or consumed, depending on both the water and nitrogen status of soils. Production of nonmethane hydrocarbons, such as isoprene, is also tied to the degree of water and nitrogen stress experienced by plants. Root production and decay, and the stabilization, chemistry and decay of humus, especially in concert with myccorhizal activity, are by far the greatest mysteries in the internal C cycle, while the sensitivity of trace gas balances to altered environmental conditions can have important feedbacks to climate change.

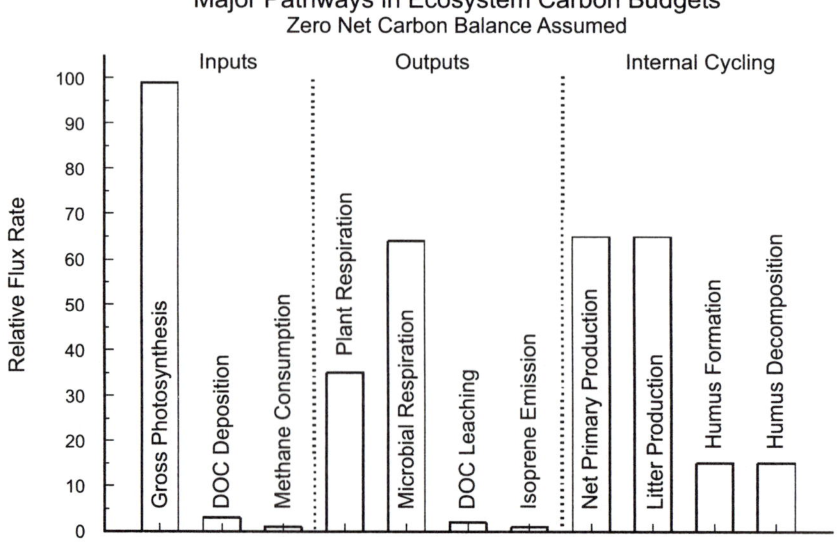

Figure 16.1 A generalized view of the relative magnitude of carbon fluxes for terrestrial ecosystems, using a temperate zone forest as an example.

Methods

Eddy correlation (or eddy covariance) methods have revolutionized the direct measurement of gross photosynthesis at the ecosystem level (Table 16.1). By providing very fine time scale measurements of total system carbon balance, this method allows the construction of relationships for net C exchange in relation to light, temperature, humidity, etc. Measurements during dark periods allow estimation of whole-system respiration as a function of temperature during light periods. By adding estimated respiration back into the measured net flux, gross photosynthesis can be estimated. While conceptually this method is equivalent to that applied to the measurement of individual leaf chambers, it avoids the severe problems associated with scaling individual leaf or chamber estimates to whole canopies.

Table 16.1 Methods used for quantifying components of the carbon cycle at the ecosystem scale and their relative flux rates (Flux), complexity of application (Comp), and precision of results (Prec).

Component	Method	Relative		
		Flux	Comp.	Prec.
Inputs				
Gross photosynthesis	Eddy correlation	High	High	High
DOC deposition	Precipitation sampling	Low	Low	High
Outputs				
Respiration				
Total	Eddy correlation	High	High	Mod.
Autotrophic	Chambers (above ground)	Mod.	High	Mod.
Heterotrophic	Chambers (soil surface)	Mod.	Mod.	High
DOC leaching	Lysimetry	Low	Mod.	Mod.
Cycling				
NPP and litter production				
Above ground	Allometrics and litter collection/clipping	Mod.	Mod.	High
Below ground	Direct sampling/ecosystem budgets	Mod.	High	Low
Litter decay and humus formation				
Above ground	Litter bags	Mod.	Low	High
Below ground	Bags/direct sampling/ecosystem budgets	Mod.	High	Low
Humus decay	Buried Bags	Mod.	Low	Mod.
	Isotopes	Mod.	High	Mod.

Eddy correlation also provides a direct measurement of total net carbon flux, or net ecosystem production (NEP, also called net ecosystem exchange, NEE), and has greatly increased the accuracy of such estimations over what could be obtained using repeated measurements of the sum of all internal pools, especially over short time periods.

Eddy correlation is, however, something of a black box method. While information on the factors controlling total system carbon fluxes can be inferred from correlations between flux rates and changing environmental conditions, partitioning contributions to fluxes between plants and soils, or roots and decomposers, requires within-system measurements. Respiration by plants and soils is generally measured using chambers which isolate that part of the system. Subsampling and spatial heterogeneity add to the complexity of these approaches and add uncertainty in extrapolating to the ecosystem scale.

Eddy correlation is both complex and expensive, which generally rules out replicated experiments. It is unlikely that we will ever be able to provide enough tower-based measurements of NEE to drive direct interpolation estimates of net carbon exchange over large areas. Significant problems remain in the measurement of fluxes during periods of atmospheric stability, especially nighttime respiration, and there are questions on the role of local flux hotspots or "chimneys" by which C fluxes would not be well-mixed and the tower values would become non-representative. For complex landscapes, there is also the "footprint" problem (known as "fetch" in an earlier generation of studies). Eddy correlation assumes that the surface conditions are homogeneous over the area that contributes to fluxes represented by the plane of the sample inlet atop the tower. In landscapes which are divided into less than 1 km^2 areas, the usefulness of the method can be questioned.

Input and output fluxes that are of minor importance to the total carbon cycle of an upland ecosystem can be critical for adjacent wetland systems. There is increasing interest in the controls on production, consumption, and retention of dissolved organic carbon (DOC) in terrestrial systems and the effects of effluxes to streams. We still do not know the extent to which DOC losses are controlled by physical/chemical versus biological processes. While undoubtedly of biological origin, DOC lost to streams is relatively inert and fluxes may be controlled as much by sorption/desorption reactions as by microbial action.

Fluxes of trace gases like methane and isoprene are among the smallest and most variable in the terrestrial C budget. Methane is generally consumed in well-aerated upland soils, and the rate of consumption increases with higher soil temperature and water-free pore space. It is suppressed by the presence of high levels of available ammonium. As soils progress from dry to wet, methane consumption declines and net methane production can occur. In wetland soils, methane flux to the atmosphere can become an important component of the total C budget, and total methane efflux is proportional to total NEE. Changes in net fluxes are driven by changes in the relative importance of gross production and consumption, neither of which is well understood in either upland or wetland systems. Chamber methods for measurement of methane fluxes have become standardized and are generally accepted. Eddy correlation techniques can be useful in the measurement of methane fluxes as long as requirements of homogeneity with the "footprint" of the tower are met.

Isoprene is mainly of plant origin, though its derivation and "purpose" are still under investigation. Isoprene release is dependent upon species and can be altered significantly by varying water and nitrogen availability. This suggests that isoprene production is extremely heterogeneous in time and space, and therefore difficult to quantify over a large area. Again, we have problems determining the small net fluxes of byproducts of aerobic respiration and synthesis.

Methods for measuring internal C fluxes are well established and generally quite accurate for aboveground processes. Estimation of fine root production remains controversial after twenty years of research, and the most accurately constrained estimates may be derived by difference from otherwise complete budgets of C and N cycling. Interactions with symbiotic mycorrhizae are a major complicating issue. The vast majority of direct physiological studies are irrelevant because they address organisms in isolation rather than in symbiosis, and in highly artificial environments. A general partitioning between root/mycorhhizal respiration (tied to immediate provision of available carbon from the plant) and free-living microbial respiration can be obtained by trenching studies in which aboveground carbon inputs are removed. These are essentially methods for dividing the black box ever more finely. We still do not understand, quantitatively, the physiology of the boxes.

As with production processes, decay processes are also well understood above ground. In fact, the decay of litter may be one of the most frequently investigated phenomena in ecosystem studies. The role of carbon fractions, nutrients, and climate in controlling foliar litter decay rates and humus formation have been summarized in equation form. Belowground, we have again the problem of delimiting root from mycorrhizae, live from dead, decomposer from symbiont. While the chemistry, activity, and decomposition of humus can be measured accurately, decay rates and CO_2 emissions cannot be predicted from chemical measurements.

Can We Close the Budget?

From an ecosystem perspective, yes we can. Eddy correlation, when carefully applied and compared with measurements of internal processes, provides as accurate an estimate of total ecosystem C balance as we are likely to obtain. From the atmospheric perspective, CO_2 budgets are relatively well defined while methane is more variable spatially and temporally, and relatively little is known about isoprene.

THE WATER CYCLE

Overview

The water cycle (Figure 16.2) is also dominated by a single input from the atmosphere, this time as wet deposition. This is partitioned into three major outputs:

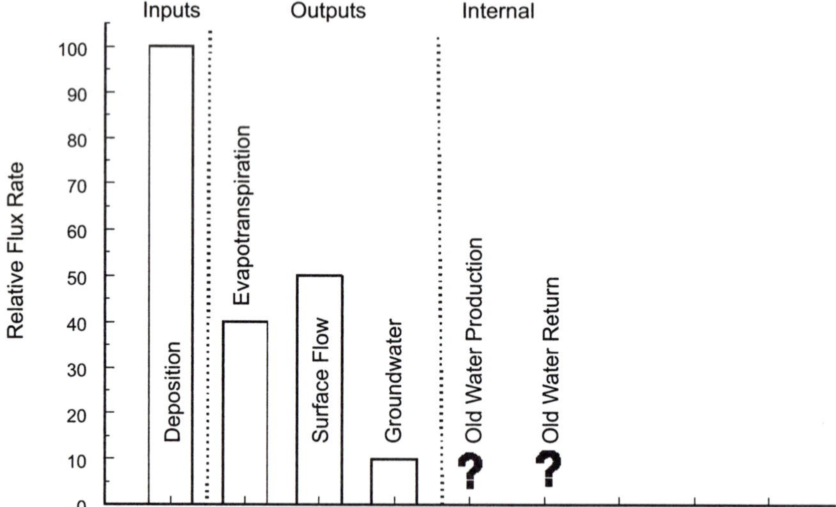

Figure 16.2 A generalized view of the relative magnitude of water fluxes for terrestrial ecosystems, using a temperate zone forest as an example.

evapotranspiration, surface runoff (streams), and inputs to groundwater. Internal pools and transfers are relatively minor, although there is increasing interest in the concepts of "old water" and "new water" as they affect the chemistry of soils, streams, and groundwater. Certain isotopic studies of "old water" suggest internal pools with resident times of many years, even in systems in which total annual throughput exceeds instantaneous storage capacity by an order of magnitude.

Methods

Precipitation may be one of the most accurately specified fluxes over most landscapes due both to standardization of methods and the existence of a large network of sampling sites (Table 16.2). Separation into rain and snow can be problematic, and inputs from fog scavenging or from cloudwater deposition at high elevations remain undersampled. Similarly, surface water drainage is also easily sampled and is again one of the most intensively sampled ecosystem or landscape parameters, both spatially and temporally.

There is a very large range in the degree of accuracy with which the two remaining hydrologic fluxes, groundwater loss and evapotranspiration (ET), can be measured directly. In systems with watertight watersheds, where groundwater loss can be assumed negligible, ET can be determined as the difference between precipitation and runoff, and the most accurate estimates of ET at the landscape scale have been obtained by this "watershed balance" method. Where groundwater losses cannot be eliminated, both groundwater loss and ET can be very difficult to estimate accurately. Eddy correlation

Table 16.2 Methods used for quantifying components of the water cycle at the ecosystem scale and their relative importance to the total cycle (Flux), complexity of application (Comp), and precision of results (Prec).

Component	Method	Relative		
		Flux	Comp.	Prec.
Inputs				
Precipitation	Precipitation sampling	High	Low	High
Outputs				
Evapotranspiration	Watershed Balance	Mod.	Low	High
	Eddy Correlation	Mod.	High	Mod.
	Tracers	Mod.	Low	Low
	Water Balance Models	Mod.	High	Mod.
Drainage	Watershed Balance	Mod.	Low	High
Internal Transfers				
Macropore/Flow Path/Storage	Tracers/Isotopes	Low	High	Low

offers the potential to estimate ET by direct measurement, and when components of the total stand energy budget are also estimated, an internal consistency check is provided. This use of the eddy correlation method has been less visible, but may prove very useful when checked against other methods of estimate. Where this method is successfully applied, and both precipitation and surface water runoff are known, then loss to groundwater can be estimated by difference. Otherwise, quantifying losses to groundwater remains a challenge. At the plot level, drainage can be estimated by measuring water disappearance from soils without significant precipitation or transpiration; however, the fate of this water at the watershed scale cannot be determined. Taking this to the next step, actual water budgets for many nonwatershed sites are routinely estimated by simple hydrologic models, as the fundamental physics are thought to be well known. Modeling will be discussed further below.

Within the soil, preferential flow paths, and the mechanisms by which water can potentially be held for some time in a system that turns over several times a year remain active areas of research and have implications for the movement of certain elements. Still, a complete understanding is not necessary for the accurate estimation of overall water, C, and N budgets at the landscape scale.

Can We Close the Budget?

Precipitation inputs have been measured and mapped for most parts of the world. ET has also been widely estimated, but direct measurements are few outside of gauged watershed systems. The physical and biological simplicity of this system may underlie the frequent use of model results in place of measurements in many studies. If we believe the models, we can close the budget in most places.

THE NITROGEN CYCLE

Overview

In contrast to water and C, inputs to the N cycle (Figure 16.3) include gaseous, hydrologic, and particulate vectors, and both inorganic and organic dissolved forms. The relative importance of these vary widely depending on vegetation type (N-fixing or not), climate (amount of wet deposition), and proximity to pollution sources, among other factors. Outputs can also occur as gaseous or dissolved forms, and again the relative importance is a function of site conditions, disturbance, and hydrologic regime, among other factors. Thus input/output balances for N are more difficult in that we must determine the net effect of several small and complex processes, any one of which might dominate on a particular site. Internal processes of net mineralization, plant uptake, and litter fall are relatively easy to measure, by comparison, although the magnitude of gross microbial fluxes (mineralization and immobilization), while significantly larger than net fluxes, are still difficult to quantify. Decomposition of litter and the production of humus are relatively easy to measure, at least for aboveground

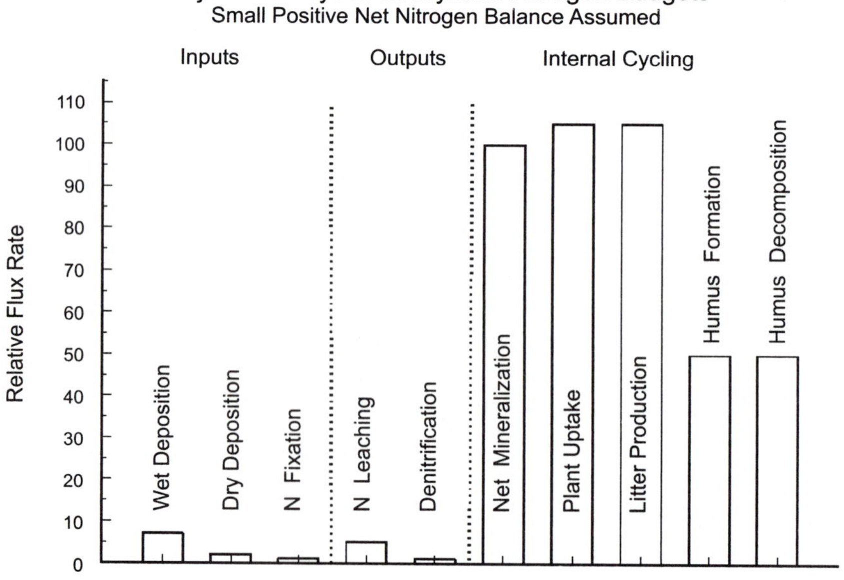

Figure 16.3 A generalized view of the relative magnitude of nitrogen fluxes for terrestrial ecosystems, using a temperate zone forest as an example.

tissues. As with C, understanding of the internal N cycle is limited by our lack of understanding of the root/mychorrhizal/soil system. For N, this uncertainty is especially important relative to processes of incorporation and stabilization of pollution-derived atmospheric N deposition.

Methods

Wet deposition inputs, derived by measurement of concentration in rain and snow, can be measured accurately, especially in comparison with other input vectors (Table 16.3). Dry deposition can be measured directly by eddy correlation. Such measurements are rare but are increasing. The more traditional method of measuring atmospheric concentrations and modeling or estimating deposition velocity is not a

Table 16.3 Methods used for quantifying components of the nitrogen cycle at the ecosystem scale and their relative importance to the total cycle (Flux), complexity of application (Comp), and precision of results (Prec).

Component	Method	Relative		
		Flux	Comp.	Prec.
Inputs				
Deposition (organic and inorganic)				
Wet	Precipitation sampling	Mod.	Low	High
Dry	Throughfall	Low	Low	Low
	Eddy correlation	Low	High	Mod.
	Deposition velocity modeling	Low	High	Low
Symbiotic N-fixation	Chronosequence mass balance	Low*	Low	Low
	Nodule activity	Low*	High	Mod.
Free-living N-fixation	Soil incubations	Low	High	Low
Outputs				
Leaching losses	Lysimetery	Mod.	Mod.	Mod.
	Watershed balance	Mod.	Low	High
Denitrification	Soil chambers and incubations	Low	High	Low
Cycling (in addition to processes under Carbon above)				
Mineralization (net)	Buried bags	High	Low	Mod.
Mineralization (gross)	Isotopes	High	High	Mod.
Uptake				
Above ground	Allometrics and litter	Mod.	Low	High
Below ground	Bags/direct sampling/ecosystem budgets	Mod.	High	Low

* Can be moderate to high in systems dominated by plants supporting symbiotic N fixation.

stable method as the effect of canopy structure, atmospheric turbulence, leaf surface condition, and stomatal conductance can all be important and are not well understood. Unlike sulfur, measuring dry deposition as throughfall enrichment does not work except in areas of high deposition, as interactions between deposition and the canopy occur, and may be variable depending on species and N status of the vegetation.

N losses in drainage water can be accurately measured in watershed studies, where the water budget is known, by measuring the concentration of N at different flow rates and times. Outside of watershed studies, concentrations in soil water can be measured in several different ways. The most common is lysimetry, and results will vary depending on the type of lysimeter used (tension versus nontension). Concentrations, collected occasionally, are then multiplied by water drainage rate to obtain flux. Drainage outside of the watershed context is usually estimated by a hydrologic model, and thus the accuracy of the N flux estimate is subject to the accuracy of the model.

Both inputs and outputs of dissolved N can be either in inorganic (DIN) or organic (DON) form. DON has recently been shown to be a significant part of the net balance of forest ecosystems in areas of low N deposition; however, few measurements of this species of N have been made. Omitting DON can result in a significant overestimation of N retention efficiency.

Gaseous exchanges, where present, can also be difficult to quantify. While N fixation by free-living organisms is low in upland systems, the presence of symbiotic fixation systems can result in very large and variable rates of N addition. Bounding the net effect of this through sequential measurements of changes in total ecosystem N content can be as accurate a method of estimation as direct measurements of the process at the root-nodule level.

Gaseous losses of N, especially those associated with denitrification, are among the most difficult of ecosystem fluxes to quantify. Gaseous oxides of N are produced by both nitrification and denitrification processes. Denitrification requires the presence of nitrate (produced only under aerobic conditions), labile carbon (consumed under aerobic conditions), and anaerobic sites. Thus, peaks of denitrification activity occur during transitional events such as wetting and drying cycles, changes in water table depth, and after N deposition or application. This is truly a "hot spot" phenomenon, and very high resolution measurements are required to capture the dynamics of the process accurately. In addition, the final product of denitrification varies. At low pH and moderately low oxygen, N_2O predominates and can be measured due to its low background concentration in the atmosphere. Under conditions where N_2 is the dominant product, traditional chamber methods cannot be used because of the large background concentration. Invasive methods for altering rates of enzyme function are required to halt the process as N_2O.

Both net nitrification and denitrification tend to be low in native ecosystems, such that gaseous losses of N tend to be highest in agricultural fields, systems which are not N limited (including many tropical systems), and in riparian zones or other locally and variably saturated soil sites. The gaseous loss of ammonia through volatilization is limited to systems with basic soils and high N availability, such as fertilized agricultural fields.

Methods and problems associated with the measurement of internal N cycles are similar to those for carbon. Net mineralization can be measured with some precision using on-site soil incubation techniques. Net nitrification, where present, can be estimated even more accurately due to the generally low level of extractable nitrate in initial soil samples used in the incubation technique. Recent studies have suggested that the gross rate of both N mineralization and nitrification are much higher (perhaps 10–20 times) than net fluxes, due to rapid and repeated recycling by soil microbes. These results have raised questions as to the source of the labile carbon, which is required in large quantities to drive these processes, especially in that microbial processes in soils are thought to be carbon (or energy) limited. Solving this puzzle is important to determine the fate of N added to ecosystems, the system's capacity to assimilate N, and to understand the carbon cycle in soils.

Plant uptake into aboveground tissues is easily measured by allometric or clipping techniques, but belowground use is unclear, as discussed for carbon. Once again, the allocation of N to root/mycorrhizal growth may be most accurately constrained as a difference calculation (net mineralization + N deposition – N leaching losses – N uptake aboveground), rather than by direct measurement.

The interaction of humus with DIN and DON in soil water needs further study, as a number of chemical interactions are very poorly understood.

Can We Close the Budget?

This is clearly the most difficult of the three budgets to close. Unlike water and carbon, there are no large gross fluxes into and out of the ecosystem. Instead, there is a large and slowly recycling internal pool maintained by a variety of relatively small fluxes, which are sensitive to a different array of controlling factors. Oddly, the situation is easiest in heavily polluted regions where inputs and losses of DIN predominate. The primary question here is how do the systems manage to retain such a large fraction of added N and what is the capacity of the retention mechanisms. The situation is more complex in agricultural systems, where fertilizer inputs can be partitioned among crop uptake, soil incorporation, denitrification, loss to groundwater, and loss to streams. In less impacted systems, the balance between gain and loss is more delicate. Changes in any one of a number of processes can shift the balance from plus to minus or back again. Overall, there are more unanswered questions in the N cycle than in those for C or water.

EXTRAPOLATING SITE-LEVEL MEASUREMENTS: INTERPOLATION WITH MODELS

Accurate measurement of a process should not be confused with understanding the mechanisms that result in the measured rate, nor with an ability to extrapolate those measurements over space or time. At this point in the study of ecosystem biogeochemistry, we tend to overvalue measurements and undervalue synthesis, understanding, and prediction. While we can carry out measurements of most ecosystem processes, our understanding of the controls over those processes and our ability to predict them are much more variable. For example, photosynthesis is perhaps the best understood biochemical process in plants, and models of photosynthesis more often suffer from too much rather than too little detail. Several models are available which can reproduce the daily carbon balance of a forest canopy without extensive calibration or "tuning" of the input parameters. In contrast, models of C allocation in plants, especially to roots and mycorrhizae, are extremely crude at this time and accurately reflect the state of our quantitative understanding of the processes involved. We have only the vaguest, general rules to apply to these processes.

Extrapolating from site measures to landscapes requires some kind of model. The degree of sophistication can vary. The simplest form is the biome-level or land-use stratified summation, which assumes that all areas of a similar type behave in the same way. Many who resist the notion of modeling altogether would like to think of this as something other than a model, but a model it is, albeit a very crude, statistical one. Additional layers of complexity can be added by further stratification, the use of correction factors, etc. However, one discovers very quickly that the complexity of the statistical model begins to approach that of a simple physiological model. The question is, then, why use a statistical model that we know will not extrapolate beyond the range of the data used to construct it (e.g., into a climate change scenario) instead of a simple, well-validated physiological model which incorporates information on the mechanisms involved, and thus may have a better chance of being more generalizable and more applicable to prediction.

These same arguments can be applied to extrapolation over time in a given place. It can be argued that using a model that captures the underlying mechanism of system response to environment has a better chance of predicting missing data points than does a straight statistical summation. Not that the model can generate missing data! Rather, when constructing, for example, monthly values from noncontinuous data, a model will do better than taking a simple average.

It may be worth demonstrating such a nonintuitive point. In a recent paper (Aber et al. 1996), a simple model of total canopy gross photosynthesis was presented and tested against daily measurements over a four-year time period obtained by eddy correlation (Figure 16.4). In this study, the model parameters were obtained directly from measured data and were not calibrated or tuned in any way. If we take the eddy correlation data as a 100% sample, run the model for each daily value, and average both of these for each 30-day period, there is quite a good, but not perfect, fit between the model and the measured estimates of flux per month. The absolute difference between the model and data, averaged monthly, is 0.72 g $C{\cdot}m^{-2}{\cdot}d^{-1}$.

Assume then that field sampling was less than continual. This can be simulated by randomly dropping individual days of eddy correlation data, recalculating the monthly mean values, and comparing them with those obtained from the complete tower

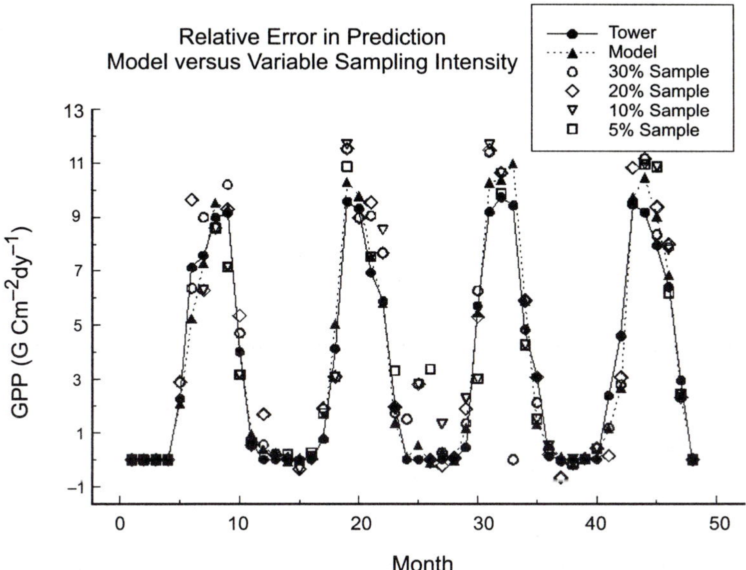

Figure 16.4 Comparison of estimates of mean daily gross photosynthesis or GPP for a mixed temperate forest at the Harvard Forest in Petersham, MA, U.S.A. produced by eddy correlation measurement and a simple model of whole canopy photosynthesis (Aber et al. 1996). Also included are mean monthly estimates obtained by removing 70–95% of the data from the eddy correlation data set.

record. As percent sampling intensity declines (fraction of data points removed increases), the error caused by taking a simple monthly average increases (Figures 16.4 and 16.5). Eventually, these errors become larger than differences between the 100% sample and the model predictions. In addition, individual months begin to have no samples in them at all, increasing the difficulty of interpolating values for those periods. To be conservative, the effects of these zero estimate months are not included in the data used in Figures 16.4–16.6. The amount of error introduced by incomplete sampling, at least for this example, increases nearly linearly as sampling intensity declines (Figure 16.6). A line fit to the values in Figure 16.6 suggests that once sampling intensity falls below 40%, it is more accurate to produce monthly estimates of mean daily GPP by driving the model with real daily climate data, than it is to take a simple mean of all measured values within that month. This cutoff value could be very different for differently structured data sets, models, and ecosystem types.

One of the principal reasons why modeling has not made greater headway as a mainline method in biogeochemistry is that it is not practiced with the same degree of rigor as applies to field investigations. A lack of consistency in the format of model papers (each should contain sections on structure, parameterization, validation, sensitivity analysis, and prediction, as well as a complete listing of all parameters and their values), and the frequent confusing of fundamentally different approaches to model

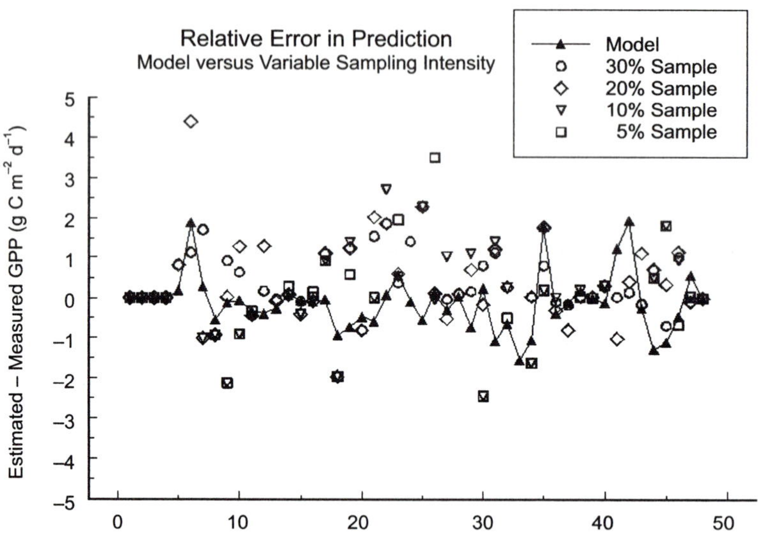

Figure 16.5 Same data as in Figure 16.4, this time expressed as differences with the 100% sample eddy correlation data set.

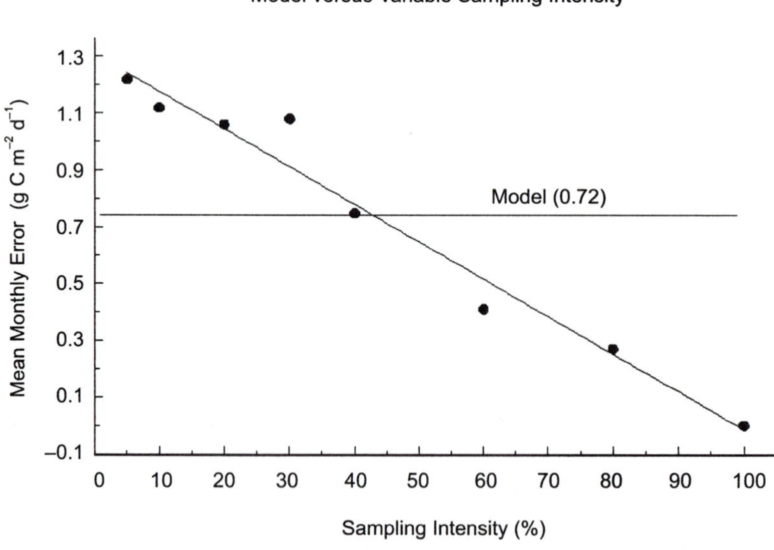

Figure 16.6 Average absolute difference between the monthly mean GPP data in Figures 16.4 and 16.5 as a function of reduction in sampling intensity (data for sampling intensities of 40%, 60%, and 80% added). The mean of monthly errors increases linearly with decreasing sampling intensity. Mean error for the model predictions is 0.72.

parameterization (calibration versus measurement) tend to obscure the relative merits of different models. This is especially evident in that modeling is the only area of research where negative degrees of freedom are allowed (indeed encouraged). Specifically, this occurs when complex models with N parameters are "fit" or "tuned" to reproduce N/10 or N/20 outputs through a generalized calibration process. Pressures to increase the complexity of models to make them more complete or "realistic" adds to the number of parameters required, many of which will not have been measured directly, and hence, need to be calibrated. In contrast, creating simple "lumped-parameter" models that capture most of the dynamics of the system under study with the fewest possible input parameters, and validating these models against large measured data sets covering several different but linked resources or processes can eliminate the negative degrees of freedom problem and eliminate the need for calibration or "tuning." This approach to modeling should be an area of active and creative research involving "modelers" and "field researchers" (with the distinction between the two hopefully diminishing through time).

EXTRAPOLATING SITE-LEVEL MEASUREMENTS OVER SPACE: ARE THERE INTERSITE INTERACTIONS?

What was said above about the value of models, and the need for rigorous applications of the modeling process, applies to spatial as well as to temporal extrapolations. However, here there is the additional potential problem of "interpixel" connections or neighborhood effects. Do the processes in a given pixel affect the processes in the adjacent pixel? This is particularly important in complex landscapes where one can assume that boundaries between very different types of ecosystems (e.g., forest and pasture, cropland and wetland) are frequent. We can consider the degree to which the adjacency problem causes the budgets over a landscape to differ from the sum of the budgets for individual pixels in two ways: (a) in terms of effect on actual balances and, (b) on the measurements of those balances.

I propose that adjacency problems are generally small, and that they may affect the measurements more than the processes.

Processes

If we assume that the pixels of interest are large enough relative to the size of the dominant plants such that processes like NPP, litter production and litter fall, shading, and interception are not subject to edge effects (e.g., pixel sizes are greater than five times the height of the dominant plants), then there are only select processes that will result in significant interpixel effects. Carbon is little affected as CO_2 exchanges with the atmosphere become rapidly well-mixed, and there is relatively little horizontal structure to concentration fields. Movements of carbon in dissolved form are a very small part of the total C flux in any system, such that water movement between systems will not affect the balance. Similar arguments can be made for nitrogen, although the potential importance of nitrate loss from certain systems in dissolved forms suggests that movement to adjacent systems with water flow may alter local balances significantly (see discussion of denitrification "hot spots" above). This may be critical from the atmospheric point of view, but may be significant for individual sites only over very long periods of time in that the local balance is the net of several relatively small exchanges, and that total pools tend to be very large relative to annual gains or losses.

Water movement between adjacent pixels is the main vector for "communication" between pixels in a complex landscape. The complexity of groundwater movement, of recharge and discharge wetlands, and of local source area flows links upland and wetland systems and creates the hot spots for certain processes. In complex landscapes, then, an accurate model of surface and groundwater hydrology may be required. A few simplified landscape hydrology models are available for expressing the position of different pixels in the landscape and the different degree of water delivery to each (see also Wood, this volume).

While ecosystem disturbance has not been discussed much here, this may be one of the principal areas in which adjacency effects modify ecosystem function. Susceptibility to fire in one pixel may affect the next through propagation of that fire. Windthrow can create edges which, in turn, increase wind stress on the edge created in an adjacent pixel, leading to further disturbance in that pixel. Any structural patterns that focus or distribute herbivore pressure, erosion, etc. will also result in adjacency effects.

Measurement

The most powerful emerging tool for direct measurement of carbon and water balances — eddy correlation — may also be the most vulnerable to adjacency effects. The requirement for relative homogeneity within the footprint of the tower becomes increasingly difficult to realize in finely divided, complex landscapes. Problems of advection between landscape patches in the environment can confound measurements by this technique. "Advection" of water between pixels through groundwater flow or lateral flow down hillslopes is also a confounding difficulty in measuring components of the water balance when using any technique that deals only with vertical flow (any one-dimensional method). As the N budget is dominated by internal recycling and measurements of gaseous exchanges are generally made using chamber techniques, measurements of N exhibit interpixel effects only to the effect that exchanges are measured by the eddy correlation technique or through horizontal water flows.

EXTRAPOLATING SITE-LEVEL MEASUREMENTS OVER SPACE: THE ROLE OF REMOTE SENSING

Other chapters in this volume address this topic in detail (Waring and Running, Prince, Mauser et al., all this volume). Here I wish to reiterate that remote sensing is the *only* available method that allows the direct, simultaneous, and contiguous measurement of ecosystem variables at fine spatial scale over large units of the landscape. Recent advances in the use of high spectral resolution instruments (imaging spectrometers) promises much greater information content than broad-band sensors and the potential for the application of laboratory spectrometric techniques for measuring chemical constituents or physiological state of the reflecting surface. This is an extremely young and rapidly evolving area of research. However, the meaning of such coarse metrics as normalized difference vegetation index (NDVI) is still unclear. Is it total foliar mass? Is it leaf area? Is it chlorophyll or nitrogen concentration? Could it be any one of these in different systems, or a combination? Is it an integrated expression of photosynthetic capacity? Different applications of the use of remote sensing to drive landscape and global models assume different meanings for this one variable. We still await a generalized understanding of broad-band signals, and are just beginning to have access to narrow-band data.

No other method approaches remote sensing for providing landscape scale data. Thus, remote sensing can be of critical importance in closing landscape budgets. However, much of the same argument regarding rigor, and especially validation, can be made for remote sensing methods as was made above for modeling. Extensive validation of the relationships derived or assumed between reflected radiation and ecosystem state and function is required.

CAN WE CLOSE THE WATER, C, AND N BUDGETS OVER COMPLEX LANDSCAPES?

If we first rephrase the question to the site level, I believe the answer is yes. Emerging and traditional techniques combined can provide cross-checked measurements of the important fluxes of these three resources, if the facilities and support are available to make those measurements.

At the landscape level, we cannot close these budgets by measurement alone. We need to develop and apply synthetic tools such as remote sensing, modeling, and GIS techniques to extrapolate site measurements accurately through space and time. At present, I believe we are more limited in our ability to close these budgets over complex landscapes by a lack of commitment to *rigorous* applications of these synthetic techniques than we are by basic site-level measurements. In general we can measure most fluxes very well, and we are still on the way up the learning curve for understanding controls and increasing prediction. Perhaps it is time to move to this next level by involving modeling as a discovery process in every study. No measurement program should be judged by the usefulness of the measurements alone, but also by how critical it is for parameterizing or validating models.

REFERENCES

Aber, J.D. 1997. Why don't we believe the models? *Bull. Ecol. Soc. Am.* **78**:232–233.

Aber, J.D., J.M. Melillo, and C.A. McClaugherty. 1990. Predicting long-term patterns of mass loss, nitrogen dynamics and soil organic matter formation from initial fine litter chemistry in temperate forest ecosystems. *Can. J. Bot.* **68**:2201–2208.

Aber, J.D., K.J. Nadelhoffer, P. Steudler, and J.M. Melillo. 1989. Nitrogen saturation in northern forest ecosystems. *BioScience* **39**:378–386.

Aber, J.D., S.V. Ollinger, and C.T. Driscoll. 1997. Modeling nitrogen saturation in forest ecosystems in response to land use and nitrogen deposition. *Ecol. Mod.* **101**:61–78.

Band, L.E. 1993. Effect of land surface representation on forest water and carbon budgets. *J. Hydrol.* **150**:749–772.

Berg, B. 1986. Nutrient release from litter and humus in coniferous soils—A mini review. *Scan. J. For. Res.* **1**:359–369.

Bugmann, H.K.M., and A.M. Soloman. 1995. The use of a European forest model in North America: A study of ecosystem response to climate gradients. *J. Biogeogr.* **22**:477–484.

Burke, I.C., T.G.F. Kittel, W.K. Lauenroth, P. Snook, C.M. Yonker, and W.J. Parton. 1991. Regional analysis of the central great plains. *BioScience* **41**:685–692.

Burke, I.C., D.S. Schimel, C.M. Yonker, W.J. Parton, L.A. Joyce, and W.K. Lauenroth. 1990. Regional modeling of grassland biogeochemistry using GIS. *Landsc. Ecol.* **4**:45–54.

Castro, M., P.A. Steudler, J.M. Melillo, J.D. Aber, and R.D. Bowden. 1995. Factors controlling atmospheric methan consumption by temperate forest soils. *Glob. Biogeochem. Cyc.* **9**:1–10.

Dise, N.B., and R.F. Wright. 1995. Nitrogen leaching in European forests in relation to nitrogen deposition. *For. Ecol. Manag.* **71**:153–162.

Feddes, R.A., M. Menenti, P. Kabat, and W.G.M. Bastiaanssen. 1993. Is large-scale inverse modelling of unsaturated flow with areal average evaporation and surface soil moisture as estimated from remote sensing feasible? *J. Hydrol.* **143**:125–152.

Galloway, J.N. 1995. Acid deposition: Perspectives in time and space. *Water Air Soil Poll.* **85**:15–24.

Hedin, L.O., J.J. Armesto, and A.H. Johnson. 1995. Patterns of nutrient loss from unpolluted, old-growth temperate forests: Evaluation of biogeochemical theory. *Ecology* **76**:493–509.

Heimann, M. 1991. The Global Carbon Cycle. Il Ciocco, Italy: North American Treaty Association.

Hilbert, D.W., A. Larigauderie, and J.F. Reynolds. 1991. The influence of carbon dioxide and daily photon-flux density on optimal leaf nitrogen concentration and root:shoot ratio. *Ann. Bot.* **68**:365–376.

Hochberg, M.E., J.C. Menaut, and J. Gignoux. 1994. The influences of tree biology and fire in the spatial structure of the West African savannah. *J. Ecol.* **82**:217–226.

Li, C., S. Frolking, and T.A. Frolking. 1992. A model of nitrous oxide evolution from soil driven by rainfall events. 2. Model applications. *J. Geophys. Res.* **97**:9777–9783.

Loaiciga, H.A., J.B. Valdes, R. Vogel, J. Garvey, and H. Schwarz. 1996. Global warming and the hydrologic cycle. *J. Hydrol.* **174**:83–127.

Martin, M.E., and J.D. Aber. 1997. Estimation of forest canopy lignin and nitrogen concentration and ecosystem processes by high spectral resolution remote sensing. *Ecol. Appl.* **7**:431–443.

Matson, P.A., and P.M. Vitousek. 1990. Ecosystem approach to a global nitrous oxide budget. *BioScience* **40**:667–672.

Nielsen, D.R, M. Kutilek, and M.B. Parlange. 1996 Surface soil water content regimes: Opportunities in soil science. *J. Hydrol.* **184**:35–55.

Noble, I.R. 1993. A model of the responses of ecotones to climate change. *Ecol. Appl.* **3**:391–403.

Pearson, S.M. 1994. Landscape-level processes and wetland conservation in the Southern Appalachian Mountains. *Water Air Soil Poll.* **77**:321–332.

Reddy, V.R., L.B. Pachepsky, and B. Acock. 1994. Response of crop photosynthesis to carbon dioxide, temperature, and light: Experimentation and modeling. *HortScience* **29**:1415–1422.

Schimel, D.S. 1995. Terrestrial biogeochemical cycles: Global estimates with remote sensing. *Remote Sens. Env.* **51**:49–56.

Schimel, D.S., B.H. Braswell, R. McKeown, D.S. Ojima, W.J. Parton, and W. Pulliam. 1995. Climate and nitrogen controls on the geography and timescales of terrestrial biogeochemical cycling. *Glob. Biogeochem. Cyc.* **10**:677–692.

Stark, J.M., and S.C. Hart. 1997. High rates of nitrification and nitrate turnover in undisturbed coniferous forests. *Nature* **385**:61–64.

Steffen, W.L., and J.S.I. Ingram. 1995. Global change and terrestrial ecosystems: An initial integration. *J. Biogeogr.* **22**:165–174.

Tietema, A., and C. Beier. 1995. A correlative evaluation of nitrogen cycling in the forest ecosystems of the EC projects NITREX and EXMAN. *For. Ecol. Manag.* **71**:143–152.

Tietema, A., W. Bouten, and P.E. Wartenbergh. 1991. Nitrous oxide dynamics in an oak-beech forest ecosystem in the Netherlands. *For. Ecol. Manag.* **44**:53–61.

van Breeman, N., and H.F.G. van Dijk. 1988. Ecosystem effects of atmospheric deposition of nitrogen in the Netherlands. *Env. Poll.* **54**:249–274.

Virginia, R.A., W.M. Jarrell, W.G. Whitford, and D.W. Freckman. 1992. Soil biota and soil properties in the surface rooting zone of mesquite (*Prosopis glandulosa*) in historical and recently desertified Chihuahuan Desert habitats. *Biol. Fert. Soils* **14**:90–98.

Wofsy, S.C., M.L. Goulden, J.W. Munger, S.-M. Fan, P.S. Bakwin, B.C. Daube, S.L. Bassow, and F.A. Bazzaz. 1993. Net exchange of CO_2 in a mid-latitude forest. *Science* **260**:1314–1317.

Wright, R.F., and N. van Breeman. 1995. The NITREX project: An introduction. *For. Ecol. Manag.* **71**:1–6.

Wu, J., R. Zhang, and J. Yang. 1996. Analysis of rainfall-recharge relationships. *J. Hydrol.* **177**:143–160.

Standing, left to right:
Will Steffen, Jim Reynolds, Jianguo "Jingle" Wu, Harald Bugmann, Pavel Kabat, Jean-Claude Menaut
Seated, left to right:
John Aber, Basil Acock, Ian Noble, Indy Burke

17

Group Report: Hydrological and Biogeochemical Processes in Complex Landscapes — What Is the Role of Temporal and Spatial Ecosystem Dynamics?

J.D. ABER and I.C. BURKE, Rapporteurs

B. ACOCK, H.K.M. BUGMANN, P. KABAT, J.-C. MENAUT, I.R. NOBLE, J.F. REYNOLDS, W.L. STEFFEN, and J. WU

INTRODUCTION

The effects of landscape complexity, including the dynamics of the landscape pattern itself, have often been neglected in studies of ecosystem functioning. Most studies of atmospheric exchange processes with the land surface and of biogeochemical cycling in complex landscapes have assumed that the pattern of the landscape (e.g., vegetation mosaic) is constant. The question has been to determine how a given landscape pattern affects flows and pools of material and energy. However, the assumption of a static landscape pattern can no longer be made. Given the speed at which humans are converting or modifying landscapes and the need to consider decadal (or longer) time scales in the context of both sustainable development and global change, the questions discussed by our group have become increasingly important. How does landscape pattern itself change? What processes drive these changes in complexity? How do such changes interact with the functioning of the landscape? How can we improve our understanding of the changing complexity of landscapes? How can we describe the interactions between process and pattern in landscapes when both are changing, on varying time scales?

The specific objectives of our discussion group were: (a) to prioritize the ecological processes that drive change in landscape pattern; (b) to review current landscape

Integrating Hydrology, Ecosystem Dynamics, and Biogeochemistry in Complex Landscapes
Edited by J.D. Tenhunen and P. Kabat

modeling, experimental, and observational approaches (with a focus on modeling); and (c) to explore ways in which these approaches can be developed towards a more complete integration with hydrological, biogeochemical, and atmospheric exchange process studies.

PATTERN AND PROCESS AT LANDSCAPE TO REGIONAL SCALES

Over the past decade, considerable interest has developed in understanding and assessing ecosystem dynamics at landscape to regional scales (e.g., Peterjohn and Correl 1984; Lajtha and Schlesinger 1988; Pielke and Avissar 1990; Aber et al. 1993, 1995; Groffman et al. 1988, 1992, 1993; Burke et al. 1990, 1991, 1997; Lindner et al. 1997). Landscapes and regions are the composite units of relevance for scientists addressing global-scale dynamics, or seeking to assess global biogeochemical and hydrologic budgets and the interactions between biospheric and atmospheric dynamics. In addition, landscape to regional scales approximate the areal extent of management and political units.

For the purposes of understanding and assessing biogeochemical, energy, and hydrological fluxes, what are the key ecosystem variables that are important to study at landscape to regional scales? There are two kinds of ecosystem attributes that we may consider here: the actual processes of interest, and the structural and functional attributes needed to calculate or assess those processes.

The key processes at landscape to regional scales include the net atmospheric exchanges of CO_2, H_2O, trace gases including N_2O, NO, CH_4, and hydrocarbons, nutrient input via wet and dry deposition, lateral transport of nutrients via the hydrologic cycle, and atmosphere–biosphere energy exchange. There are also ecosystem processes that are of primary interest to managers at landscape to regional scales. Many of these are summarized in Daily (1997) and include natural resource production such as timber, grain, forage, and livestock yield, which are closely associated with primary and secondary productivity.

Structural attributes over the landscape or region that are necessary to determine these key processes include topography, soil mineralogy, vegetation composition, leaf area index, surface roughness, albedo, primary productivity, and respiration, insofar as they influence net ecosystem production (both primary productivity and respiration), evapotranspiration, percolation, overland flow, and microbial processes responsible for trace gas fluxes. In addition, information regarding land-use management practices such as harvesting history, cultivation, fertilization, and crop types may be crucial for assessing large-scale ecological functioning (Cohen et al. 1996; Burke et al. 1991). These criteria apply to the study of individual sites or plots as undertaken in traditional ecological or ecosystem research. However, placing these sites in complex landscapes implies that transfers across plot or unit boundaries are important and that they affect the dynamics of all units.

Landscapes are four-dimensional systems with considerable complexity in the spatial domain: in addition to within-plot dynamics, as measured in traditional studies, placing these sites in complex landscapes assumes that transfers across plot or unit boundaries is an important process that affects the dynamics of the unit. The vertical transfer of materials is important in integrating landscapes with the atmosphere and broader biogeochemical cycles, adding a third dimension. Finally, the representation of landscapes as dynamic, with the status and distribution of the basic landscape units changing over time, defines the fourth dimension. Considering the full range of interactions and feedbacks between processes, landscape units, and adjacent atmospheric and hydrospheric systems, the complexity of this approach becomes apparent.

Adjacency, Self-organization, and Thresholds

One of the major tasks that distinguishes landscape ecology from other environmental disciplines, and which lends complexity to its practice, is understanding and representing the phenomena that arise from and are associated with the adjacency of the landscape elements. One implication of the recognition of adjacency effects is the potential for positive feedbacks affecting these transfers, resulting in increasingly complex and nonrandom distributions across the landscape. The tendency toward self-organization creates important breakpoints in the dynamics and distribution of ecological characteristics. These three issues will be discussed below.

Adjacency

Adjacency is important wherever there are significant transfers of material, energy, or information between adjacent patches. Adjacency itself is scaled differently for different processes. The transport of materials by mass flow is usually between directly connected patches (e.g., Peterjohn and Correll 1984), while for the transport of propagules, adjacency is defined by a wider neighborhood (e.g., Coffin and Lauenroth 1989). Spatially correlated events, such as fires, disease outbreaks, and landslides, are important phenomena in which adjacency is significant (e.g., Turner and Romme 1994).

Adjacency has usually been incorporated in models of landscapes by adopting pixelated, or less commonly, polygonal representations of the horizontal structure (Costanza et al. 1990; Keane et al. 1996). More formal mathematical representations, such as fields and diffusion gradients, have sometimes proved useful for investigating components of complex systems, but have proved to be of limited value in more comprehensive models (Wu and Levin 1994, 1997). Both the pixelated and polygonal representations usually require some form of classification of the landscape elements into a discrete number of groups. The issue of general classification schemes of landscape elements has not been fully resolved in landscape ecology or modeling (see section on **LANDSCAPE FUNCTIONAL TYPE**).

There are no models that have successfully incorporated all of the important biological (e.g., propagule dispersal, animal movement), biogeochemical, and disturbance flows in a single system. The main challenge is that different processes are optimally represented at different temporal and spatial scales. Integrating across a variety of scales within a single model remains a challenge awaiting solution.

For some tasks, a complete representation of all processes at optimal resolutions may not be essential. For example, in many of the interactions with the landscape, atmospheric processes integrate over scales of several kilometers (Pielke et al. 1997), and finer–scale representations of structure and function (e.g., biogeochemical processes) may not be essential. Models are now being developed that take advantage of this observation, such that at regional scales, mesoscale circulation models are being linked to ecosystem simulation models to represent the interactions among landcover, biogeochemical, and atmospheric processes (Pielke et al. 1993). At larger scales, the same types of frameworks are being developed through linkages of the DGVMs (Dynamic Global Vegetation Models) and General Circulation Models (GCMs). However, simplification is not straightforward. Most processes are nonlinear and are affected by numerous stochastic (and possibly chaotic) phenomena, so great care must be taken in averaging over smaller scales.

Self-organization

Self-organization in a landscape occurs when a process or event establishes a pattern in the landscape, which is then reinforced by subsequent processes induced by that pattern (Holling et al. 1996). For example, in semi-arid systems, a sparse and patterned distribution of shrubs establishes localized patches into which nutrients are concentrated by plant uptake from the surrounding bare ground and by deposition in the patch as litter, as well as by entrapment of airborne particles (Charley and West 1977; Burke et al. 1989; Schlesinger et al. 1990). Increasing redistribution of resources from plant interspaces to under shrubs may lead to shrub-associated dune formation and the persistence of shrubs in the same location in the landscape (Schlesinger et al. 1990). Another example is the introduction of pattern into an initially homogeneous landscape by disturbances such as fire or landslide (Romme and Despain 1989; Turner and Romme 1994). Subsequent distributions of plants and changes in structural components of the resulting patches can increase the chances for another disturbance of the same type (e.g., a moist forest is converted to a grassland that dries more rapidly), further reinforcing the initial disturbance-derived pattern. Thus, the original patch distribution tends to persist for very long periods. Self-organizing disturbance can also take the form of moving fronts of vegetation decline and regeneration such as in the fir waves of high elevation temperate mountain systems, in which a disturbed edge increases wind-stress and desiccation along this edge, leading to continued mortality along that edge, and the development of a mortality "wave" (Sprugel 1984). The effect of self-organization on the study of landscapes is that point models which do not include adjacency effects will not capture a primary process controlling the distribution of pattern and function over those landscapes, and hence will not provide an accurate representation of landscape phenomena.

Thus, there are three general classes of conditions that tend to lead to large adjacency effects and self-organization in landscapes: (1) places with dynamic geomorphology (e.g., landslides, river channels), (2) where limiting resources exhibit high mobility between patches (semi-arid systems), and (3) where disturbance creates subsequent pattern in vegetation which increases the chance of repeated or continuing disturbance (fire and fir wave examples). Self-organization may be present in most systems but is not an obvious factor in all of them. Strong environmental gradients tend to override the tendency towards self-similarity of adjacent patches but self-reinforcing patterns may still be present. Where a process tends to rehomogenize the landscape (e.g., very extensive fires or droughts), there will be little opportunity for self-organization to arise.

Important feedbacks can also be introduced through effects of landscape pattern on atmospheric processes. Mesoscale climate models have shown that nonhomogenous distributions of vegetation types within a landscape (e.g., fields or burned areas adjacent to forests) can result in very significant increases in the total amount of precipitation received in a landscape area, as well as highly nonhomogeneous distribution of that precipitation (Pielke and Avissar 1990). The general effect of this feedback is to increase precipitation in the drier parts of the landscape, providing a negative feedback on the inhomogeneity of soil moisture. There may be potential in these landscape level studies of vegetation–atmosphere interactions to understand the effects of different patterns of land use on mesoscale climate such that, for example, patterns of clearing in tropical forests could be designed to minimize local climatic effects. The DGVMs mentioned above are designed to increase the realism of these feedbacks for GCM-scale modeling.

Thresholds

Essentially, self-organizing processes will tend to maintain landscape units within a specific condition until a sufficiently destabilizing disturbance pushes that patch into a different state. Thus, patches that appear stable over long periods of time may be "pushed" into a very different state by a single disturbance event. Self-organizing processes may then tend to maintain that new state. The boundary between these states is the threshold condition.

Some landscapes are mosaics of patches representing different sides of a threshold. For example, savannahs can switch between grass- and tree-dominated states, and discrete disturbance events can move patches between these two states. In general, patterns generated by ecotones between systems dominated by different physiognomic groups of plants are candidates for important threshold effects (e.g., forest–prairie border, alpine treeline, taiga–tundra border).

Major disturbance events may also push otherwise stable landscape units across thresholds and into new states (e.g., Zobel and Antos 1997). Examples could include major erosional events, volcanoes, extreme fires, and altered groundwater depth. Some of the most severe threshold effects are associated with human land use. Conversion of land from forests to croplands and pastures as well as from agricultural land to urban-suburban and industrial land are extreme and often nonreversible (due as much to social as to ecological constraints). Our ability to predict future changes in patterns of human use of the landscape is very limited, and feedbacks between altered land use and climate models are not fully recognized. For example, extensive conversion of forest land to crops and pastures could alter land-surface albedo in a way that would lead to large climate changes, both locally and globally (Pielke and Avissar 1990; Pielke et al. 1991). Such changes could be greater than the effects of global-scale changes in radiative forcing by increased trace gas concentrations.

Indicators of Landscape State

Given the difficulties of understanding and modeling the complex interactions resulting from adjacency, self-organizing, and threshold effects, short-term analyses and monitoring of landscape status may require the development of simple, robust indicators. If landscapes are concrete entities, they can be characterized by indicators that summarize a larger or more complex set of information or that capture some of the emergent properties. Such indicators should ideally be readily measurable, repeatable, should be directly related to important ecological processes or conditions, and should integrate over space and/or time. Examples might include river nitrate concentrations, albedo or leaf area index, biological diversity, and other tangible and measurable indicators. A number of theoretial variables have also been suggested, such as ecotone length, indices of degree of fragmentation/contagion and connectivity, and pattern analysis (Li and Reynolds 1997); a key potential area of development is the connection between these two types of landscape indicators. As models of landscape dynamics develop, measurements of indicator variables can provide validation data for those models which span over large spatial scales.

Indicators have little value *per se* and should relate to a specific question/problem, either scientific or practical (e.g., management issues). They will be mainly considered as comparative tools, either between landscapes or between varying states of a given landscape with time. They can be simple "descriptors" of extent (e.g., total area, number of units) or more complex ones such as geomorphology and hydrological network. They can also be "integrators" and relate to structural, functional, or dynamic properties of the landscape. Some originate from ground-based measurements (e.g., water yield and nitrate concentration in streams, [Bormann and Likens 1979]) and some from space-borne instruments (e.g., albedo or foliar nitrogen concentration [Reich et al. 1997; Martin and Aber 1997]), which call for statistical methods comparing sequential scene measures.

Some indicators may be more difficult to obtain than others. As an example, for landscape architecture, such indicators may vary from ecotone length to indices of degree of fragmentation/contagion and connectivity, including pattern analysis. Though simply measured, some can derive from previous process studies and then be interpreted in relation to cross-boundary transfers or retention phenomena (sources and sinks, e.g., sediment load, chemistry of stream water).

Indicators of disturbances may constitute early warning signals (e.g., river nitrate concentration as in Cole et al. [1993], Caraco et al. [this volume], Howarth et al. [1996]) or reflect thorough changes in structure and/or function (e.g., water DOC/DON concentration, species biodiversity). Preferably, indicators should be cross-linked, thus leading to more thorough interpretation. Indicators that reflect known patterns of change in ecosystems, substantiated by experimental studies, will be most valuable.

There are other types of indicators, including those that represent system state but may not be easily measured over landscapes, and do not integrate spatially. These include biodiversity, the presence or absence of rare species in general, or of species that signify ranges of environmental conditions (e.g., the disappearance of lichens from tree bark in polluted areas). There are also measurements that are difficult to make but do integrate spatially, such as eddy covariance (e.g., Wofsy et al. 1993) or aircraft-borne measurement of total landscape gas fluxes.

LANDSCAPE FUNCTIONAL TYPES

The complexity inherent in connected landscapes suggests that a simplifying heuristic is needed in developing manageable conceptual and computer models. The most commonly proposed heuristic has been hierarchy theory (see Reynolds and Wu, this volume; Noble, this volume). Hierarchy theory is rich in conceptual approaches and provides useful prescriptions for breaking complex landscape systems into manageable levels and scales; however, to date its utility has been limited to generating broad guidelines rather than a specific and rigorous integrative structure.

Hierarchies can be described in terms of both spatial distribution and processes (O'Neill et al. 1986; Wu and Loucks 1995). For spatial distribution, a fundamental concept is that of the landscape patch or landscape functional type (LFT). An analogous concept, that of plant functional types (PFTs), has been subject to considerable research efforts and much debate over the past years (cf. Chapin et al. 1996; Woodward and Cramer 1996; Smith et al. 1997). According to the PFT concept, plants can be grouped or classified in a variety of ways, some of which are based on functional differences (e.g., photosynthetic pathways = C_3 vs. C_4) or structural ones (e.g., growth form = trees, grass, herbs). Such classifications (or simplifications) are valuable for identifying general patterns, for reduction of complexity, and for integration across different disciplines. For example, in the latter instance, PFTs are the basis of many global vegetation models. While the concept of PFTs is appealing, many approaches and philosophies have been identified, and there are a number of methodological approaches to the PFT concept (cf. Gitay and Noble 1997). Consequently, a large variety of different PFT schemes have been proposed for various purposes (Woodward and Cramer 1996; Smith et al. 1997).

In an analogous way, the concept of landscape structural and functional types (LFTs) has been proposed (see Reynolds and Wu, this volume). Like PFTs, the definition and physical meaning of an LFT will vary depending on the processes under study and the questions being addressed. In general, an LFT is an identifiable component of a landscape that represents an integrated assemblage of biological and physical entities which exchange and deliver mass and energy (via atmospheric couplings, inputs to rivers, etc.). Different LFTs operate at different characteristic scales in space and time; they are patch mosaics when viewed at finer scales and relatively homogeneous units when viewed from coarser scales. Thus, many if not most landscapes can be represented as hierarchies of patch mosaics or LFTs (Coffin and Lauenroth 1989; Wu and Loucks 1995; Wu and Levin 1994, 1997; Reynolds and Wu, this volume). Although heterogeneity occurs across scales, LFTs can be used to provide a hierarchical structure to complex landscapes so that we can focus only on a limited number of discrete scales (hierarchical levels) with insignificant loss of information. Such a simplification is necessary because (a) it is impractical to measure all processes of consideration continuously across space and time, (b) understanding is conversely related to complexity, and (c) it has been suggested that both physical and ecological processes tend to operate at characteristic scales and thus a hierarchical approach with LFTs should facilitate properly matching scales and integrating ecological with physical processes across complex landscapes.

Recognition of the boundaries of patches is not a well-developed technique. For structural characteristics, spatial statistics of two-dimensional surface distributions have been used. There have been few attempts to determine LFTs based on processes, although the identification of a given area of "homogeneous" landscape is inherent in many techniques. An example is eddy covariance, where the assumption of homogeneity within the footprint of the tower is critical, and one for which rigorous analyses are not available. Most plot-based methods also assume homogeneity within the plot; however, this is generally more easily supported for 1–10 m units than for the landscape. One method for calculating the size of the LFT for biogeochemistry studies is through the spatial divergence term (see Raupach et al., this volume). Essentially, the ratio of the divergence of the horizontal flux of mass or energy to the divergence of the vertical flux, or the vertical flux plus internal production and consumption terms, should not exceed an established critical value. Thus LFTs for nitrogen-cycling studies would be much smaller within a given landscape than LFTs for carbon balance. Regardless of the size of the LFT or the method used to determine it, land-surface parameters must be aggregated to the size of this unit. The effect of using an inappropriate patch size in a landscape-scale study of hydrologic or biogeochemical processes is unknown; however, in general, using too large a patch size will lead to errors in systems where nonlinear processes operating at sub-patch scales are important.

In studies where remote sensing data layers form an important part of the information content of the work, pixels instead of patches tend to be the fundamental landscape units used. Pixels may represent an invariant basic spatial unit that can change state through time. It is possible to combine adjacent pixels with similar characteristics into larger polygons which can then be used with landscape pattern statistics, thus combining continuous and discrete methods of spatial analysis. Mathematical treatments are better developed for discrete than for continuous descriptions of the landscape.

In addition to hierarchies of structure and distribution, complex systems can be thought of in terms of hierarchies of processes. A method for structuring models to reflect this hierarchy of processes is described in the next section.

INTEGRATING ECOSYSTEM DYNAMICS WITH HYDROLOGY AND BIOGEOCHEMISTRY AT LANDSCAPE SCALES

Ecological Models Applied at Landscape to Regional Scales

In this chapter, three types of models are distinguished that have been applied to study landscape-scale dynamics of ecosystems: patch scale models and DGVMs, both of which typically ignore interactions between neighboring patches or "pixels," and landscape models *sensu strictu*, which take these interactions into account.

Patch Models

Patch scale models that have been applied at landscape to regional scales can be classified broadly into two families: gap models and biogeochemistry models.

Gap models. The so-called "gap models" (Shugart 1984), first presented over 25 years ago (Botkin et al. 1972), simulate the competitive dynamics between individuals of different species on patches that have an area of 100–1000 m^2. These patches are supposed to be small enough so that all trees can be assumed to interact with each other. Patch-level descriptions of environmental conditions defining the light, nutrient, temperature, and water regimes of the site are used to drive establishment, growth, and mortality routines (Shugart 1984). Initially conceived as a simple four-dimensional description of the species' niche, gap models have increased considerably in complexity over time (cf. review by Bugmann et al. 1996).

While the treatment of vertical canopy structure (e.g., Botkin et al. 1972; Prentice et al. 1993), internal nutrient cycling (Pastor and Post 1986; Aber et al. 1982), and vertical water fluxes (e.g., Solomon 1986; Prentice et al. 1993) is treated in some detail in most gap models, the interactions between patches, i.e., the horizontal flows of energy, matter, and/or information, are neglected in all models except those originating from ZELIG (Urban et al. 1991). Gap models have been fairly successful in simulating species succession over time for individual "sites" (or, in the ZELIG case, small areas); however, these models are too detailed to be run with full spatial coverage for areas larger than a few dozen km^2 (e.g., a landscape of 10 km^2 is composed of 2000 patches of 500 m^2). Instead, some gap models have been applied to study regional-scale patterns of ecosystem types based on the assumption that the study area can be divided into fairly large, homogeneous units, e.g., a 10×10 km rectangular grid as done by Lindner et al. (1997), or a polygon-based approach derived from a geographical information systems (GIS) analysis of the model's driving variables (Bugmann et al. 1999).

Biogeochemical models. Biogeochemical models are usually structured as "box-and-flow" models describing the fluxes of energy, carbon, nutrients, and water, thereby assuming that the structure of the ecosystem is static (e.g., Parton et al. 1987; Running and Coughlan 1988). The major processes determining the rates of transfer of material between the compartments are formulated based on physiological principles, involving submodels for photosynthesis, allocation, stomatal conductance, etc.

Similar to gap models, most biogeochemistry models treat the vertical dimension in some detail, but they do not consider horizontal adjacency effects. As a matter of fact, these models have no specific areal extent and thus have been used at the plot, watershed, regional, and global level. At scales larger than the plot, they are frequently run in conjunction with a GIS to define the number and areal extent of homogeneous landscape units (cf. Pierce and Running 1995), and to produce spatially explicit outputs (Aber et al. 1995; Burke et al. 1997). Differences in vegetation composition between landscape units are summarized as changes in physiological parameters, generally representing plant functional types such as evergreen versus deciduous, tree versus grass, etc. (e.g., Nemani and Running 1996). Leaf area index (LAI) is a central driver of ecosystem processes in these models, and some biogeochemistry models aim at predicting LAI internally (e.g., Schimel et al. 1994), whereas others use LAI as an input variable derived from remote sensing (e.g., Running and Coughlan 1988).

Dynamic Global Vegetation Models (DGVMs)

Currently under development for use with GCMs, DGVMs combine descriptions of structural, physiological, and successional dynamics. In most of them, the equations are derived by considerable simplifications of existing patch-scale models, although one of the current DGVMs employs a full gap type approach. All DGVMs include biogeochemistry at the individual plant level, and most include a disturbance scheme. The vegetation is described in terms of plant functional types (e.g., C_3 vs. C_4 grasses, broadleaved vs. needleleaved trees). All climatically suitable PFTs compete for resources within a pixel, and competition rules of varying degree of sophistication are used to determine the resulting mixture of PFTs, which then is mapped as a "biome type" for that pixel.

DGVMs do not treat adjacency effects, partly due to the large size of the pixels used in global-scale applications (typically, $0.5° \times 0.5°$). The two major challenges facing this approach are: (a) to determine the number of PFTs required and to model how their ratio changes in the landscape, and (b) how to incorporate realistic descriptions of disturbances at these large spatial scales. By increasing their spatial resolution and perhaps also the number of PFTs that are distinguished, it is possible in principal to apply these models at continental to regional scales.

Landscape Models

Landscape models, as defined here, explicitly address the relationship between spatial patterns and ecosystem processes, taking into account the phenomena of adjacency, self-organization, and thresholds. Most of these models are based on a gridded description of the landscape (cf. Wu and Levin 1994, 1997) and treat spatially correlated processes such as lateral flows of resources, propagules, and disturbances. The structure of these models is quite variable. Examples include Markov models and cellular automata (e.g., Gardner et al. 1996), combined Markov-cellular automata modeling (e.g., Li and Reynolds 1997), other rule-based approaches (e.g., Noble and Gitay 1996), and gap and biogeochemistry models with connections between adjacent patches (e.g., Urban et al. 1991; Leadley et al. 1996). In some cases, a hierarchy of models is used in which parameters for the simpler, low-resolution models are derived from complex models of higher-resolution phenomena (e.g., Grant and French 1990; cf. Luan et al. 1996).

In contrast to the other approaches, most landscape models treat vertical ecosystem structure only marginally but emphasize the horizontal dimension. Most of the studies using landscape models were largely theoretical, studying abstract landscapes (cf. Noble, this volume). A central challenge here is to evaluate these findings with respect to the patterns and processes in real landscapes. There are some examples of models that address this issue; however, these case studies deal mainly with water transport and employ simple routing routines and retention times within landscape units (e.g., Costanza and Maxwell 1991; Ostendorf and Reynolds 1993; Vorosmarty 1996, 1997).

Integrating Landscape-scale Models of Hydrology, Ecosystem Dynamics, and Biogeochemistry

Historically, studies of the hydrology, ecosystem dynamics, and biogeochemistry of landscapes have proceeded largely independently of each other. In each discipline, it was assumed that the variables considered by the other disciplines could be treated as constant boundary conditions. Consequently, there are only a few examples of landscape- to regional-scale studies that transgressed these disciplinary boundaries. For example, Band et al. (1993) linked a biogeochemistry model with a hydrological model for a forested watershed; over the past years, the distinction between ecosystem dynamics (gap) models and biogeochemistry models has become increasingly blurred (e.g., Pastor and Post 1986; Korol et al. 1995; Keane et al. 1996; Friend et al. 1997) because it was recognized that ecosystem structure and function cannot be treated separately, but must be considered within a single model framework. However, we are not aware of a functioning model integrating all three disciplines.

Achieving an integrated model of hydrology, ecosystem dynamics, and biogeochemistry for a given landscape could profit greatly from methodological advances like model modularity and object-oriented design, which are discussed below.

Modularity of Landscape Models

The processes of ecosystem dynamics, hydrology, and biogeochemistry are currently modeled at different temporal and spatial scales. The weather systems that drive hydrology occur at global or continental scales, ecosystems typically occupy areas of a few square kilometers, and much biogeochemistry is driven by soil microorganisms and soil conditions that vary over meters. Temporal scales are similarly variable. Previously it has been difficult to combine models over such a wide range of scales, but recently some appropriate tools have become available.

Most existing models of landscape processes have been written in procedural languages such as FORTRAN. They have described the processes operating in the system, and the state of the system has been represented by the state variables. However, many modelers are now turning to object-oriented designs (Coad and Yourdon 1990; Acock and Reynolds 1997). These designs identify the objects in the system and encapsulate in the objects both the state variables and the processes appropriate to the object. They enhance the modularity and genericness of the code, thus opening the model to contributions from many authors (Reynolds and Acock 1997). Figure 17.1 shows one possible outline for a landscape model with an object-oriented design.

With object-oriented designs, it is convenient to think in anthropomorphic terms about what the objects "know" about themselves and what processes they "know" how to perform. The objects can communicate by requesting information from each other. Thus it is possible to have objects at large spatial and temporal scales aggregating information from objects at smaller scales. For example, a landscape object might "know" that it is composed of patches and that some of these are urban, some vegetation, and some water. It might also "know" how local weather is affected by these patches, how materials move between the patches, and how to aggregate water evaporated from all the patches.

Patches can be further subdivided into vegetation types, plant types, and even individual plants if so desired (Figure 17.1). These different objects can step through simulated time at different rates as appropriate to our knowledge of the processes. Thus it is possible to develop a structure of intercommunicating objects that cover all scales of interest in the landscape. Each object can, and must, be tested individually. Such a structure will be most useful if it contains objects at all organizational levels used by the modelers who might be expected to contribute to a landscape model. It is also important that objects pass information up and down the organizational hierarchy without jumping levels. So, for example, if a plant object needs solar radiation information to calculate transpiration, and that information is a state variable in the patch, it should pass the request for the information through "Plant_type" and "Vegetation_type" to

Figure 17.1 Example object-oriented design for a landscape model.

The principal steps required to accommodate all processes discussed in the position papers are shown. Each step consists of Object_name.function_call. Successive subdivisions of the Landscape object are indented to show the hierarchy of spatial aggregation. Any or all of the steps in this structure could be expanded to give more detail and introduce additional objects.

each run:	
Simulation_controller.run	
Landscape.map_patches	Map by composition, configuration, flow paths.
Landscape.sequence_patches	Sequence based primarily on elevation.
Landscape.read_weather	Read input weather file.
Landscape.step_time	
each time step:	
Landscape.calculate_weather	Calculate effects of topography, veg. height, etc.
Landscape.sudden_change	Fire, disease, clear cutting, etc.
Land_patch.change	Change first, next, or all patches affected.
Landscape.spread_change	Propagate fire, disease to next patch. Iterate.
Landscape.call_patches_in_sequence	
Patch.step_time	Call first patch in elevation sequence
for land patches:	
Land_patch.read_weather	Get patch weather from landscape.
Land_patch.material_from_adjacent_patches	
Get materials.	
Vegetation.photosynthesize	
[Plant_type.photosynthesize	
[Plant.photosynthesize, etc. to desired level.]	
Vegetation.transpire	
Vegetation.grow	
Vegetation.species_change	Track succession dynamics
Soil.water_flux	
Soil.chemistry	
Land_patch.trace_gasses_emit	
Land_patch.expand_contract	Change map of patches.
for water patches:	
Water_patch.calculate_flow_rate	
Water_patch.sedimentation	Calculate change in sediment load
Vegetation.nutrient_uptake	
Water_patch.chemistry	
Water_patch.expand_contract	
Landscape.material_flow_between_patches	Iterate back to Patch.step_time.

the patch. Only by observing these constraints can we preserve the flexibility to integrate all contributions.

When water is the principal medium moving materials between the patches, it makes sense to calculate first changes in patches at the highest elevations in the landscape, move materials to lower patches if appropriate, then calculate changes in the lower patches. The patches are therefore sequenced for calculation according to water flow paths in the landscape. In object-oriented programming languages (OOL), it is possible to send a message like "calculate" to each patch in the sequence and have each patch know what type it is (urban, vegetation, water, etc.), making the calculations appropriate for that type of patch. This is done by using some features peculiar to OOLs (inheritance, virtual functions, and function overriding) that are not available in procedural languages (Cox 1986; Meyer 1988; Wegner 1990; Wirfs-Brock et al. 1990; Booch 1991). Thus the use of OOLs and an object-oriented design (OOD) for landscape models has distinct advantages over traditional alternatives.

PATCHES, POLYGONS, AND PIXELS

Any model that integrates all the landscape processes must be able to handle data on spatial distribution. Remotely sensed data comes from satellites as pixels, the data layers in GIS are either vector or raster representations, and other data may be collected as lists of point data with latitude and longitude coordinates. The simplest way to accommodate all these types of data in landscape models is to use a grid to represent the landscape surface, and to calculate change at each point (intersection) on the grid. This is not as computationally intensive as it might first appear. If we have already identified the patches in the landscape, we can make our calculations once for each patch and apply the results to all the points within that patch. In effect we are assuming that the square around the point is uniform. If a datum falls within the square, it is assumed to apply to the point. Using a grid of points makes it easier to represent the increase and decrease in area of the patches, e.g., when one vegetation type gradually encroaches on the area occupied by another. This encroachment is represented in the model by transferring points from one patch to another. Pixels are similar to patches because they will normally cover several points, and the model can be related to the remotely sensed data by aggregating over the points covered by each pixel. Polygons are more difficult to fit over the landscape and are rigid, i.e., cannot represent expansion and shrinkage of patches.

MODELING OVER MANY LEVELS OF ORGANIZATION

Using an OOD allows us to cover more than three levels of organization in a landscape model (Figure 17.1) when processes farther away from the level of interest have penetrating influences. An OOD allows us to treat some processes coarsely while treating others in great detail. However, there may be practical considerations such as run time. If it is desired, for whatever reason, to reduce the number of levels of organization represented in the model, this can be done by replacing the more detailed levels with empirical equations. The procedure is to run the more detailed levels of the model alone and record their responses to the input variables (e.g., Reynolds et al. 1993). Then the responses are captured in empirical equations such as multiple regression equations or neural nets. It is often possible to capture more of the behavior of objects in this way than to generate the empirical equations directly from experimental observations.

The Modeling Process

It is useful to revisit the role of simulation modeling in ecosystem and landscape to regional-scale science. Modeling represents a significant activity in these disciplines, partly because empirical analysis is difficult at these large scales, partly because simulation modeling can integrate some of the complexity inherent at these scales, and partly because of the need by policy-makers for sensitivity analyses at regional scales.

Ecosystem and landscape studies follow the general paradigm for all scientific research (Figure 17.1). The process of modeling is one of several activities that comprise synthesis. When this is the case, the set of hypotheses represented by the model must be tested against field data not used in the construction of the model or derivation of parameters ("validation"). Validation is a critical step in testing and documenting the value of any model, and it is important to use validation data that are as independent as possible from the data used in building the model.

We often learn the most when the models fail; however, such failures are rarely reported, as they are difficult to publish. This, in combination with pressure to produce a result, often leads to "tuning" or over-calibration of the model, a process in which the many input parameters in a complex model are changed until the "right" answer is obtained. Although the tuning can possibly provide insights into the model itself, these insights are rarely generally applicable to ecological systems. Instead we report on an overly fitted model whose predictions are not robust. When models fail to reproduce a validation data set, the analysis of the reasons for this failure could lead to increased understanding about the underlying ecological processes, and thus it often would increase the robustness of model-based sensitivity studies conducted under scenarios of environmental change.

Model testing at the landscape scale poses considerable, specific problems. One is the difficulty of conducting experiments at this scale, for both financial and logistical reasons, although "natural experiments" (e.g., fires or windthrow) sometimes provide opportunities for testing model behavior under contrasting conditions. It may also be possible to do experiments for only parts of a landscape, e.g., in small watersheds. Extensive transects along environmental gradients can also be useful for testing models under systematically changing conditions. The power of remote sensing to provide high resolution spatial information over large areas can be used to generate validation data sets when the error parameters of the algorithm used to convert radiance to a biophysical property are known. The limited availability of data at large scales still constitutes a major limitation to developing robust models that are aimed at assessing the sensitivity of landscapes to direct and indirect anthropogenic environmental changes.

Several examples of model validation at the landscape to regional scales have been published. These can be divided into two broad categories: (a) studies that evaluated model behavior at several to a large number of points or plots within a landscape, e.g., with respect to soil organic matter, net primary production (NPP), or the water balance and (b) spatially explicit comparisons of the simulation results with maps derived from large regional data sets, or comparisons of simulated variables like NPP with variables derived from remote sensing data, such as cumulative NDVI. For example, regional runoff maps produced by the U.S. Geological Survey and regional forest production data produced by many national forest services offer the potential for validation of NPP (Burke et al. 1991) and water yield data (Aber et al. 1995) at the landscape scale.

Given the scarcity of landscape-scale data sets, it is important to develop protocols for the standardization of methods and interlaboratory comparisons for analyses, so that data sets from other research groups and other geographical areas can be used for model testing. This is doubly important because few experiments can be performed at this scale. Some large-scale ecological studies, such as those funded by NASA in the grasslands, boreal forest, and the Amazon (Sellers et al. 1992; Sellers 1995; Hall and Sellers 1995), provide data-layer-rich spatial data sets for use in model parameterization and validation.

The power of modeling approaches in ecology could be strengthened considerably if a more rigorous modeling paradigm was applied. Today, many modeling activities in ecology are considered to be of little value by a large number of field-oriented researchers. In this context, it would be desirable to establish more rigorous criteria for reviewing modeling papers, increased expectations for validation, tighter standards for calibration, and especially a more complete presentation of the modeling process in the reviewed literature, including reports on those instances where models fail. This could increase the value of modeling as a tool for ecological analyses.

Problems in Publishing and Distributing Models

Publishing ecosystem and landscape models within the framework provided for experimental work can be difficult. Full and accurate presentation of models within peer-reviewed journals, including a complete description of model structure, a complete listing of all variables used in the model and the values assigned to each parameter, as well as the source of each value, can require substantial amounts of space and is seen as uninteresting material by some editors. It is also true that models can be most useful and informative when they fail; however, papers presenting poor agreement between measurements and model predictions are difficult to publish.

There are examples of disciplines in which a central modeling repository is provided by the professional association (e.g., groundwater hydrology modeling). With the proliferation of models at the landscape scale and the difficulty in publishing models in a format that meets the needs of the community, it is time for one of the organizations in the global change community to provide this repository and publication function.

Such a service could consist of a web-based information system in which a model code could be stored along with standardized data sets and associated outputs.

Minimal standards for documentation should be established, and the repository should be available to the entire scientific community at no cost. In association with this repository, an electronic journal should be established that would provide for peer review and "publication" of papers in electronic form only.

REFERENCES

Aber, J.D., C. Driscoll, C.A. Federer, R. Lathrop, G. Lovett, J.M. Melillo, P. Steudler, and J. Vogelmann. 1993. A strategy for the regional analysis of the effects of physical and chemical climate change on biogeochemical cycles in northeastern (U.S.) forests. *Ecol. Mod.* **67**:37–47.

Aber, J.D., J.M. Melillo, and C.A. Federer. 1982. Predicting the effects of rotation length, harvest intensity and fertilization on fiber yield from northern hardwood forests in New England. *For. Sci.* **28**:31–45.

Aber, J.D., S.V. Ollinger, C.A. Reich, P.B. Federer, M.L. Boulden, D.W. Kicklighter, J.M. Melillo, and R.G. Lathrop, Jr. 1995. Predicting the effects of climate change on water yield and forest production in the northeastern United States. *Clim. Res.* **5**:207–222.

Acock, B., and J.F. Reynolds. 1997. Introduction: Modularity in plant models. *Ecol. Mod.* **94**:1–6.

Band, L.E., P. Patterson, R. Nemani, and S.W. Running. 1993. Forest ecosystem processes at the watershed scale: Incorporating hillslope hydrology. *Agr. For. Meteorol.* **63**:93–126.

Booch, G. 1991. Object-oriented Design with Applications. Redwood City, CA: Benjamin-Cummins.

Bormann, F.H., and G.E. Likens. 1979. Pattern and Process in a Forested Ecosystem. New York: Springer.

Botkin, D.B., J.F. Janak, and J.R. Wallis. 1972. Some ecological consequences of a computer model of forest growth. *J. Ecol.* **60**:849–872.

Bugmann, H., M. Lindner, P. Lasch, M. Flechsig, B. Ebert, and W. Cramer. 1999. Scaling issues in forest succession modelling. *Clim. Change*, in press.

Bugmann, H.K.M., Y. Xiaodong, M.T. Sykes, P. Martin, M. Lindner, P.V. Desanker, and S.G. Cumming. 1996. A comparison of forest gap models: Model structure and behaviour. *Clim. Change* **34**:289–313.

Burke, I.C., T.G.F. Kittel, W.K. Lauenroth, P. Snook, C.M. Yonker, and W.J. Parton. 1991. Regional analysis of the central great plains, sensitivity to climate variability. *BioScience* **41**:685–692.

Burke, I.C., W.K. Lauenroth, and W.J. Parton. 1997. Regional and temporal variation in net primary production and nitrogen mineralization in grasslands. *Ecology* **78(5)**:1330–1340.

Burke, I.C., W.A. Reiners, and D.S. Schimel. 1989. Organic matter turnover in a sagebrush steppe landscape. *Biogeochemistry* **7**:11–31.

Burke, I.C., D.S. Schimel, C.M. Yonker, W.J. Parton, L.A. Joyce, and W.K. Lauenroth. 1990. Regional modeling of grassland biogeochemistry using GIS. *Landsc. Ecol.* **4**:45–54.

Chapin, F.S., M.S. Bretharte, S.E. Hobbie, and H.L. Yhong. 1996. Plant functional types as predictors of transient responses of arctic vegetation to global change. *J. Veg. Sci.* **7**:347–358.

Charley, J.L., and N.E. West. 1977. Micro-patterns of nitrogen mineralization activity in soils of some shrub-dominated semi-desert ecosystems of Utah. *Soil Biol. Biochem.* **9(5)**:357–365.

Coad, P., and E. Yourdon. 1990. Object-oriented Analysis. Englewood Cliffs, NJ: Prentice Hall.
Coffin, D.P., and W.K. Lauenroth. 1989. Disturbances and gap dynamics in a semiarid grassland: A landscape-level approach. *Landsc. Ecol.* **3(1)**:19–27.
Cohen, W.B., M.E. Harmon, D.O. Wallin, and M. Fiorella. 1996. Two decades of carbon flux from forests of the Pacific Northwest. *BioScience* **46**:836–844.
Cole, J.J., B.L. Peierls, N.F. Caraco, and M.L. Pace. 1993. Nitrogen loading of rivers as a human-driven process. In: Humans as Components of Ecosystems, ed. M.J. McDonnell and S.T.A. Pickett, pp. 163–174. New York: Springer.
Costanza, R., and T. Maxwell. 1991. Spatial ecosystem modelling using parallel processors. *Ecol. Mod.* **58**:159–183.
Costanza, R., F.H. Sklar, and M.L. White. 1990. Modeling coastal landscape dynamics. *BioScience* **40**:91–107.
Cox, B. 1986. Object-oriented Programming: An Evolutionary Approach. Reading, MA: Addison-Wesley.
Daily, G.C., ed. 1997. Nature's Services: Societal Dependence on Natural Ecosystems. Washington, D.C.: Island Press.
Friend, A.D., A.K. Stevens, R.G. Knox, and M.G.R. Cannell. 1997. A process-based, terrestrial biosphere model of ecosystem dynamics (HYBRID v3.0). *Ecol. Mod.* **95**:249–287.
Gardner, R.H., W.W. Hargrove, M.G. Turner, and W.H. Romme. 1996. Climate change, disturbances and landscape dynamics. In: Global Change and Terrestrial Ecosystems, ed. B.H. Walker and W.L. Steffen, pp. 149–172. International Geosphere-Biosphere Programme Book Series, vol. 2. Cambridge: Cambridge Univ. Press.
Gitay, H., and I.R. Noble. 1997. What are functional types and how should we seek them? In: Plant Functional Types: Their Relevance to Ecosystem Properties and Global Change, ed. T.M. Smith, H.H. Shugart, and F.I. Woodward, pp. 3–19. Cambridge: Cambridge Univ. Press.
Grant, W.E., and N.R. French. 1990. Response of alpine tundra to a changing climate: A hierarchical simulation model. *Ecol. Mod.* **49**:205–227.
Groffman, P.M., C.W. Rice, and J.M. Tiedje. 1993. Denitrification in a tallgrass prairie landscape. *Ecology* **74**:855–862.
Groffman, P.M., J.M. Tiedje, D.L. Mokma, and S. Simkins. 1992. Regional scale analysis of denitrification in north temperate forest soils. *Landsc. Ecol.* **7**:45–53.
Groffman, P.M., J.M. Tiedje, G.P. Robertson, and S. Christensen. 1988. Denitrification at different temporal and geographical scales: Proximal and distal controls. In: Advances in Nitrogen Cycling in Agricultural Ecosystems, ed. J.R. Wilson. Wallingford, U.K.: CAB International.
Hall, F.G., and P.J. Sellers. 1995. First International Satellite Land Surface Climatology Project (ISLSCP) Field Experiment (FIFE) in 1995. *J. Geophys. Res.* **100:**383–395.
Holling, C.S., G. Peterson, P. Marples, J. Sendzimir, K. Redford, L. Gunderson, and D. Lambert. 1996. Self-organization in ecosystems: Lumpy geometries, periodicities and morphologies. In: Global Change and Terrestrical Ecosystems, ed. B.H. Walker and W.L. Steffen, pp. 346–384. Cambridge: Cambridge Univ. Press.
Howarth, R.W., G. Billen, D. Swaney, A. Townsend, N. Jaworski, K. Lajtha, J.A. Downing, R. Elmgren, N. Caraco, T. Jordan, R. Berendse, J. Freney, V. Kudeyarov, P. Murdoch, and Z. Zhao-liang. 1996. Regional nitrogen budgets and riverine N and P fluxes for the drainages to the North Atlantic ocean: Natural and human influences. *Biogeochemistry* **35**:75–139.
Kareiva, P., and U. Wennergren. 1995. Connecting landscape patterns to ecosystem and population processes. *Nature* **373**:299–302.

Keane, R.E., P. Morgan, and S.W. Running. 1996. FIRE-BGC — A mechanistic ecological process model for simulating fire succession on coniferous forest landscapes of the northern Rocky Mountains. U.S.D.A. Forest Service Research Paper INT–RP–484. U.S. Dept. of Agriculture.

Knapp, A.K., J.T. Fahnestock, S.P. Hamburg, L.B. Statland, T.R. Seastedt, and D.S. Schimel. 1993. Landscape patterns in soil-plant water relations and primary production in tallgrass prairie. *Ecology* **74**:549–560.

Korol, R.L., S.W. Running, and K.S. Milner. 1995. Incorporating intertree competition into an ecosystem model. *Can. J. For. Res.* **25**:413–424.

Lajtha, K., and W.H. Schlesinger. 1988. The biogeochemistry of phosphorus cycling and phosphorus availability along a desert soil chronosequence. *Ecology* **69**:24–39.

Leadley, P.W., H. Li, B. Ostendorf, and J.F. Reynolds. 1996. Road-related disturbances in an arctic watershed: Analyses by a spatially explicit model of vegetation and ecosystem processes. In: Landscape Function and Disturbance in Arctic Tundra, ed. J.F. Reynolds and J.D. Tenhunen, pp. 387–415. Ecological Studies Series 120. Berlin: Springer.

Li, H., and J.F. Reynolds. 1997. Modeling effects of spatial pattern, drought, and grazing on rates of rangeland degradation: A combined Markov and cellular automaton approach. In: Scaling of Remote Sensing Data for Geographical Information Systems, ed. D.A. Quattrochi and M. Goodchild, pp. 211–230. Chelsea, MI: Lewis Publ.

Lindner, M., H. Bugmann, P. Lasch, M. Flechsig, and W. Cramer. 1997. Regional impacts of climatic change on forests in the state of Brandenburg, Germany. *Agr. For. Meteorol.* **84**:123–135.

Luan, J., R.I. Muetzelfeldt, and J. Grace. 1996. Hierarchical approach to forest ecosystem simulation. *Ecol. Mod.* **86**:37–50.

Martin, M.E., and J.D. Aber. 1997. Estimation of forest canopy lignin and nitrogen concentration and ecosystem processes by high spectral resolution remote sensing. *Ecol. Appl.* **7**:431–443.

Meyer, B. 1988. Object-oriented Software Construction. Englewood Cliffs, NJ: Prentice Hall.

Nemani, R., and S.W. Running. 1996. Implementation of a hierarchical global vegetation classification in ecosystem function models. *J. Veg. Sci.* **7**:337–346.

Noble, I.R., and H. Gitay. 1996. A functional classification for predicting the dynamics of landscapes. *J. Veg. Sci.* **7**:329–336.

O'Neill, R.V. 1988. Hierarchy theory and global change. In: Scales and Global Change, ed. T. Rosswall, R.G. Woodmansee, and P.G. Risser, pp. 29–45. New York: Wiley.

O'Neill, R.V., D.L. DeAngelis, J.B. Waide, and T.F.H. Allen. 1986. A Hierarchical Concept of Ecosystems. Princeton: Princeton Univ. Press.

Ostendorf, B., and J.F. Reynolds. 1993. Relationships between a terrain-based hydrologic model and patch-scale vegetation patterns in an arctic tundra landscape. *Landsc. Ecol.* **8**:229–237.

Parton, W.J., D.S. Schimel, C.V. Cole, and D.S. Ojima. 1987. Analysis of factors controlling soil organic matter levels in Great Plains grasslands. *Soil Sci. Soc. Am. J.* **51**:1173–1179.

Pastor, J., and W.M. Post. 1986. Influence of climate, soil moisture, and succession on forest carbon and nitrogen cycles. *Biogeochemistry* **2**:3–28.

Peterjohn, W.T., and D.L. Correll. 1984. Nutrient dynamics in an agricultural watershed: Observations on the role of a riparian forest. *Ecology* **65**:1466–1475.

Pielke, R.A., and R. Avissar. 1990. Influence of landscape structure on local and regional climate. *Landsc. Ecol.* **4**:133–155.

Pielke, R.A., G. Dalu, J.S. Snook, T.J. Lee, and T.G.F. Kittel. 1991. Nonlinear influence of mesoscale land use on weather and climate. *J. Clim.* **4**:1053–1069.

Pielke, R.A., T.J. Lee, J.H. Copeland, J.L. Eastman, C.L. Ziegler, and C.A. Finley. 1997. Use of USGS-provided data to improve weather and climate simulations. *Ecol. Appl.* **7**:3–21.

Pielke, R.A., D.S. Schimel, T.J. Lee, T.G.F. Kittel, and X. Zeng. 1993. Atmosphere-terrestrial ecosystem interactions: Implications for coupled modeling. *Ecol. Mod.* **67**:5–18.

Pierce, L.L., and S.W. Running. 1995. The effects of aggregating sub-grid land surface variation on large-scale estimates of net primary production. *Landsc. Ecol.* **10**:239–253.

Prentice, I.C., M.T. Sykes, and W. Cramer. 1993. A simulation model for the transient effects of climate change on forest landscapes. *Ecol. Mod.* **65**:51–70.

Rebele, F. 1994. Urban ecology and special features of urban ecosystems. *Glob. Ecol. Biogeogr. Lett.* **4**:173–187.

Reich, P.B., M.B. Walters, and D.S. Ellsworth. 1997. From tropics to tundra: Global convergence in plant functioning. *Proc. Natl. Acad. Sci. USA* **94**:13,730–13,734.

Reynolds, J.F., and B. Acock. 1997. Modularity and genericness in plant and ecosystem models. *Ecol. Mod.* **94**:7–16.

Reynolds, J.F., D.W. Hilbert, and P.R. Kemp. 1993. Scaling ecophysiology from the plant to the ecosystem: A conceptual framework. In: Scaling Processes between Leaf and the Globe, ed. J. Ehleringer and C. Field, pp. 127–140. New York: Academic.

Romme, W.H., and D.G. Despain. 1989. Historical perspective on the Yellowstone fires of 1988. *BioScience* **39(10)**:695–699.

Running, S.W., and J.C. Coughlan. 1988. A general model of forest ecosystem processes for regional applications. I. Hydrologic balance, canopy gas exchange and primary production processes. *Ecol. Mod.* **42**:125–154.

Schimel, D.S., B.H. Braswell, E.A. Holland, R. McKeown, D.S. Ojima, T.H. Painter, W.J. Parton, and A.R. Townsend. 1994. Climatic, edaphic, and biotic controls over storage and turnover of carbon in soils. *Glob. Biogeochem. Cyc.* **8**:279–293.

Schlesinger, W.H., J.F. Reynolds, G.L. Cunningham, L.F. Huenneke, W.M. Jarrell, R.A. Virginia, and W.G. Whitford. 1990. Biological feedbacks in global desertification. *Science* **247**:1043–1048.

Sellers, P.J., F.G Hall, and G. Asrar. 1992. An overview of the First International Satellite Land Surface Climatology Project (ISLSCP) field Experiment (FIFE). *J. Geophys. Res.* **97**:18,345–18,371.

Sellers, P., F. Hall, H. Margolis, R. Kelly, D. Baldocchi, G. den Hartog, J. Cihlar, M.G. Ryan, B. Goodison, P. Crill, K.J. Ranson, D. Lettenmaier, and D.E. Wickland. 1995. The Boreal Ecosystem-Atmosphere Study (BOREAS): An Overview and Early Results from the 1994 Field Year. *Bull. Am. Meteorol. Soc.* **76**:1549–1577.

Shugart, H.H. 1984. A Theory of Forest Dynamics. The Ecological Implications of Forest Succession Models. New York: Springer.

Smith, T.M., H.H. Shugart, and F.I. Woodward, eds. 1997. Plant functional types: Their relevance to ecosystem properties and global change. International Geosphere-Biosphere Programm Book Series, vol. 1. Cambridge: Cambridge Univ. Press.

Solomon, A.M. 1986. Transient response of forests to CO_2-induced climate change: Simulation modeling experiments in eastern North America. *Oecologia* **68**:567–579.

Sprugel, D.G. 1984. Density, biomass, productivity, and nutrient-cycling changes during stand development in wave-regenerated balsam fir forests. *Ecol. Monogr.* **54**:165–186.

Turner, M.G., and W.H. Romme. 1994. Landscape dynamics in crown fire ecosystems. *Landsc. Ecol.* **9(1)**:59–77.

Urban, D.L., G.B. Bonan, T.M. Smith, and H.H. Shugart. 1991. Spatial applications of gap models. *For. Ecol. Manag.* **42**:95–110.

Vorosmarty, C.J., K. Sharma, B. Fekete, A.H. Copeland, J. Holden, J. Marble, and J.A. Lough. 1997. The storage and aging of continental runoff in large reservoir systems of the world. *Ambio* **26**:210–219.

Vorosmarty, C.J., C.J. Willmott, B.J. Choudhury, A.L. Schloss, T.K.Stearns, S.M. Robeson, and T.J. Dorman. 1996. Analyzing the discharge regime of a large tropical river through remote sensing, ground-based climatic data, and modeling. *Water Res.* **32**:3137–3150.

Wegner, P. 1990. Concepts and paradigms of object-oriented programming. *Oops Mess.* **1**:7–87.

Wirfs-Brock, R., B. Wilkerson, and L. Wiener. 1990. Designing Object-oriented Software. Englewood Cliffs, NJ: Prentice Hall.

Wofsy, S.C., M.L. Goulden, J.W. Munger, S.-M. Fan, P.S. Bakwin, B.C. Daube, S.L. Bassow, and F.A. Bazzaz. 1993. Net exchange of CO_2 in a mid-latitude forest. *Science* **260**:1314–1317.

Woodward, F.I., and W. Cramer. 1996. Plant functional types and climatic change: Introduction. *J. Veg. Sci.* **7**:306–430.

Wu, J., and S.A. Levin. 1994. A spatial patch dynamic modeling approach to pattern and process in an annual grassland. *Ecol. Monogr.* **64(4)**:447–464.

Wu, J., and S.A. Levin. 1997. A patch-based spatial modeling approach: Conceptual framework and simulation scheme. *Ecol. Mod.* **101**:325–346.

Wu, J., and O.L. Loucks. 1995. From balance-of-nature to hierarchical patch dynamics: A paradigm shift in ecology. *Qtly. Rev. Biol.* **70**:439–466.

Zobel, D.B., and J.A. Antos. 1997. A decade of recovery of understory vegetation buried by volcanic tephra from Mount St. Helens. *Ecol. Monogr.* **67**:317–344.

Author Index

Subject Index

Index compiled by Indexing Specialists, Hove, East Sussex